S0-AVN-731

Topics in Applied Physics Volume 29

Topics in Applied Physics Founded by Helmut K. V. Lotsch

Hydrogen in Metals II

Application-Oriented Properties

Edited by G. Alefeld and J. Völkl

With Contributions by

G. Alefeld B. Baranowski H. Brodowsky
T. Schober B. Stritzker H. Wenzl Ch. A. Wert
E. Wicke H. Wipf R. Wiswall H. Wühl

With 162 Figures

Springer-Verlag Berlin Heidelberg New York 1978

sep/ae PHYS

Professor Dr. *Georg Alefeld*
Dr. *Johann Völkl*

Physik-Department der Technischen Universität München,
D-8046 Garching, Fed. Rep. of Germany

ISBN 3-540-08883-0 Springer-Verlag Berlin Heidelberg New York
ISBN 0-387-08883-0 Springer-Verlag New York Heidelberg Berlin

Library of Congress Cataloging in Publication Data. Main entry under title: Hydrogen in metals. (Topics in applied physics; v. 28–29) Includes bibliographical references and indexes. Contents: v. 1. Basic properties.—v. 2. Application-oriented properties. 1. Metals—Hydrogen content. I. Alefeld, G., 1933— II. Völkl, J., 1936— TH690.H97 669′.94 78-4487

Monophoto typesetting, offset printing and bookbinding: Brühlsche Universitätsdruckerei, Lahn-Gießen
2153/3130-543210

Preface

Progress in solid-state sciences results from many effects: the preparation of new materials, the development of new experimental methods, or increase in technological interest. All three aspects are valid for research on hydrogen in metals.

Although these systems are not completely new, advanced preparation techniques yielding well-defined samples not only made the results on diffusion, solubility, and phase transitions, for example, more and more reproducible, but also stimulated the application of new methods like neutron scattering and Mössbauer effect. The technological interests range from old problems like hydrogen embrittlement to applications of metal-hydrogen systems in fission and fusion reactors, fuel cells, as energy storage systems, etc. The concept for future energy transport and supply known as "hydrogen economy" also caused appreciable increase in research on interaction of hydrogen with metals.

Due to this broad interest, this group of materials presently is studied by many scientists affiliated with various disciplines such as physics, chemistry, physical chemistry, metallurgy and engineering. Therefore, research results are scattered throughout many different journals or conference reports.

The books *Hydrogen in Metals I. Basic Properties*, and *Hydrogen in Metals II. Application-Oriented Properties* contain reviews written by experts in the field. Those topics which have experienced the greatest progress over the recent years have been selected. Areas for which comprehensive review papers exist have been omitted. Although the editors have coordinated arrangement and content of the various contributions, they have purposely not eliminated all diverging points of view which necessarily exist in a rapidly expanding area of research.

The great number of subjects made it necessary to divide the contributions into two volumes, the first one devoted more to basic properties, the second one more to application-oriented properties. It is evident that ambiguities in this subdivision were unavoidable.

The editors are grateful to the authors for being very cooperative, for fast response to changes, and for being on time. The editors are also grateful to the publishers and their staff, and Dr. H. Lotsch for encouragement.

Munich, March 1978 *G. Alefeld · J. Völkl*

Contents

Contents of Hydrogen in Metals I

Basic Properties (Topics in Applied Physics, Vol. 28)

Contributors

Alefeld, Georg

Physik-Department der Technischen Universität München,
D-8046 Garching, Fed. Rep. of Germany

Baranowski, Bogdan

Polska Akademia Nauk, Instytut Chemii Fizycznej,
Kasprzaka 44/52, 01-224 Warszawa, Polen

Brodowsky, Horst

Institut für Physikalische Chemie, Universität Kiel,
D-2300 Kiel, Fed. Rep. of Germany

Schober, Tilman

Institut für Festkörperforschung, Kernforschungsanlage Jülich,
D-5170 Jülich 1, Fed. Rep. of Germany

Stritzker, Bernd

Institut für Festkörperforschung, Kernforschungsanlage Jülich,
D-5170 Jülich 1, Fed. Rep. of Germany

Wenzl, Helmut

Institut für Festkörperforschung, Kernforschungsanlage Jülich,
D-5170 Jülich 1, Fed. Rep. of Germany

Wert, Charles A.

Department of Metallurgy and Mining Engineering,
University of Illinois, Urbana, IL 61801, USA

Wicke, Ewald

Westfälische Wilhelms-Universität, Institut für Physikalische Chemie,
D-4400 Münster, Fed. Rep. of Germany

Wipf, Helmut

Physik-Department der Technischen Universität München,
D-8046 Garching, Fed. Rep. of Germany

Wiswall, Richard

Department of Energy and Environment,
Brookhaven National Laboratory, Upton, L.I. NY 11973, USA

Wühl, Helmut

Institut für Festkörperforschung, Kernforschungsanlage Jülich,
D-5170 Jülich 1, Fed. Rep. of Germany

1. Introduction

G. Alefeld

With 6 Figures

Applications of metal-hydrogen systems extend into many areas. The following list, which certainly is not complete, demonstrates the great variety of immediate or potential applications:

1) Purification of hydrogen, hydrogen filters [1.1, 2].

2) Hydrogen embrittlement, powder metallurgy.

3) Metal hydrides applied as moderator, reflector, shielding, or control materials in nuclear reactors [1.3].

4) Isotope separation using isotope-dependent properties of metal hydrides.

5) The fusion reactor, e.g., interaction of H with the plasma-containing wall, permeation of H, extraction of tritium out of lithium, etc.

6) Hydrogen technology, water splitting, methane-water splitting, hydrogen pipelines, etc. [1.4].

7) Electrodes for fuel cells or batteries [1.5].

8) Hydrogen storage for automotive propulsion [1.6] or electric utility load leveling [1.7].

9) Thermal compressor for H using metal-hydrogen systems [1.8, 9].

10) Hydrogen-Brayton cycle with metal hydrides as compressor [1.10].

11) Heat storage using metal hydrides [1.11, 12].

12) Metal-hydride heat pumps for home heating, especially for use of solar heat [1.9, 13, 14].

13) Energy cascading, i.e., topping cycles for power-generation stations using metal-hydrogen systems [1.15].

In the preceding Topics volume [Ref. 1.16, Chap. 1], it was pointed out that metal-hydrogen systems quite often have played the role of prototypes for certain basic physical properties. The same is partly true for some of the applications, namely those listed above under Items 9)–13). Besides the potential use of metal hydrides, the understanding of the energy storage or energy conversion cycles, using metal-hydrogen systems, also allows the understanding of the general underlying principles and thus a generalization to other working substances.

It is not the intention of this volume to cover each of these subjects in a separate chapter, but to provide information needed if applications are intended. As in [1.16], only those areas in which the largest advances have been made in the past few years are included. A certain ambiguity in deciding whether a chapter is more relevant to basic properties or to applications was unavoidable, thus, for example, the chapter on diffusion coefficients [Ref. 1.16, Chap. 12] could as well have been included in this volume. The arrangement of the chapters in this volume is as follows.

The first four chapters deal mainly with thermodynamics and phase diagrams, the following two with electronic properties and the last one with trapping effects. Chapter 2 by *Schober* and *Wenzl* contains a collection of the phase diagrams of H in V, Nb, and Ta. Chapter 3 by *Wicke* and *Brodowsky* summarizes the knowledge about the most intensively studied system, the Pd–H system. Chapter 4 by *Baranowski* contains the recent results on new hydride phases produced by the high-pressure technique. Then, Chapter 5 by *Wiswall* summarizes the knowledge needed for use of metal-hydrogen systems for energy storage or energy conversion. This chapter also summarizes some of the potential applications. *Stritzker* and *Wühl* present in Chapter 6 the state of knowledge about superconducting metal-hydrogen alloys. In Chapter 7 *Wipf* describes the recent experimental progress in electrotransport of protons in metals. The final Chapter 8 by *Wert* on the interaction of hydrogen with other defects presents observations which are in themselves interesting, but which also yield parameters that finally may help to understand and prevent embrittlement.

As already mentioned, the methods listed under Items 9)–13) have partly model characteristics and are partly considered for real applications. The absorption-heat pump using two metal hydrides with different vapor pressures may be considered as an example for the well-known "resorption process" [1.17]. The main drawback of the metal hydrides for use of solar energy and especially for home heating is the high price of the materials and the low heat conduction of the metal-hydride powders which therefore require large heat-transfer surfaces. Investment costs are a less restrictive argument for the application listed under Item 13). Since the applications of metal hydrides in a topping cycle for power generation have not yet been published[1], we shall give a short introduction to the essential underlying ideas.

1.1 Metal Hydrides as Heat Transformer for Energy Cascading

It is well known that besides the irreversibilities in the combustion process the main losses in power-generating stations are caused by heat flowing irreversibly from a high temperature level to that temperature level at which it can be used in the Clausius-Rankine cycle with water. An improvement in the efficiency of a power-generating plant could be achieved if this heat transport from a technically unusable to a technically usable temperature level could be performed reversibly, that means without loss of availability, i.e., the ability to be converted to work. Therefore, if heat from the temperature level T_2 (e.g., 1000 °C) is transformed *reversibly* into heat at the temperature level T_1 (e.g.,

[1] Presented at the Meeting of the Austrian Physical Society, Leoben, Oct. 1977 and at the 2nd World Hydrogen Energy Conference, Zurich, Aug. 1978.

300 °C for vaporization of water or 560 °C for superheating), the amount of heat must increase at the lower temperature level according to the Carnot factors of the two temperature levels T_2 and T_1. It is easy to show that with two heat reservoirs at different temperatures, a reversible transport of heat without performing work is not possible. However, this is possible if three or more heat reservoirs are used. Assuming $T_2 > T_1 > T_0$, the first and second laws of thermodynamics, namely

$$Q_2 + Q_1 + Q_0 = 0\,, \tag{1.1}$$

$$\frac{Q_2}{T_2} + \frac{Q_1}{T_1} + \frac{Q_0}{T_0} = 0\,, \tag{1.2}$$

have a solution if the heat exchange occurs according to the following scheme:

$$Q_2 < 0 \quad Q_1 > 0 \quad Q_0 < 0\,, \tag{1.3}$$

i.e., heat is transported from the reservoirs with the temperatures T_2 and T_0 to the reservoir with the temperature T_1.

The ratio Q_2/Q_1 amounts to

$$\frac{|Q_2|}{|Q_1|} = \frac{\dfrac{T_1 - T_0}{T_1}}{\dfrac{T_2 - T_0}{T_2}}\,. \tag{1.4}$$

The first and the second laws of thermodynamics therefore allow reversible heat transformation from the temperature level T_2 to the temperature level T_1 with an increase in the amount of heat according to the Carnot factor if at least one further temperature reservoir is involved. The increase of heat at the level T_1, which may be a temperature level adapted to the Clausius-Rankine cycle, results from heat which was originally at the temperature level T_0 with low or no availability. The loss of availability of the heat Q_2 transformed from T_2 to T_1 is recovered in the availability of that amount of heat Q_0 at the temperature T_1 which has been raised from T_0 to T_1. The previous concept of the reversible heat transformation is due to *Nesselmann* [1.18] who gave a thermodynamic formulation of much older ideas by *Schneevogl* [1.19], and *Altenkirch* and *Tenckhoff* [1.20] for the transformation of heat. These authors noted that an absorption-heat pump is one possible technical instrument which can transform heat (in principle reversibly) from one temperature level to another. Their ideas did not receive much attention, apparently since no "refrigerant/absorbent pair" usable for high-temperature heat-pumps was suggested.

In addition to others [1.15], metal-hydrogen systems may be used as high-temperature heat-transformer. Alloys for high-temperature storage of hydrogen, in contrast to those of the low- and medium-temperature regime, have not been studied intensively in the past. We shall therefore demonstrate the

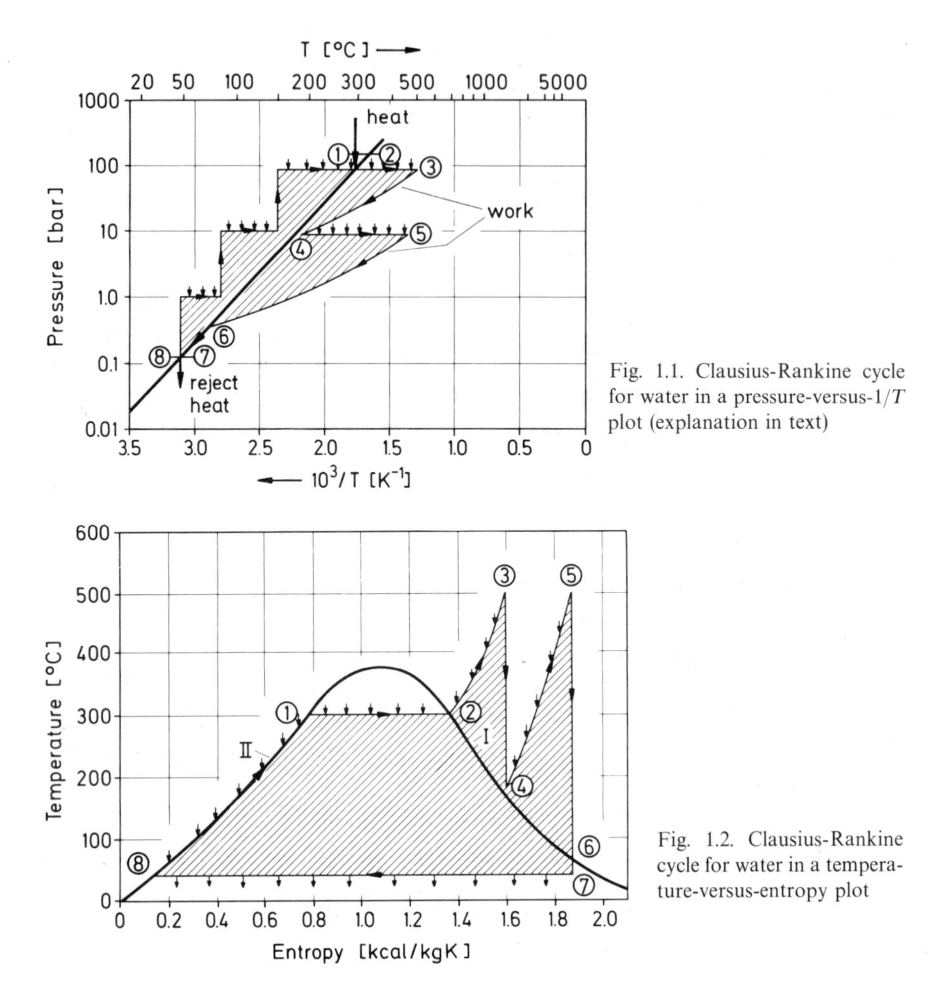

Fig. 1.1. Clausius-Rankine cycle for water in a pressure-versus-$1/T$ plot (explanation in text)

Fig. 1.2. Clausius-Rankine cycle for water in a temperature-versus-entropy plot

principle of the method using two alloys with well-known properties, namely Mg_2Ni [1.21] and $LaNi_5$ [1.9]. The latter alloy may as well be replaced by $CaNi_5$ or FeTi. It should be pointed out that this combination is by no means optimum for a topping cycle to the Clausius-Rankine cycle with water. Nevertheless, the virtues of the method can be demonstrated by this combination.

Absorption-heat pumps are usually presented in a plot $\ln p \rightarrow 1/T$, whereas for the Clausius-Rankine cycle the temperature-entropy diagram is most often used. To demonstrate combination of a metal-hydrogen topping cycle with the Clausius-Rankine water cycle, both processes should be presented in comparable diagrams. We shall use both the temperature-entropy and the pressure-temperature diagram. Figure 1.1 displays the Clausius-Rankine cycle in a pressure-temperature diagram, whereas Fig. 1.2 shows the Clausius-Rankine cycle in the familiar temperature-entropy plot. Corresponding process points are labelled by corresponding numbers. The arrows indicate addition of heat if

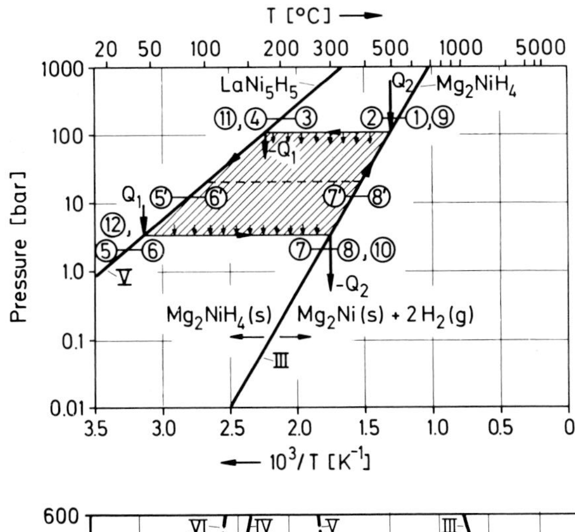

Fig. 1.3. Heat-transformation cycle in a pressure-versus-temperature plot. Working fluid (absorbens): H_2; absorber: Mg_2Ni; resorber: $LaNi_5$ (explanation in text)

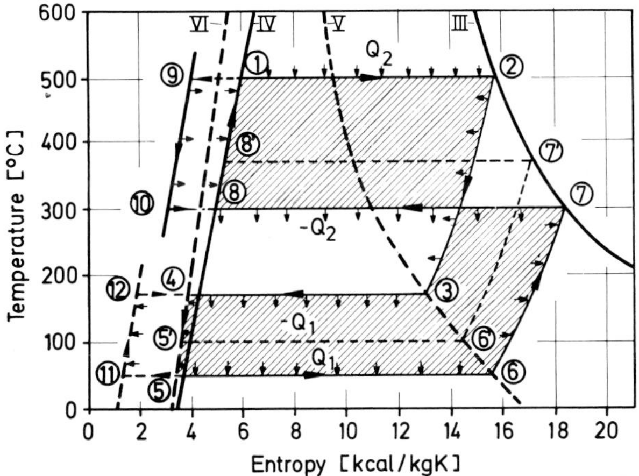

Fig. 1.4. Heat-transformation cycle in a temperature-versus-entropy plot. Working fluid (absorbens): H_2; full heavy lines: Mg_2NiH_4 system (absorber); dashed heavy lines: $LaNi_5H_5$ system (resorber), (explanation in text)

the arrowhead ends at the cycle or rejection of heat if the end of the arrow starts from the cycle. The process data are chosen as follows: $p_{nax} = 90$ bar, $T_{max} = 500$ °C, $T_{min} = 45$ °C, reheating at 9 bar. Figure 1.3 shows the pressure-versus-temperature relations for the two alloys Mg_2NiH_4 and $LaNi_5H_5$. The heat-pump cycle shown in Fig. 1.3 is also displayed semiquantitatively in Fig. 1.4 in a temperature-entropy diagram. Lines III/IV and V/VI correspond to the Lines I (saturated vapor line) and II (saturated liquid line) in Fig. 1.2. Figure 1.4 is constructed as follows: In the existing temperature-entropy diagram for hydrogen, first the vapor-pressure curve III of hydrogen in equilibrium with Mg_2NiH_4 using the data of Fig. 1.3 is plotted. The horizontal distance to Line IV represents the heat of vaporization of 1 kg H_2 out of Mg_2NiH_4 divided by T [K]. Lines V and VI are constructed in the same way for the $LaNi_5H_5$ system. The numbers in Figs. 1.3 and 1.4 label corresponding states of the systems.

For the following we assume that the hydrogen pressure may not exceed 100 bar, which corresponds to a maximum temperature of 500 °C for the Mg_2NiH_4 system. The heat-transformation process with four temperature levels proceeds as follows (Figs. 1.3 and 1.4):

①–②: Production of hydrogen with 100 bar at 500 °C, heat of vaporization $Q_2 = 15.5$ kcal/mole H_2. ②–③: Isobaric cooling of the hydrogen gas to 175 °C. ③–④: Absorption of hydrogen by $LaNi_5$, $Q_1 = 7.5$ kcal/mole H_2. ④–⑤: Drop of the temperature of the saturated $LaNi_5H_5$ to 45 °C. ⑤–⑥: Production of hydrogen gas at 45 °C and about 3 bar, $Q_1 = 7.5$ kcal/mole H_2. ⑥–⑦: Isobaric heating to 300 °C. ⑦–⑧: Reabsorption of the hydrogen by Mg_2Ni, $Q_2 = 15.5$ kcal/mole H_2. ⑧–①: Rise of the temperature of Mg_2NiH_4 to 500 °C.

In Fig. 1.4 the cycles of the hydrogen-free alloys are also indicated schematically. ⑨–⑩: Drop of the temperature of Mg_2Ni from 500 to 300 °C. ⑪–⑫: Rise of the temperature of $LaNi_5$ from 45 to 175 °C. The arrows indicate the possibility of internal heat exchange.

The usefulness of the heat pump as a topping cycle may be demonstrated by applying it for the following purpose (for simplicity the temperature differences ΔT at heat exchangers are ignored). We assume that 500 °C or higher heat is available. The heat required at 300 °C for the evaporization ①–② in Fig. 1.2 therefore usually flows irreversibly from 500 to 300 °C. With the heat pump of Figs. 1.3 and 1.4 the heat Q_2 added at 500 °C is available at 300 °C to provide the heat of evaporization for the Clausius-Rankine cycle. But, in addition, the amount of heat Q_1 which previously had a temperature of 45 °C is now available at 175 °C. The process data in Figs. 1.1 and 1.2 for the reheat branch ④–⑤ have been chosen such that steam, produced at 175 °C with 9 bar using the extra amount of heat Q_1, can be fed into the Clausius-Rankine reheat branch at point ④. Therefore, the heat Q_1 can be used completely at the temperature level of 175 °C, i.e., without loss of availability. Assuming that by regenerative heat exchange (either internally or by heat exchange with the Clausius-Rankine cycle) the temperature cycling of the absorbers and the hydrogen can be performed reversibly, we can estimate the improvement by comparing the availability w of Q_2 and Q_1 at the four temperature levels 500, 300, 175, and 45 °C:

$$w_{500} = Q_2\left(1 - \frac{318}{773}\right) = Q_2 \cdot 0.58$$

$$w_{300} = Q_2\left(1 - \frac{318}{573}\right) = Q_2 \cdot 0.45$$

$$w_{175} = Q_1\left(1 - \frac{318}{448}\right) = Q_1 \cdot 0.29$$

$$w_{45} = Q_1\left(1 - \frac{318}{318}\right) = Q_1 \cdot 0.00\,.$$

Inserting $Q_2 = 15.5$ kcal/mole, $Q_1 = 7.5$ kcal/mole one finds correctly

$$w_{500} + w_{45} = w_{300} + w_{175} .$$

Since $w_{175} = 0.32 w_{300}$, a 32% increase of the efficiency compared to the normal irreversible heat flow from 500 to 300 °C has been achieved. But the improvement can go even beyond this value for the following reason.

The regenerative heating of the Clausius-Rankine cycle without a topping cycle is only possible for the part ⑧–① in Fig. 1.2. In Figs. 1.3 and 1.4 an alternative process path ⑤–⑥–⑦–⑧ is shown. These process data would be used if heat is required at 375 °C instead of at 300 °C. This alternative demonstrates that such a topping cycle in principle allows transformation of heat from the high temperature to any value between 300 and 500 °C. Therefore, such a topping cycle provides also for the superheating processes ②–③ and ④–⑤ in Figs. 1.1 and 1.2 the heat at the temperature level required. How much of the regenerative heating of the water steam in the superheating step ②–③ and ④–⑤ (Fig. 1.2) is provided by heat from processes ⑦–⑧ (Figs. 1.3 and 1.4) or by heat exchange with hydrogen in the process ②–③ (Figs. 1.3 and 1.4) must be left to optimization considerations and will not be considered here. In any case it can be concluded that by applying this topping cycle the effective Carnot temperature can be raised close to 500 °C. The Carnot efficiency therefore reaches $\approx 60\%$; with a process efficiency of about 0.8 a total efficiency close to 50% seems possible. Again it should be pointed out that the alloys discussed represent by no means the optimal choice for efficiency increase. Alloys with lower equilibrium pressures (e.g., TiV_4, ZrV, ZrNi, etc.), so that higher temperatures T_2 for the upper heat input are possible may allow a further increase of the efficiency. Due to lack of applications, a systematic study of the high-temperature hydrogen-storage alloys has not yet been performed. There is a very good chance that suitable stable alloys can be tailored. The ultimate limitation in temperature and pressure will be given by the embrittlement of the containers and heat exchangers.

An alternative topping cycle is shown in Figs. 1.5 and 1.6. In this case the high pressure hydrogen produced at 500 °C (①–②) is first expanded in a hydrogen turbine (②–③), then isobarically reheated to 300 °C (③–④) and reabsorbed by the alloy (④–⑤). The isentropic expansion may stop at 300 °C and then proceed quasiisothermal to ④. If the hydrogen turbine tolerates higher temperatures, the gas may be superheated at ② and subsequently expanded to ④. In any case, similar to the preceding heat-transformation process, the heat Q_2 originally available at 500 °C is now available at 300 °C; but instead of producing the heat Q_1 at 175 °C, work has been generated in the process ②–③ (Figs. 1.5 and 1.6). Clausius-Rankine processes of this type using a more-component system have already been discussed, for example, by *d'Arsonval* [1.22] for water as absorber and SO_2 as working fluid, by *Koenemann* [1.23] for liquid $ZnCl_2 \cdot NH_3$ as absorber and NH_3 as working fluid, and was even used by *Honigmann* [1.24] in 1881 to run a steam engine between Jülich and Aachen using the combination H_2O/NaOH. Recently such

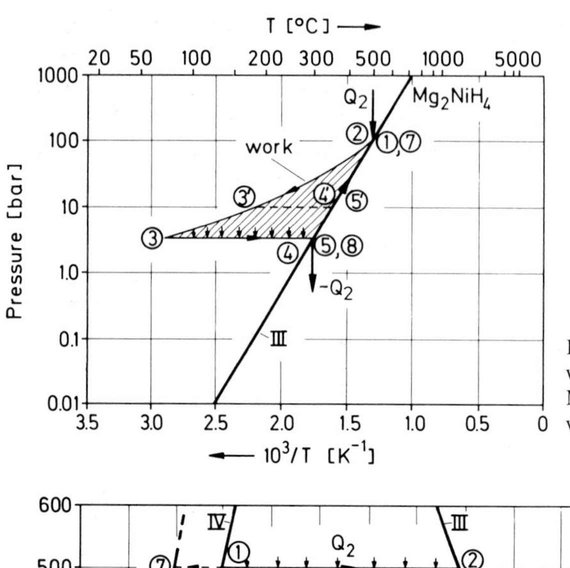

Fig. 1.5. Clausius-Rankine cycle with H_2 as working fluid and Mg_2Ni as absorber in a pressure-versus-$1/T$ plot

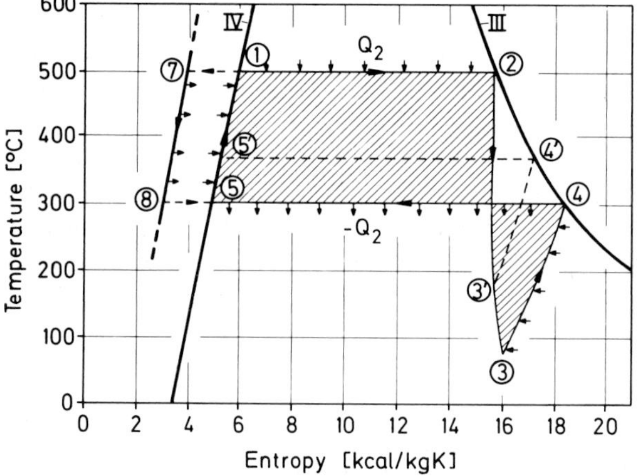

Fig. 1.6. Clausius-Rankine cycle with H_2 as working fluid and Mg_2Ni as absorber in a temperature-versus-entropy plot

a process using metal hydrides was also discussed by *Gruen* et al. [1.14] and *van Mal* [1.9] in combination with home heating and production of electricity using solar energy. The alternative path ③'–④'–⑤' (Figs. 1.5 and 1.6) shows that, depending on how much work is extracted in the expansion ②–③, the heat Q_2 or part of it can be recovered at temperatures between 300 and 500 °C (or also lower than 300 °C if the expansion proceeds further than $p=3$ bar). Therefore, regenerative heating of the subsequent Clausius-Rankine cycle of Fig. 1.2 over the whole temperature range is possible in this case as well. For the work-producing topping process of Figs. 1.5 and 1.6, only one metal as absorber and less heat-exchanger surface is needed compared to the heat-transformation process (Figs. 1.3 and 1.4), whereas the non-work producing topping process (Figs. 1.3 and 1.4) does not require a hydrogen turbine which may be endangered not only by hydrogen embrittlement but also by errosion due to a small amount of metal powder still contained in the hydrogen stream. An evaluation which of the two processes as topping cycle is more economical certainly needs more detailed considerations. The increase of the efficiency of a power-generating plant by a topping cycle would not only reduce the amount of primary fuel required per kWh, but also reduce investment costs, which partly compensate for the extra investment for the heat transformer. An increase of the efficiency by 50%, e.g., from 33 to 50%, reduces the amount of waste heat per kWh and thus the required cooling-tower capacity by a factor of 2. Also the expensive investment for pollution reduction (e.g., coal scrubbers for SO_2 removal) is reduced by a factor of 0.67 (per unit work produced).

The virtue of the metal-hydrogen topping cycles as compared to other two-component topping cycles [1.15] can be stated as follows:

1) The reaction speed is very large.

2) It can be anticipated that suitable alloys with long-time stability will be available. Since heat transfer occurs mostly at changing temperatures, it is not necessary to use hydrides which have a flat plateau for the pressure versus concentration; even homogeneous solutions, like those in niobium or vanadium, can be considered as working media. In contrast to NH_3 [1.23], for example, the working gas H_2 is stable in the temperature regime considered.

3) The vapor pressure of the metals can be ignored compared to that of H_2, which is important not only for technical purposes but also for high efficiency of the heat-pump cycle.

A disadvantage of metal hydrides is the low heat conduction of the powders. Yet for a large-scale operation the fluidized-bed technique seems not unfeasible. Finally, the problem of hydrogen embrittlement is restrictive in respect to material selection for containers and heat exchangers.

References

1.1 T. Graham: Phil. Trans. Roy. Soc. (London) **156**, 399 (1866)

1.2 T. Graham: Proc. Roy. Soc. (London) **16**, 422 (1868) or Phil. Mag. **36**, 63 (1868) or C. R. Acad. Sci. **66**, 1014 (1868) or Ann. Chim. Phys. **14**, 315 (1868)

1.3 e.g. W. Mueller, P. Blackledge, G. Libowitz: *Metal Hydrides* (Academic Press, New York, London 1968)

1.4 For details see: Proceedings: The Hydrogen Economy, Miami Energy Conf. THEME (Univ. of Miami, Coral Gables, Florida 1974)

1.5 H. Ewe, E. W. Justi, K. Stephan: Energy Conversion **13**, 109 (1973)

1.6 R. L. Meijer: Philips Techn. Rev. **31**, 169 (1970)

1.7 e.g. J. M. Burger, P. A. Lewis, R. S. Isler, F. J. Salzano: Proc. 9th Intersoc. Energy Conv. Eng. Conf. ASME (New York 1974) p. 428

1.8 H. H. van Mal: Intern. Meeting on Hydrogen in Metals, Jül-Conf-6 (Vol. II), 739 (1972); Chem. Ing. Tech. **45**, 80 (1973)

1.9 H. H. van Mal: Phil. Res. Repts. Suppl. (1976) No. 1

1.10 J. R. Powell, F. J. Salzano, W.-S. Yu, J. S. Milan: Science **193**, 314 (1976)

1.11 G. Libowitz: Proc. 9th Intersoc. Energy Conv. Eng. Conf. ASME (New York 1974) p. 322

1.12 G. Alefeld: Energie **27**, 180 (1975)

1.13 J. G. Cottingham: Informal Report BNL-19914, Brookhaven Nat. Lab. (1975)

1.14 D. M. Gruen, R. L. McBeth, M. Mendelsohn, J. M. Nixon, F. Schreiner, I. Sheft: Proc. 11th Intersoc. Energy Conv. Eng. Conf. A.I.Ch.E., Vol. 1 (New York 1976) p. 681

1.15 G. Alefeld: Rationelle Energienutzung durch Wärmespeicherung, VDI Bericht **288** (VDI-Verlag, Berlin 1977)
1.16 G. Alefeld, J. Völkl (eds.): *Hydrogen in Metals I: Basic Properties*. Topics in Applied Physics, Vol. 28 (Springer, Berlin, Heidelberg, New York 1978)
1.17 W. Niebergall: *Sorptions-Kältemaschinen*, Handbuch der Kältetechnik, Band 7, Hrsg. R. Plank (Springer, Berlin, Göttingen, Heidelberg 1959)
1.18 K. Nesselmann: *Zur Theorie der Wärmetransformation*, Bd. XII (Wiss. Veröff. Siemens-Konz. 1933) p. 89; see also:
K. Nesselmann: *Die Grundlagen der Angewandten Thermodynamik* (Springer, Berlin, Göttingen, Heidelberg 1950) p. 185
1.19 F. Schneevogl: Verfahren zur Umformung von Wärmegefällen in gleichwertige, jedoch technisch leichter verwertbare Wärmegefälle, Patent DRP 196746 (1905)
1.20 E. Altenkirch, B. Tenckhoff: Absorptionskältemaschine zur kontinuierlichen Erzeugung von Kälte und Wärme oder auch Arbeit, Patent DRP 278076 (1911)
1.21 J. J. Reilly, R. H. Wiswall: Inorg. Chem. **7**, 2254 (1968)
1.22 J. A. d'Arsonval: Revue des cours scientifiques de la France et de l'étranger, Sept. 17, 1881
1.23 E. Koenemann: Ein neues Zweistoffverfahren zur Krafterzeugung. Trans. World Power Conf., Vol. 5 (VDI-Verlag, Berlin 1930)
1.24 See, for example, W. Pauer: *Energiespeicherung* (Theodor Steinkopf Dresden, Leipzig 1928);
A. Riedler: Die Honigmannsche Dampfmaschine mit feuerlosem Natronkessel, Zeitschr. VDI Vol. **XXVII**, 47 (1883);
M. F. Gutermuth: Untersuchungen an Honigmannschen feuerlosen Natronkesseln, Zeitschr. VDI Vol. **XXVIII**, 10 (1884)

2. The Systems NbH(D), TaH(D), VH(D): Structures, Phase Diagrams, Morphologies, Methods of Preparation

T. Schober and H. Wenzl

With 48 Figures

2.1 Background

Vanadium, niobium, and tantalum crystals can dissolve large quantities of hydrogen on interstitial sites. Alloys containing two hydrogen atoms per metal atom have been prepared. Various disordered and ordered arrangements of hydrogen are possible with homogeneous phases existing over an appreciable range of composition between dilute solutions and highly concentrated hydrides. The NbH, TaH, and VH phase diagrams are schematically presented in Fig. 2.1. A very important feature is the high mobility of hydrogen at and below room temperature [*Völkl* and *Alefeld* [2.2] (Ref. [2.3], Chaps. 8, 9, 12)].

Disorder-order phase transitions of the hydrogen interstitials are correlated with structural transitions of the matrix metal. The different unit cells can be generated by relatively small distortions of the bcc cell of the pure metal. Therefore, all phases can be called interstitial solid solutions or interstitial alloys. These binary metal-hydrogen systems behave as quasi-monocomponent systems of hydrogen existing in the form of gaseous, liquid, and solid phases in "interstitial space" (*Alefeld* [2.4]).

Hydrogen interstitials are sources of crystal lattice distortions which induce lattice parameter changes proportional to the hydrogen concentration (Fig. 2.2). Therefore, phase transitions are often correlated with coherency stresses. A dependence of phase separation on the shape of the sample may result [Ref. 2.3, Chap. 2].

The unusual properties of these metal hydrides have stimulated many scientists to investigate the phenomena and the related elementary processes in these systems. In addition, preparation of samples is relatively easy, which facilitates experimental studies.

In materials technology one of the most dangerous effects is the deterioration of mechanical properties due to embrittlement of initially ductile metals by hydrogen. Brittle fracture of Ta containers used in the chemical industry is intimately related to the fast diffusion and ordering of hydrogen interstitials (Chap. 9). Hydrogen embrittlement is also observed in Nb, an important technical superconductor (NbTi [2.8]). Nb is also a possible construction material in fusion reactor technology [2.9]. Hydrides can be applied in energy technology for storing the energy carrier hydrogen (Chap. 5), for storing heat (*Alefeld* [2.10]), for purifying hydrogen by semipermeable membranes (PdAg-cells) or by absorption (*Klatt* et al. [2.11]).

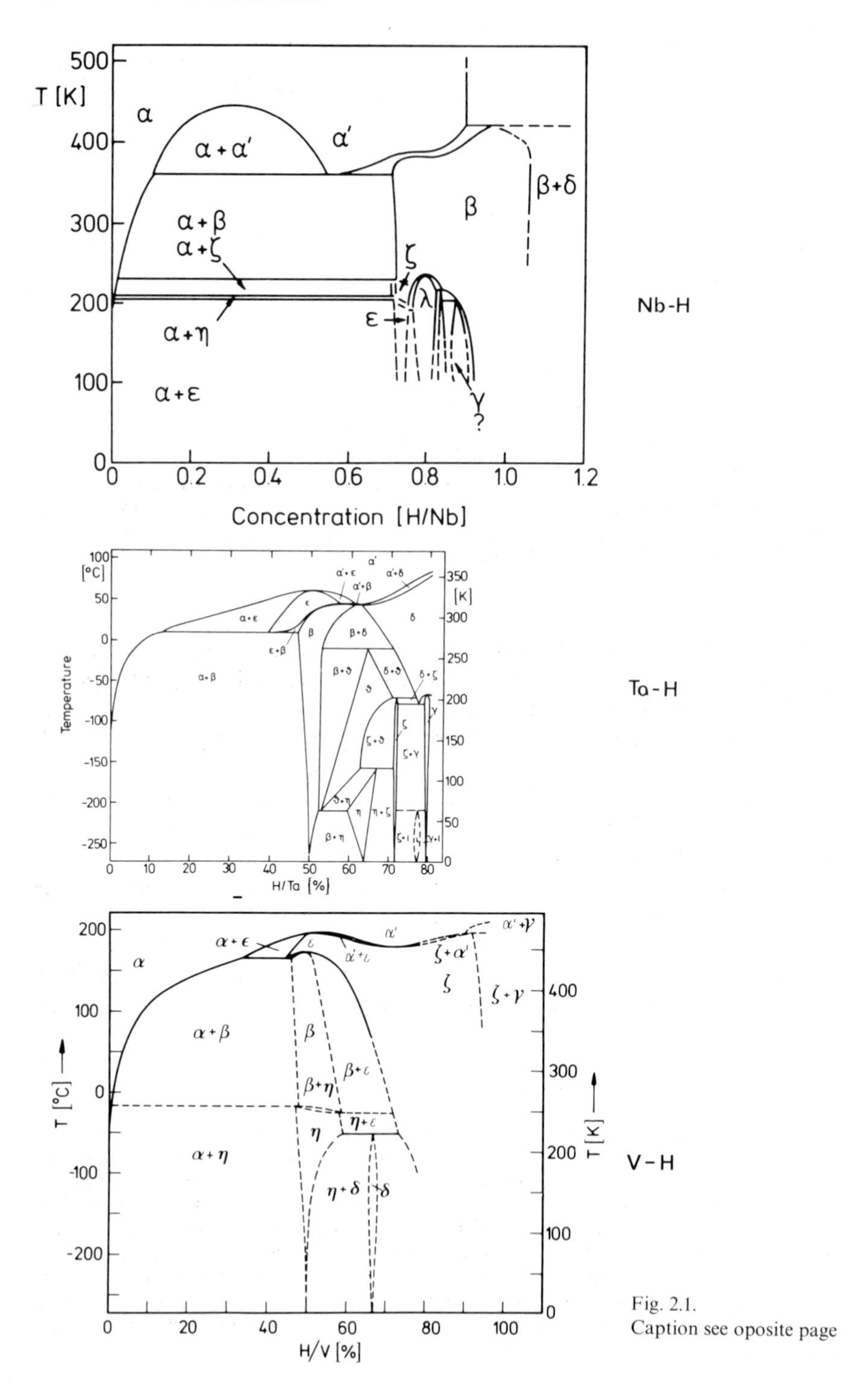

Fig. 2.1.
Caption see oposite page

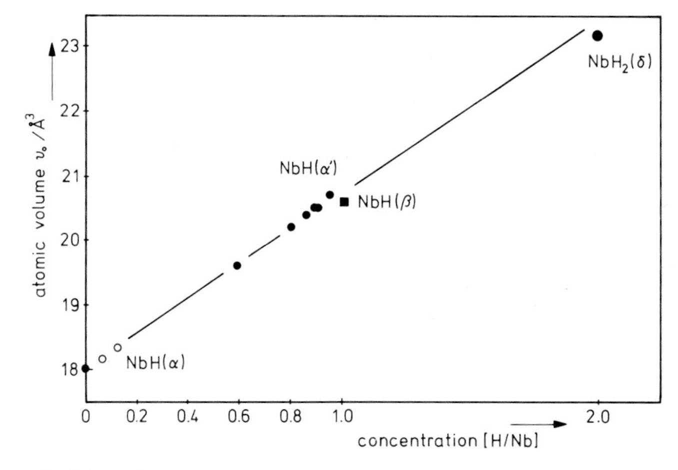

Fig. 2.2. Volume per Nb atom v_0 at about 300 K as a function of hydrogen concentration as determined by x-ray diffraction ($\Delta v_0/v_0 \approx 0.15\,c$). A similar linear relation holds for the V-H system in the range $0 \leqq c \leqq 2$ (H$\Delta v_0/v_0 \approx 0.17\,c$) and for the Ta-H system known in the range $c \leqq 0.8$ (H$\Delta v_0/v_0 \approx 0.16\,c$) [2.5–7], see also [Ref. 2.3, Chap. 3]

Embrittlement and other property changes of V, Nb, and Ta systems by hydrogen are representative of most metals. Therefore, these special systems can be considered as model substances which are useful for a general understanding of the phenomenological and atomistic behavior of metal-hydrogen systems.

The present chapter is organized as follows: In Section 2.2 we describe the various structures. In Section 2.3 the concentration and temperature range of the different structures or phases are presented in the form of binary phase diagrams. Solid hydrogen as a border phase is not discussed. Coherent phase separation, spinodals, critical behavior, etc., are excluded (see [Ref. 2.3, Chap. 2]). Section 2.4 is devoted to the discussion of phase morphology. In Section 2.5 standard methods of sample preparation and characterization are described. More unusual methods like ion implantation are discussed by *Stritzker* and *Wühl* in Chapter 6.

◀ Fig. 2.1. Schematic phase diagrams of the NbH, TaH and VH systems based in part on the work at KFA Jülich. These systems form exothermal solutions of H (or D) from the gas phase at elevated temperatures (>300–450 K) [2.1]. Below these temperatures, the disordered solutions of H (α, α' phase) transform to various ordered interstitial phases which in turn may undergo further phase transitions at even lower temperatures. In Nb and V an ordered phase with composition MH_2 was found. At 0 K, only fully ordered phases are expected. Note the exceptional behavior of the NbH system where the $\alpha+\beta$ two-phase region extends at room temperature to concentrations of $\sim$70%. Also, a miscibility gap between disordered bcc phases α and α' may be found in that system at temperatures above the disorder-order phase boundary lines. The experimental data points are presented in Section 2.3

2.2 Structures

2.2.1 Methods of Structure Determination

X-*ray diffraction*. Despite the small scattering amplitude of hydrogen, it is also a powerful tool in hydride structure investigations. It is well suited for precision lattice parameter measurements of MH_x alloys (M=Nb, Ta, V, etc.). The important information contained in lattice parameter data of disordered alloys as a function of concentration is the trace of the dipole force tensor P_{ij} as discussed in detail in [Ref. 2.3, Chapter 3] (Fig. 2.2). High precision x-ray diffractometers can clearly resolve the narrowly spaced peaks occurring upon ordering in M-H alloys, thus giving information on the deformation of the metal lattices due to ordering (*Pick* [2.5, 6]). X-ray diffraction fails, however, to see the hydrogen sublattice formed in ordered M-H compounds due to the weak scattering by the H-atoms as compared with that of the heavy metal atoms. It is therefore of little value when the structure of the H sublattice or the lattice positions of H-atoms are investigated.

Neutron diffraction has the advantage that there is only a small variation of the coherent neutron scattering amplitude with atomic number Z. Hence, assemblies of light elements such as hydrogen sublattices in ordered M-H compounds may easily be seen by neutron diffraction. With standard diffractometers, no high angular resolution may be obtained. Typical widths of diffraction lines are 1° at low θ angles. Predominantly, powder diffraction has been carried out on hydrides with the advantages of 1) negligible extinction effects and, therefore, 2) the possibility of precisely measuring absolute intensities. The latter point is essential for the unambiguous determination of hydride structures. Single-crystal diffractometry on M hydrides (which, of course, would often be preferable) is virtually impossible since 1) ordered hydrides have a domain structure with many variants and 2) the preparation of large suitable specimens is not an easy task. Due to the much lower incoherent scattering cross section for deuterium as compared with hydrogen, deuterides have been preferentially investigated by n diffraction. We bear in mind, however, that there may be significant differences between the M-H and M-D phase diagrams and structures.

Transmission electron microscopy (TEM) has unique advantages over x-ray and n diffraction. It allows simultaneous imaging and diffraction of a small area, typically 2 μm in diameter, when selected area diffraction is used (SAD). Given a domain size greater than 2 μm (which is often the case), SAD may be carried out on one domain—or single crystal—of the ordered hydride phase. Furthermore, the scattering of electrons by hydrogen is sufficiently large to allow for an observation of superstructure reflections. Thus, single-crystal diffraction patterns of the hydrogen sublattice may be obtained. Also, the three-dimensional reciprocal lattices of hydride structures may be determined by recording successive two-dimensional sections at distinct tilting angles of the goniometer. The technical and fundamental difficulties with this method, however, are manifold. First, no accurate intensity measurements are possible. Second, the preparation of well-defined, homogeneously electropolished samples is difficult. Third, often hydrogen loss and plastic relaxation occur in the thinned areas of the sample. Fourth, dimensional resolution in reciprocal space is relatively poor, of the order of 0.5%.

2.2.2 Niobium Hydrides and Deuterides[1]

Phase α and α'. Phases α and α' are disordered solutions of H in bcc Nb with low and high hydrogen concentrations, respectively [2.5, 6, 12]. The lattice parameters of α and α' were found to increase linearly with concentration and are expressed by (2.1) [2.5, 6] (see also [Ref. 2.3, Chap. 3])

$$\Delta a/a = (4.72 \pm 0.25) \cdot 10^{-4}/\%\ \mathrm{H/Nb}\,. \tag{2.1}$$

A neutron diffraction study (*Somenkov* et al. [2.13]) revealed a very weak noncubic (011) reflection in the α' phase, the origin of which remains unclear. In contrast, similar experiments by *Pick* [2.5, 6] resulted only in the normal bcc reflections. We have no real reason to doubt that phase α' is a disordered bcc solution of H in Nb. *Carstanjen* and *Sizmann* [2.14] found by channeling that deuterium occupies tetrahedral sites.

The β phase is an ordered interstitial solid solution of hydrogen. Early x-ray work [2.12, 15] demonstrated that its structure was fc orthorhombic. Two precision x-ray investigations [2.5, 6, 16] confirmed that result and showed the linear dependence of the fco lattice parameters *a*, *b*, *c* with concentration H/Nb. We now confine our attention to the H sublattice structure which was found and analyzed by *Somenkov* et al. [2.13, 17, 18]. In that n diffraction work, additional superlattice reflections from hydrogen ordering were observed and their intensities were measured quantitatively. It was concluded that the H-atoms occupy tetrahedral positions at well-defined spacings. A model of the NbH β phase is shown in Fig. 2.3a. The unit cell in grey may be described by the orthorhombic base vectors $\boldsymbol{a}$, $\boldsymbol{b}$, and $\boldsymbol{c}$, where $a \approx b \approx a_0 \cdot \sqrt{2}$ and $c \approx a_0$. The hydrogen sublattice may be also described by translations T_1 and T_2 (*Hauck* [2.19, 20]). A confirmation of the reciprocal lattice found by *Somenkov* et al. [2.13] was obtained by *Schober* et al. [2.21] through selected area electron diffraction (=SAD). The latter work was performed on a single domain of the β phase and, therefore, represents a true single-crystal measurement.

The ζ phase was found by TEM upon cooling of β phase areas to between -45 and $-65\,^\circ$C (*Schober* [2.22–24]). Its reciprocal lattice was identical to the one previously observed for β-Ta_2H (see below). Thus it may be concluded that ζ-NbH and β-Ta_2H are isomorphic. An illustration of the structure is given in Fig. 2.3b. The H sublattice may be described by translation T_1 alone [2.19, 20].

[1] Hydrides and deuterides are treated together here since no apparent structural differences have been reported.

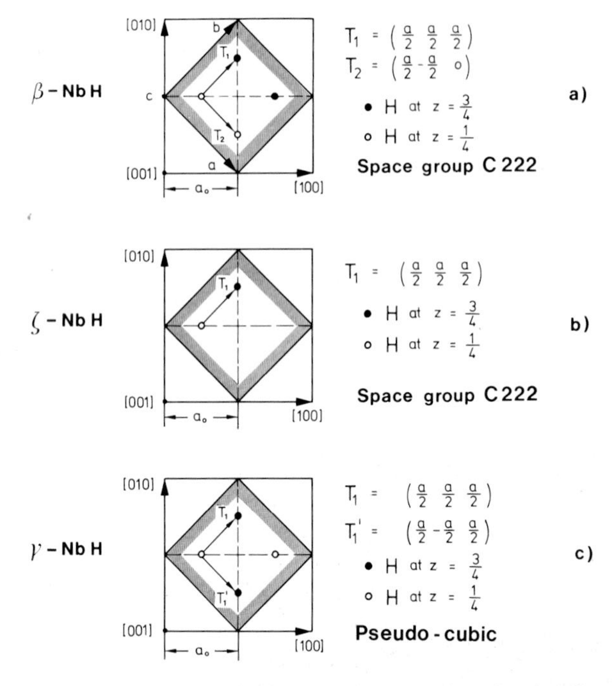

Fig. 2.3a–c. The structure of NbH phases. For the sake of clarity, the metal atoms are not shown. (a) The orthorhombic β phase. (b) The orthorhombic ζ phase. Strictly speaking we expect also a filling of the right-hand sites in the unit cell. The vacancy concentration in the two hydrogen planes, however, must be different. (c) Proposed model for the structure of the γ-phase [2.19, 20]

ζ arises from β by ordering of the "H vacancies". This gives rise to an additional periodicity along $[1\bar{1}0]_c$ (c = cubic) which is twice the distance occurring in the bcc structure. Hence, half-order superlattice reflections are observed along $[1\bar{1}0]_c$ on (001) reciprocal lattice planes. These additional reflections on (001) are the distinctive features of the ζ phase with respect to β-NbH. Recent independent TEM work by *Pesch* [2.25] confirmed the existence of phase ζ.

The ε phase structure was first found by *Somenkov* et al. [2.13, 17, 18] and has the composition Nb_4H_3. It is described by a cell with parameters $a \approx b \approx 2a_0 \cdot \sqrt{2}$, $c \approx a_0$ and is completely ordered. An illustration of the structure is seen in Fig. 2.4. Comparison with Fig. 2.3 shows that the ε phase unit cell is obtained by appropriately mixing and joining elements of the β and ζ unit cells. The same phase ε was found by TEM using SAD techniques [2.23, 24]. This conclusion was reached since the reciprocal lattice obtained by TEM seemed to be identical with the one found by n diffraction. A (001) section of the reciprocal lattice as obtained by electron diffraction is shown in Fig. 2.5 [2.24].

The γ phase, first described by *Pick* [2.5, 6], is a pseudocubic, high-concentration, low-temperature phase of the approximate composition

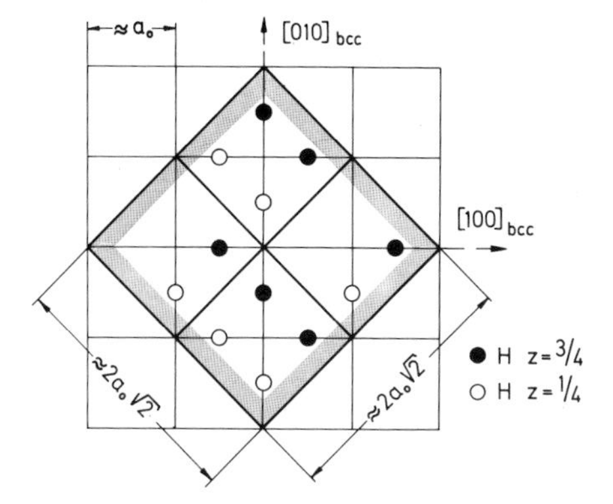

Fig. 2.4. The orthorhombic unit cell of the ε phase of H in Nb (*Somenkov* et al. [2.13, 17, 18]). $a \approx b \approx 2a_0 \cdot \sqrt{2}$, $c \approx a_0$

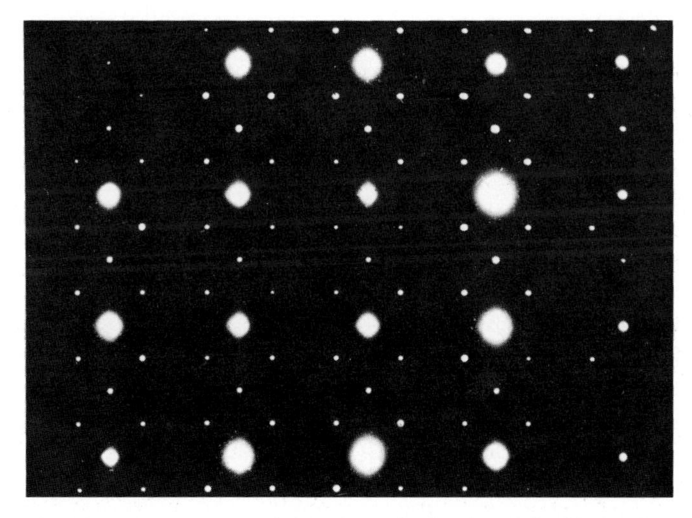

Fig. 2.5. (001) SAD pattern of the ε phase of H in Nb (*Schober* [2.23, 24]). The strong spots lying on a square grid are the Nb matrix reflections. The weaker spots are the H superstructure reflections

$NbH_{0.9}$. It is stable below 200 K and is visible in x-ray diffraction through the disappearance of the orthorhombic reflections of the β phase. The structure as proposed by *Hauck* [2.19, 20] is presented in Fig. 2.3c.

The λ phase was recently found in DTA work by *Welter* [2.26]. It has an approximate concentration of 80%. No structural information is available on phase λ. Recent metallographic work on the $\beta \rightarrow \lambda$ transition indicated that λ is an ordered phase with a domain structure [2.27]. It has been speculated that there exist still further ordered low-temperature phases with larger unit cells than that of ε [2.18]. Although it seems an attractive possibility, we wish to point out the great experimental difficulties in finding and analyzing such structures.

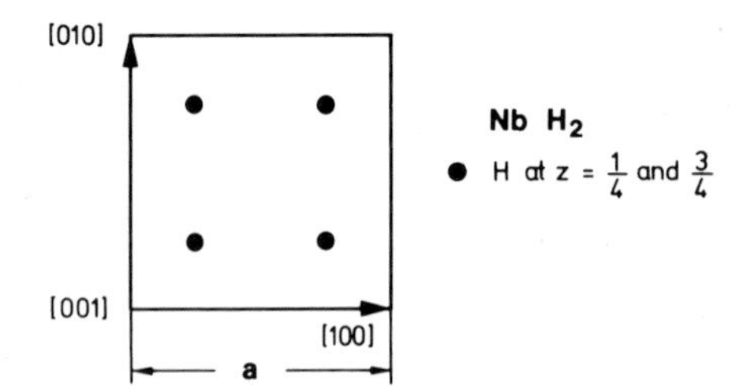

Fig. 2.6. The δ phase (NbH_2). The structure is of the CaF_2 type. The metal atoms form a fcc lattice and are omitted in the drawing

The δ phase has been known for many years. It was shown to have a fcc structure corresponding to composition NbH_2 [2.28, 29]. It has a grey, metallic appearance and is very brittle. Values of its metal atom lattice parameters vary from 0.4536 nm [2.29] to 0.4556 ± 0.0002 nm [2.23]. The crystal structure presumably is of the CaF_2 type where the H-atoms occupy tetrahedral sites. An illustration is provided in Fig. 2.6. We finally note that further incorporation of hydrogen could occur on octahedral sites of the metal lattice, which would result in compositions above H/Nb=2. Such structures have not been observed, i.e., there is no evidence for a phase NbH_3. Such trihydride phases are indeed encountered in the light rare earth-hydrogen systems.

2.2.3 Tantalum Hydrides and Deuterides[2]

The α phase is a disordered solution of H in bcc Ta [2.30, 31] even up to very high concentrations. The linear dependence of the lattice parameter on concentration c=H/Ta is given by (2.2) [2.32]

$$a = 0.3306 + 0.015 \cdot c \ \text{(nm)}. \tag{2.2}$$

This relation is valid at room temperature up to $c \approx 0.25$. No measurements have been reported for higher concentrations at elevated temperatures. The tetrahedral site occupancy was demonstrated by *Antonini* and *Carstanjen* [2.33].

Phase β designates the ordered low-concentration phase which in previous work was assigned the composition Ta_2H. The TaH phase diagram (see next section) gives the range of composition of β versus temperature. Numerous x-ray diffraction studies gave conflicting results as to the metal atom structure of β. We consider the work by *Ducastelle* et al. [2.32], who found an orthorhombic metal lattice structure to be very reliable. Similar results were obtained by *Stalinski* [2.34].

Focusing our attention on the H sublattice, we now consider the neutron work of *Somenkov* et al. [2.30, 31]. The structure found in that work is shown in Fig. 2.7. Hydrogen occupies (as in Nb) tetrahedral sites within the ortho-

[2] Again, no structural differences were reported between Ta hydrides and deuterides.

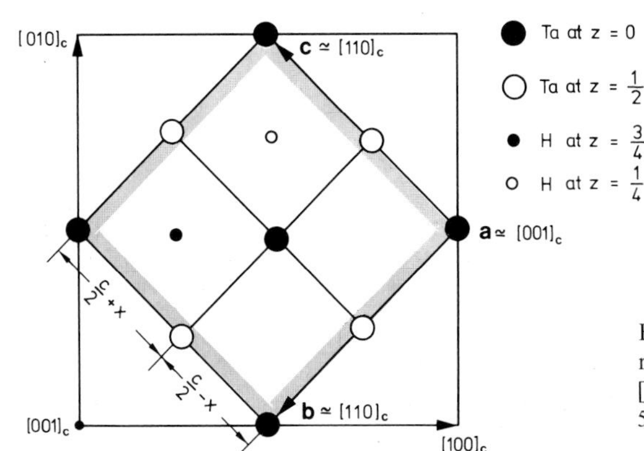

Fig. 2.7. The structure of the orthorhombic TaH β phase (*Somenkov* [2.30, 31]) at the concentration of 50%

rhombic unit cell having base vectors ***a***, ***b***, and ***c***. Only alternate $(\bar{1}10)$ planes are occupied with H-atoms. The spacing of these planes increases by x, whereas the adjacent planes are "compressed" by the amount x. In the above study, the absolute value of the displacement of nearest $(\bar{1}10)$ planes $x = 0.017 \cdot a_0 \cdot \sqrt{2}$, whereas a recent theoretical calculation yielded a value of $0.02 \cdot a_0 \cdot \sqrt{2}$ (*Khachaturyan* and *Shalatov* [2.35]). The above considerations strictly apply only to alloys of stoichiometry Ta_2H. Deviations from stoichiometry may be either achieved by adding H-atoms in appropriate tetrahedral positions in the adjacent $(\bar{1}10)$ planes, or by removing H-atoms from the filled $(\bar{1}10)$ planes (*Asano* et al. [2.36]). Additional firm evidence for the reciprocal lattice model of the Ta_2H structure as proposed by *Somenkov* et al. [2.30, 31] came from a recent TEM study of *Wanagel* et al. [2.37]. We again note the lack of intensity measurements in electron diffraction work which makes the distinction between different physical structures having identical reciprocal lattices impossible. In a quite different TEM study, the present authors investigated the relative lengths of vectors ***a***, ***b***, and ***c*** in Fig. 2.7 [2.38]. The work was carried out by precisely measuring length ratios in the reciprocal lattice of the Ta_2H structure. Thus, the overall deformation of the Ta_2H unit cell with known relative orientation of the H sublattice was obtained. Given the geometry in Fig. 2.7, the following axial ratios were determined [2.38]:

$$c/b = 1.011 \quad c/a = 1.4015 \quad \text{hence} \quad b/a = 1.387\,. \tag{2.3}$$

It follows that the lattice expands more along $[\bar{1}10]_c$ than along $[\bar{1}\bar{1}0]_c$, which may be a consequence of the repulsion between adjacent hydrogen-rich $(\bar{1}10)$ planes. Similarly, the 1.9% increase of ***a*** with respect to ***b*** is attributable to the elastic softness of Ta along $\langle 100 \rangle_c$. We wish to emphasize that such a study of length ratios of the hydride unit cell base vectors is impossible in standard neutron diffraction for lack of resolution and because of occurrence of the domain structure. Likewise, x-ray diffraction fails to provide this information

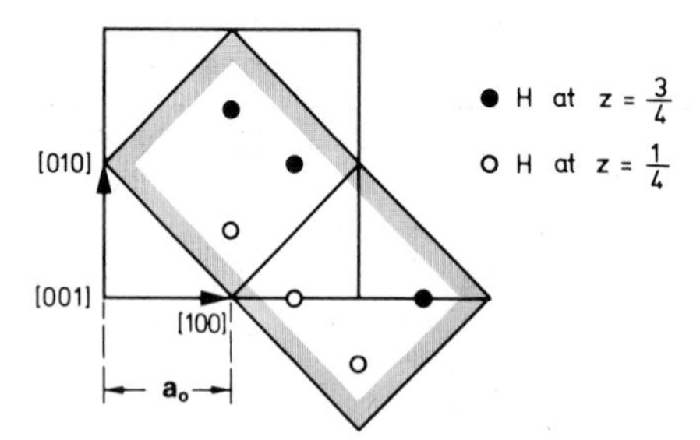

Fig. 2.8. The unit cell of the fully ordered orthorhombic Ta_4D_3 phase (*Somenkov* [2.18, 40]). The metal atoms are not shown

due to the invisibility of the hydrogen sublattice and the domain structure, although it is superior to TEM techniques with regard to resolution.

Asano [2.39] succeeded recently in obtaining single-crystal x-ray diffraction patterns of the β phase using the precession method. These patterns displayed superlattice reflections due to the systematic displacements of the Ta atoms in the ordered β phase. The observed distortions were in full agreement with the results in our study [2.38].

The γ phase. Following *Asano* et al. [2.36], we refer to the low-temperature phase observed in that work as γ. The existence of this low-temperature phase is considered firmly established in the light of our own DTA measurements [2.105]. However, in the latter work, the γ phase is centered around the concentration of 80%.

The ζ phase. Somenkov et al. [2.18, 40] described a phase of the assumed composition Ta_4H_3. The unit cell has base vectors $a \simeq a_0 \cdot \sqrt{2}$, $b \simeq 2\sqrt{2} \cdot a_0$, $c \simeq a_0$ and belongs to space group I 222 [2.18]. The phase is shown in Fig. 2.8 and referred to as ζ. Phase ζ was found below $-23\,^\circ\mathrm{C}$ in the Russian work [2.18] and below approximately $-20\,^\circ\mathrm{C}$ in our DTA work [2.105].

Some evidence for Somenkov's reciprocal lattice of the ζ-Ta_4D_3 structure [2.40] was recently obtained in TEM work by *Schober* [2.27]. Also, additional evidence for at least two ordered low-temperature phases was seen in that work. Furthermore, a pseudocubic phase was detected at low temperatures similar to the γ-NbH phase [2.27].

Phase δ, as it was called by *Asano* et al. [2.36], is the high-concentration phase TaH_{1-x} which has a structure identical with β-NbH [2.30, 40] (see Fig. 2.3a). Formally, the δ phase was formed by filling up the allowed tetrahedral sites of the β phase unit cell with the same probability. [We note that if it were not the same probability, the extra reflections typical of the β phase would appear on (001).]

The ε phase. Finally, *Asano* et al. [2.36] proposed the existence of still another low-concentration, high-temperature phase which was labelled β_2 in that work. This new phase, referred to as ε here, was also found in a recent metallographic and DTA study at KFA Jülich [2.105]. The structure of ε still has to be clarified, although there is some metallographic evidence that ε is a tetragonal phase.

Table 2.1. Summary of the V-H and V-D phases

	Vanadium hydrides	Relevant references		Vanadium deuterides	Relevant references
α_H	Disordered bcc solution α'_H = high-concentration	[2.41, 42]	α_D	Disordered bcc solution 90% tetrahedral occupancy, α'_D = high-concentration phase	[2.44, 52, 53]
β_H	V_2H; monoclinic, octahedral O_z occupancy, $c_0/a_0 = 1.1$	[2.42–48]	β_D	V_2D, monoclinic, octahedral O_z occupancy, $c_0/a_0 \simeq 1.1$ isomorphic with β_H	[2.41, 44, 52, 54]
γ_H	VH_2, fcc dihydride phase, $a_0 \simeq 0.424$ nm	[2.29, 41]	γ_D	VD_{1-x}, low-temperature fully ordered phase orthorhombic unit cell twice as large as in δ_D. Mainly tetrahedral occupancy	[2.52, 54]
δ_H	V_3H_2, low-temperature ordered phase. Monoclinic structure, presumably octahedral occupancy	[2.49–51]	δ_D	VD_{1-x}, low-temperature partially ordered phase, mainly tetrahedral occupancy, orthorhombic structure as in β-NbH	[2.52–54]
ε_H	V_2H, exists between 175 and 197 °C; probable structure: tetrahedral, all octahedral O_z sites are filled with probability 1/2. No similar deuteride observed. The hydride VH_{1-x} (x = 0.1–0.2) is referred to here as ζ_H	[2.42, 43]	ε_D	VD_2, a fcc dideuteride phase	[2.55]
η_H	V_2H, low-temperature (speculative!) modification of β_H	[2.194]	ζ_D	New low-temperature modification. Presumably orthorhombic, $c \simeq 75\%$	[2.195]

2.2.4 Vanadium Hydrides and Deuterides

A distinction is made here between hydride and deuteride phases and phase diagrams. The situation is even more complicated due to the simultaneous occurrence of tetrahedral and octahedral sites for the H- or D-atoms. Thus, the V-H and V-D systems and structures are distinctly different from their Nb-H(D) or Ta-H(D) counterparts. A summary of the pertinent V-H and V-D phases is given in Table 2.1.

The V-H System

Maeland [2.41] determined the linear variation of the lattice parameter of the disordered bcc *phase* α_H to be

$$\Delta a/a = (5.73 \pm 0.3) \cdot 10^{-4}/\text{at.\% H/V} . \qquad (2.4)$$

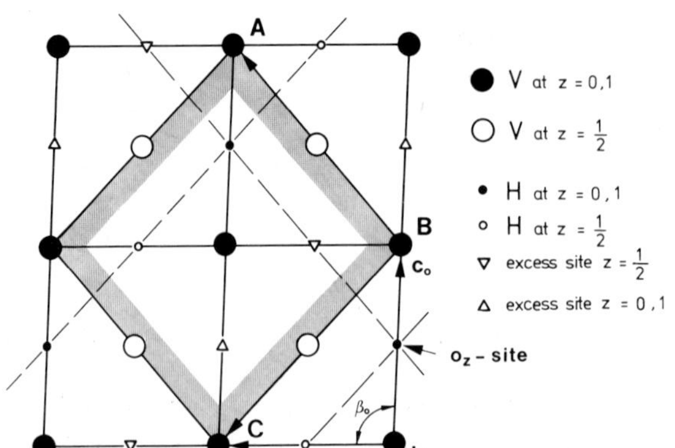

Fig. 2.9. The monoclinic unit cell of the V_2D-(β_D) phase. Deuterium atoms occupy octahedral sites

Phase β_H is an ordered interstitial solution of hydrogen in vanadium. Because of its isomorphism [2.43, 45] with the deuterium phase β_D we shall discuss its monoclinic structure in detail in Section 2.3.6.

The high-temperature phase ε_H. Recently *Asano* et al. [2.42] and *Fukai* and *Kazama* [2.56] used x-ray, NMR resistivity, and calorimetric measurements to obtain evidence for an ordered high-temperature phase. Neutron diffraction on the ε_H phase in the temperature range from 175 to 220 °C showed that the superlattice reflections of β_H had disappeared (*Asano* et al. [2.43]). It was concluded that ε_H is partially ordered and that the hydrogen is distributed over all O_z sites (full circles and triangles in Fig. 2.9) with the probability 1/2. [An extension of this model to higher H concentrations leads to the superstructure ζ_H(VH) [2.43] probably identical in symmetry with ε_H. The metal atom structure of ζ_H should be bct with H at the O_z sites $(0, 0, \frac{1}{2})$ and $(\frac{1}{2}, \frac{1}{2}, 0)$.]

The low-temperature phase δ_H. Evidence for a low-temperature phase transition was obtained by the investigators using resistivity, x-ray, and electron diffraction measurements [2.49, 50, 57]. *Asano* and *Hirabayashi* [2.51] made a detailed calorimetric and structural study of δ_H and concluded that it has the composition V_3H_2. In their work, the $\delta_H \rightarrow \beta_H$ phase changes were observed as peaks in the specific heat curves [2.51]. Using the electron diffraction results of *Wanagel* et al. [2.49], the same authors proposed that δ_H is monoclinic as shown in Fig. 2.10. [The unit vectors are $\boldsymbol{A} = (a_0^2 + b_0^2)^{1/2}$, $\boldsymbol{B} = a_0$, $\boldsymbol{c} = (4a_0^2 + c_0^2)^{1/2}$ with $\beta = 77°$; $\boldsymbol{a}_0$ and $\boldsymbol{c}_0$ are the unit vectors of the pseudo-bct metal atom structure.] Regarding the question of octahedral or tetrahedral occupancy, *Asano* and *Hirabayashi* [2.51] concluded on the basis of the strong pseudo-tetragonality of V_2H at room temperature that hydrogen also occupies octahedral sites in the δ_H phase.

The η_H *phase*. This phase was tentatively introduced in recent DTA-work [2.194] on the basis of weak but reproducible thermal effects in the vicinity of -20 °C. It is too early to decide whether η_H can be established as a new VH-phase. There is the possibility that the above authors have observed a

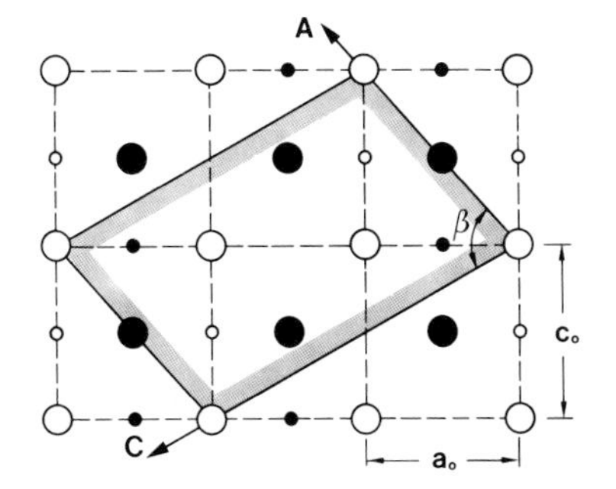

V: ○ y = 0 ● y = ½

H: ∘ y = 0 • y = ½

Fig. 2.10. Monoclinic structure of δ_H-VH phase (*Asano* et al. [2.51]). The composition is V_3H_2

parasitic phase transition in the V-O-H system. In this case phase η_H and the corresponding lines in the VH-phase diagram in Fig. 2.25a could be omitted.

The dihydride phase $\gamma_H(VH_2)$. The phase VH_2 was found to have an fcc structure ($a_0 \approx 0.424$ nm) [2.29, 41]. It arises due to the inability of the ζ_H phase to incorporate hydrogen beyond the concentration of 100%. The metal atom structure must transform to fcc to provide equivalent tetrahedral sites. Presumably, VH_2 is isomorphic with NbH_2 and has a CaF_2 structure.

The V-D System

Concerning the disordered bcc phase α_D, it follows from the work of *Asano* and *Hirabayashi* [2.52] that

$$\Delta a/a = (5.65 \pm 0.3) \cdot 10^{-4}/\text{at.\% D/V} . \tag{2.5}$$

Neutron diffraction work by *Somenkov* et al. [2.44] on α_D with $c = 50\%$ gave a lattice parameter of 0.313 nm. Also, the surprising result was obtained that only 90% of the interstitial atoms occupy tetrahedral; the rest occupy octahedral sites. Similar conclusions were reached by *Chervyakov* et al. [2.53] working with a 80% specimen, having an x-ray lattice parameter of 0.3157 ± 0.0002 nm. Here, an octahedral occupancy of 7% was found for deuterium.

The β_D *(or* V_2D*) phase* has been extensively investigated [2.44, 52, 54]. The crystal structure is monoclinic[3] as shown in Fig. 2.9. The deuteride unit cell was described by vectors $\boldsymbol{A} = (a_0^2 + c_0^2)^{1/2}$, $\boldsymbol{B} = b_0$, and $\boldsymbol{C} = (a_0^2 + c_0^2)^{1/2}$. At the composition V_2D, two deuterium atoms (D's) in the shaded unit cell occupy octahedral sites at $z = 0$ and $z = 1/2$. Refining the model somewhat, *Somenkov* et al. [2.44] concluded that 95% of the D's occupy octahedral sites, while the rest occupy tetrahedral sites thus producing a model of mixed octa-tetrahedral structure. The large c_0/a_0 ratio of approximately 1.1 in the β_D phase is surprising [2.41, 52]. The pseudo-tetragonality is thus much higher than in comparable Ta and Nb hydrides. *Asano* et al. [2.42] addressed themselves to the problem of deviation from the stoichiometry V_2D. In the hypostoichiometric case, they proposed a statistical occupancy on O_z sites (full circles in Fig. 2.9). In contrast, hyperstoichiometry is achieved by filling randomly the excess sites marked by triangles in Fig. 2.9.

[3] To a first approximation, the structure is tetragonal [2.41].

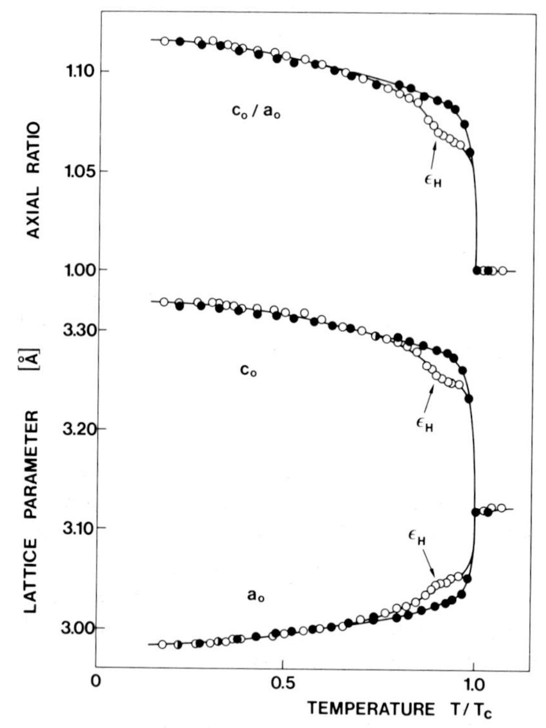

Fig. 2.11. Temperature dependence of the lattice parameters of vanadium hydrides and deuterides β_H(○) and β_D (●) (*Asano* et al. [2.42])

Westlake et al. obtained the following lattice parameter values for β_D [2.54]:

$$A \approx C = 0.446\ \text{nm}, \qquad B = 0.300\ \text{nm} \qquad \beta = 95.5° .$$

The β_D phase is stable to 425 K only, whereas 473 K was reported for β_H [2.41][4]. The 95 to 5% octa-tetrahedral occupancy by D was not disputed by *Westlake* et al. [2.54], as it provided a better explanation of their results. From the quadrupole splitting of the deuterium and the vanadium resonance lines in single crystal vanadium foils, *Arons* et al. [2.45] concluded that octahedral sites are occupied in the β_D phase, whereas the tetrahedral sites are dominant in the δ_D phase.

Regarding the isomorphism of β_H and β_D, *Wanagel* et al. [2.52] presented evidence for a monoclinic structure in their TEM study of β_H. Their reciprocal lattice was also found to be identical with the one of β_D derived by *Somenkov* et al. [2.44]. The absence of intensity measurements in TEM did not allow for a distinction between tetrahedral or octahedral sites. In either case one would obtain the same reciprocal lattice. *Cambini* et al. [2.47] and *Takano* et al. [2.48] obtained additional evidence for a monoclinic structure of β_H using TEM. In

[4] More exact transition temperatures are 406 K for β_D and 470 K for β_H [2.194, 195]

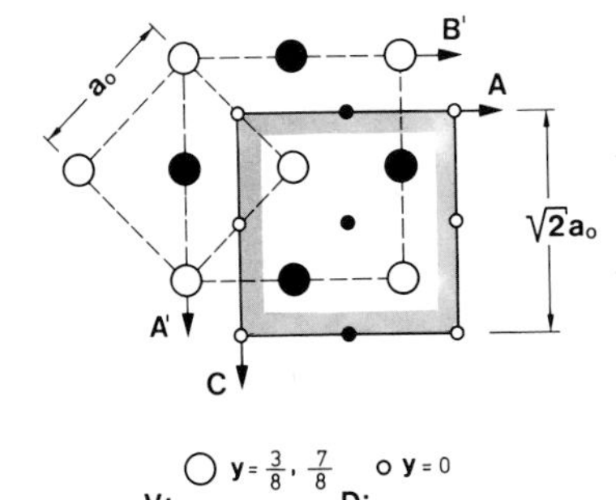

Fig. 2.12. The γ_D structure [2.52]. Here, $A=C=\sqrt{2}\cdot a_0$ and $B=2a_0$ ($a_0=0.3143$ nm). The dashed lines depict unit cells of the metal lattice and phase δ_D

order to investigate possible structural differences between β_H and β_D, *Asano* et al. [2.42] measured the lattice parameters a_0 and c_0 and therefore c_0/a_0 of both phases as a function of normalized temperature T/T_c. Here, T_c is the first-order transition temperature where $\beta \rightarrow \alpha$. On the basis of identical curves (see Fig. 2.11) in the important temperature range $0.2 \leqq T/T_c \leqq 0.75$, it was tentatively concluded that β_H and β_D are isomorphic. More substantial evidence for this isomorphism was recently presented in the neutron diffraction work on V_2H by *Asano* et al. [2.43].

The low-temperature deuteride phases γ_D *and* δ_D. As discussed previously, V-D alloys with D/V$=0.75$ are in the α phase at room temperature and display predominantly tetrahedral occupancy. If such alloys are cooled below approximately 205 K, the superstructure reflections of a new phase, the δ_D *phase*, appear in neutron diffraction (*Asano* and *Hirabayashi* [2.52]). Consistent with previous Russian work [2.53], an orthorhombic cell with $\boldsymbol{A}=\boldsymbol{B}=\sqrt{2}\cdot a_0$ and $\boldsymbol{C}=a_0$ ($a_0=0.3156$ nm) and space group P_{nnn} was found. The D's occupy regularly the tetrahedral sites as previously illustrated for the case of β-NbH (see Fig. 2.3a). A minority of D's was thought to occupy other tetrahedral sites [2.52], whereas *Chervyakov* et al. [2.53] postulated that 10% were in octahedral sites. At any rate, δ_D is very similar in structure to the ordered phases $TaD_{0.8}$ and $NbD_{0.8}$. The α_D-δ_D transition was also found in the neutron and x-ray experiments by *Westlake* et al. [2.54], who left the octahedral/tetrahedral question unresolved.

Further cooling of δ_D below approximately 150 K finally resulted in the appearance of *phase* γ_D [2.52, 54]. γ_D is produced through further ordering of the vacancies within the D sublattice and again has a composition near V_4D_3. An illustration of γ_D is provided in Fig. 2.12. It is seen that γ_D is obtained from δ_D by doubling the unit cell along the c axis. Here $A=C=\sqrt{2}\cdot a_0$ and $B=2a_0$ ($a_0=0.3143$ nm) with space group P_{cc2} [2.52].

Westlake et al. [2.54] described the same phase transformation $\delta_D \rightarrow \gamma_D$. Their crystal structure of γ_D seems to be at variance, however, with the above

Table 2.2. Character of phase transitions between matrix structures. 1 indicates transition which must be first order, 2 transition which can be second or first order

	bcc	fco	tetr	fcc
bcc	×	2	2	1
fco	2	×	2	2
tetr	2	2	×	2
fcc	1	2	2	×

structure of *Asano* and *Hirabayashi* [2.52]. Thus, further work is necessary to substantiate the proposed model of the latter authors.

We finally note that a concentration of 100% may be achieved for V-D alloys, i.e., the compound VD, which is in the δ_D phase below a certain temperature. Obviously, this δ_D structure is markedly different from the corresponding stoichiometric structure VH. Also, the order-disorder temperatures are very far apart for these analogous hydrides and deuterides. The disparities may be attributed to the differences in vibrational entropy [2.58] or to tunneling effects [2.59].

The phase ζ_D was introduced in recent DTA-work [2.195]. Presumably, it is also orthorhombic and exists between 75 and 83%; the corresponding temperature range is -62 to -39 °C.

The dideuteride phase VD_2 *(or* ε_D*)* presumably has the same structure as VH_2. *Hardcastle* and *Gibb* [2.55] found a lattice parameter of 0.427 nm. *Mair* [2.196] reported a value of 0.42555 nm.

2.3 Phase Diagrams

In this section we present phase diagrams using temperature T and hydrogen concentration c as variables to describe boundary lines $T(c)$ between the various phases discussed in Section 2.2.

Experimentally, these phase boundaries were determined by measuring transitions between the structures and changes of properties during the phase transitions. Due to the "self-sealing" of samples below 300 °C (see Sect. 2.5) most experiments were performed by changing the temperature at constant average concentration of hydrogen. According to the high mobility of hydrogen also at low temperatures (*Völkl* and *Alefeld* [2.2]; Chap. 12, Ref. 2.3) thermodynamic equilibrium of and between phases inside the samples can be reached in a reasonable time even at the temperature of liquid nitrogen. Transitions between phases of different matrix structure can occur continuously or discontinuously according to the "Landau rules" (*Boccara* [2.60], *Somenkov* [2.18]). Table 2.2 summarizes the results of these rules for transitions between the structures discussed in Section 2.2.

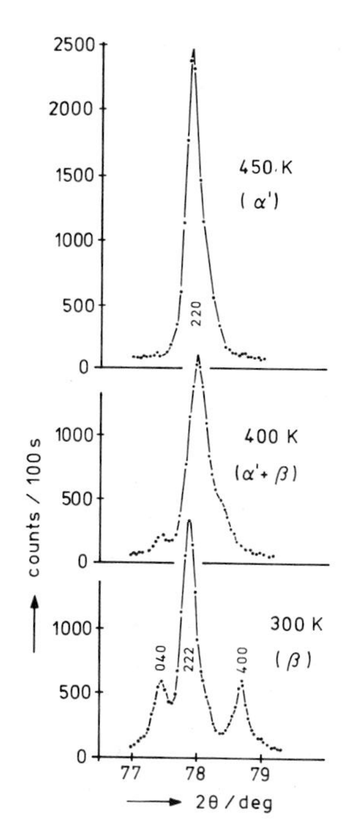

Fig. 2.13. X-ray rocking curve on powder sample of $NbH_{0.9}$ at three different temperatures. The (220) Bragg peak of the bcc α' phase (450 K) splits into three peaks of the fco β phase on cooling

2.3.1 Examples of Phase Transitions

a) Structural phase transitions of the matrix can be revealed directly by Bragg diffraction of x-rays. Figure 2.13 shows an example of x-ray powder diffraction of NbH at different temperatures (*Pick* [2.6]). This change of structure is correlated with a disorder-order phase transition of hydrogen. The spacing of the hydrogen superstructure planes and the temperature dependence of the degree of ordering can be determined by measuring the intensity of neutron or electron diffraction. Figure 2.14a shows an example for thermal neutron diffraction (*Alefeld* et al. [2.61]), Fig. 2.5 for electron diffraction. In Fig. 2.14a the high-temperature superstructure neutron intensity drops to zero in a narrow temperature range where the order-disorder transition between β and α' phase occurs. At low temperatures a phase transition between two different superstructures is indicated by the sudden change of the intensity. The temperature dependence of the superstructure line in Fig. 2.14a within the homogeneous β phase indicates the onset of disordering due to generation of thermal defects in the ordered hydrogen arrangement. These defects probably consist of "hydrogen Frenkel defects" which are hydrogen interstitials occupying forbidden tetrahedral sites, thus leaving allowed sites vacant.

b) The analysis of thermal effects during phase transitions has been used extensively by the authors and elsewhere to determine phase boundary lines in the phase diagram. Phase transitions are clearly visible as sharp DTA peaks

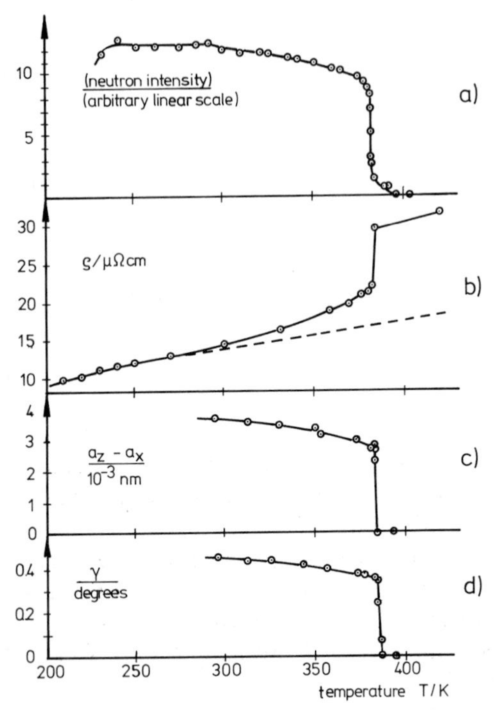

Fig. 2.14. (a) Neutron intensity of superstructure Bragg peak $(\frac{1}{2}\ \frac{3}{2}\ 1)_c$ (see Fig. 2.3) on $NbH_{0.83}$ as a function of temperature. (b) Electrical resistivity $\varrho(T)$ on $NbH_{0.83}$ as a function of temperature. (c) The "tetragonal distortion" $a_z - a_x$ of the bcc lattice was determined from the splitting of x-ray diffraction peaks as indicated in Fig. 2.13. The difference in the cube dimensions a_z in direction [001] and a_x in direction [100] is the result of ordering of hydrogen on half of the z sites in the β phase (see Fig. 2.3). The "tetragonality" $(a_z - a_x)$ of the β phase structure decreases with increasing thermal disorder by Frenkel defect formation. At the first-order phase transition $\beta - \alpha'$ the tetragonality drops to zero. The experiments were performed by *Pick* [2.5, 6]. (d) The angle γ characterizes the distortion of the original bcc cell along [110] due to the ordered "chains" of H along [110] (see Fig. 2.3). The effect of thermal disorder increasing with temperature can also be observed in the temperature dependence of γ. At the $\beta - \alpha'$ transition γ drops to zero. The experiments were performed by *Pick* [2.5, 6]

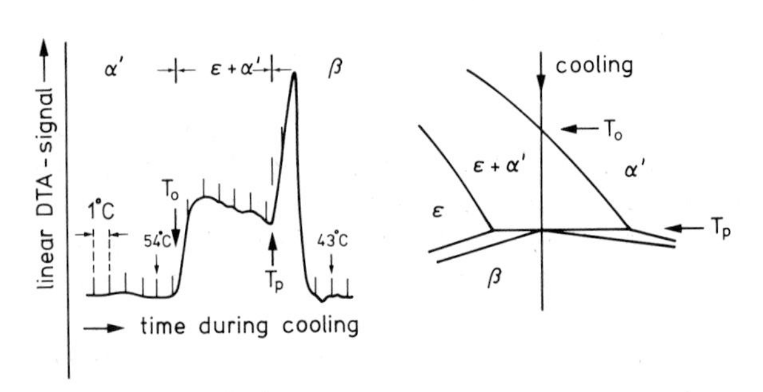

Fig. 2.15. Low-speed DTA cooling curve (0.1 °C/min) of a 59% TaH specimen. The path in the phase diagram is schematically shown to the right. The two phase $(\varepsilon - \alpha')$ region extends over several degrees. A strong signal is seen at the onset of the peritectoid reaction (T_p). Short vertical lines are temperature markings

when one given phase transforms to another phase, or two other phases. Examples are the eutectoid decomposition of a phase upon cooling into two neighbor phases, or the peritectoid transformation of a phase upon heating resulting again in two neighboring phases. Less pronounced effects are obtained when alloys already containing two phases are heated (or cooled) until one of the phases has disappeared. Figure 2.15 shows an example on

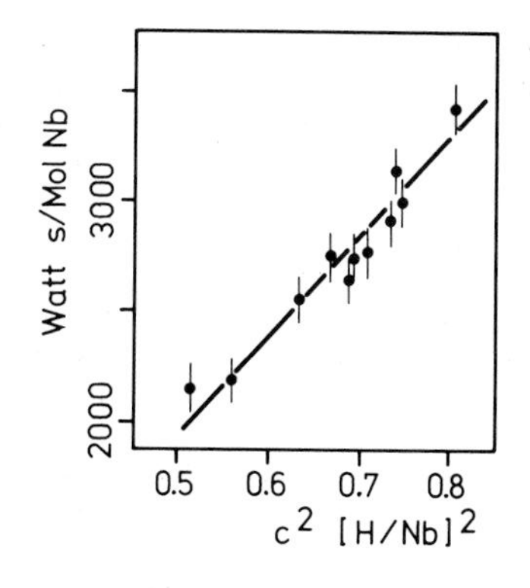

Fig. 2.16. Heat consumed during the phase transition $\beta-\alpha'$ in NbH as a function of hydrogen concentration c

$TaH_{0.59}$. A clear thermal effect at the transition temperature between β-NbH and ζ-NbH (as revealed by TEM [2.23, 24], see Sect. 2.2) was recently observed by *Welter* [2.62]. Figure 2.16 presents experimental results on the heat consumed during the β-α' transition in NbH (*Welter* [2.62]).

c) The electrical resistivity reacts quite sensitively to phase transitions. The solubility limit at low concentrations was often determined by this method (e.g., [2.63–65]). Figure 2.14b shows an example. The electrical resistivity also indicates formation of thermal defects in the ordered hydrogen lattice within a homogeneous phase (Fig. 2.14, *Welter* [2.62]).

Several anomalies are superimposed on the normal temperature dependence of the resistivity due to phonon scattering (dashed line):

1) The steep rise at about 386 K is caused by the order-disorder transition β-α'.

2) Within the β phase the resistivity increases according to the relation $d\varrho/dT = d\varrho/dT$ (phonons) + const. $\exp(-E/kT)$. Presumably, the exponential term is related to formation of additional defects in the hydrogen superstructure. Assuming formation of Frenkel defects [2.66], the formation energy is $2E \approx 0.6$ eV. The electrical resistivity of the β phase NbH is nearly the same as that of pure Nb (15 μΩ cm at 300 K) due to the ordering effect.

3) Small changes of slope of the curve at low temperatures indicate a phase transition, which also shows up in part a) at the same temperature.

d) Observation of well-polished surfaces of samples as a function of temperature by optical microscopy reveals phase transitions as shown in Fig. 2.35 (*Schober* [2.27]). The transformation of cubic into noncubic phases is accompanied by formation of domains which show up in polarization contrast even if the corresponding distortions of the initially cubic elementary cells are less than 1%. Thus this method can compete in sensitivity with the x-ray structure analysis. In addition, it provides information about the phase morphologies, the phase boundaries and the kinetics in a region of a few tens of nm below the surface of the sample. Also, phase transitions give rise to changes of the surface topology of samples, which can be analyzed by interference contrast (*Schober* and *Linke* [2.67, 68].

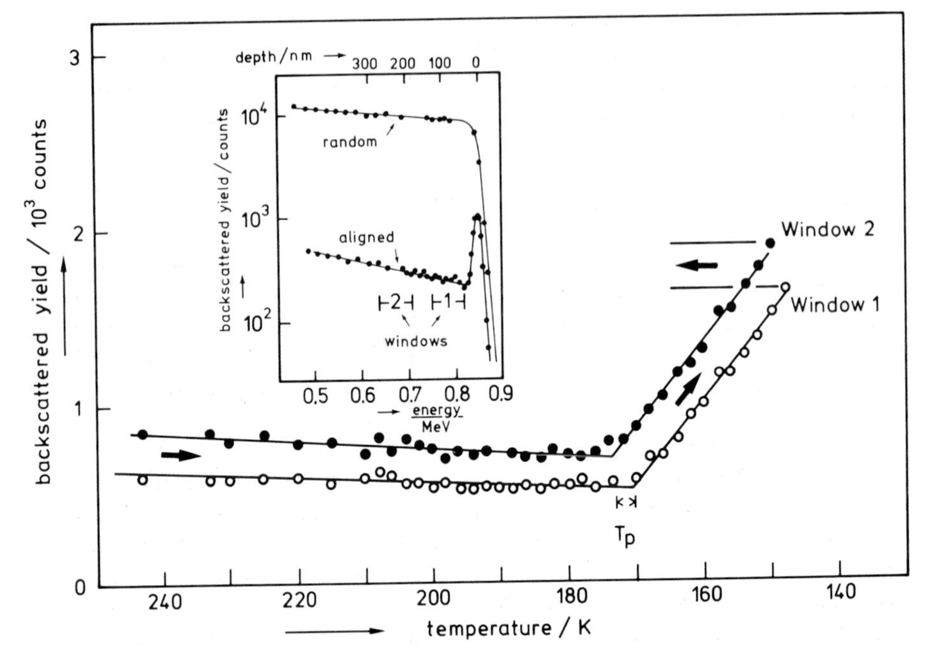

Fig. 2.17. The insert shows aligned and random spectra of 1 MeV He backscattered from {100} Nb [2.72]. The aligned yield is that obtained when the crystal axis is perpendicular to the beam. The peak at the high energy end is due to enhanced scattering from atoms displaced, for example, by surface oxide layer. A measure of crystal perfection is obtained from the ratio of aligned yield just below the surface peak to random yield at the same depth. Electronic windows are pre-set at the depth shown. The main part of the figure presents values of the aligned backscattering yield during cooling on $NbH_{0.33\,at.\%}$. The yield initially decreases slowly with temperature due to the gradual decrease in atomic thermal vibrational amplitude, then increases abruptly due to the lattice distortion associated with precipitation of hydrogen in the form of hydride particles with large hydrogen concentration

e) Spinodals have been determined by using the Gorsky effect (*Völkl* [2.70], *Tretkowski* [2.71]). From these experiments the critical point T_c of the NbH(α-α') transition between bcc phases with different concentrations of disordered hydrogen was evaluated (see Fig. 2.18).

f) Methods of nuclear physics can also be used to reveal phase transitions. Figure 2.17 presents results of Rutherford backscattering measurements on dilute NbH (*Whitton* et al. [2.72]). Enhanced backscattering of channeled He ions is observed as the lattice distorts due to formation of precipitates during cooling below the solubility limit.

g) Many other properties change in the course of phase transitions, e.g., internal friction, sound velocity and damping [2.73], NMR[5] [2.45, 193], superconductivity (Chap. 6, and [2.69]), ductility (Chap. 9), the dislocation

[5] An extensive review is given by *Cotts* (Chap. 9, Ref. 2.3)

density (increased by internal plastic deformation due to coherency stresses) [2.23, 24, 69, 74], magnetic susceptibility [2.75], etc. (see [2.76]). If readily available, these methods are able to provide valuable additional information about phase transitions and phase boundaries in the phase diagram.

2.3.2 Nomenclature; Isotope, Impurity Effects

We shall discuss only hydrogen reactions with pure metals typically of 99.99% atomic purity (see Sect. 2.5). Impurities may drastically influence the solubility limits at low temperatures (e.g., [2.64, 77–79]), and also triple point temperatures (compare Fig. 2.18 with the *Walter* and *Chandler* NbH diagram [2.12]).

Most experiments were performed with hydrogen or deuterium. Isotope effects are most pronounced in vanadium. Tritium-metal systems have not yet been investigated in detail. Increasing activity in the analysis of metal-tritium systems can be expected, e.g., in connection with the experimental test of predictions for isotope effects and in connection with fusion reactor technology.

As usual in metal-hydrogen systems the hydrogen concentration $c = n_H/n_M$ is determined by the ratio of the number of hydrogen atoms n_H and the number of metal atoms n_M in a sample. The following relation is valid between c and the more usual concentration values used in other binary systems $n_H/(n_H + n_M)$:

$$n_H/n_M \equiv c = 1/[(n_H + n_M)/n_H - 1] .$$

Different phases have been given consecutive letters of the Greek alphabet as they were determined in the course of time. Thus, corresponding structures in the V-H, V-D, Nb, and Ta systems may have different names. Because it seems unlikely that all existing phases have already been found, we did not change the nomenclature to avoid confusion in comparison with the original publications.

2.3.3 The NbH System

Figure 2.18 presents the phase diagram of the NbH system, Fig. 2.19 the solubility limits in the α phase at low concentrations. The boundary lines between the phases have been determined by various methods as indicated in the preceding subsection. The structures were discussed in Section 2.2; phase morphologies will be described in Section 2.4.

Remarks

1) The classical phase diagram of *Walter* and *Chandler* [2.12] only qualitatively describes the phase boundaries between α, α', and β phase above 300 K.

2) The temperature dependence of the terminal concentration c_α in Fig. 2.19 can be characterized by $c_\alpha = c_\alpha^0 \exp(-\Delta H_p/kT)$ with $c_\alpha^0 = 5.35$, enthalpy gain by

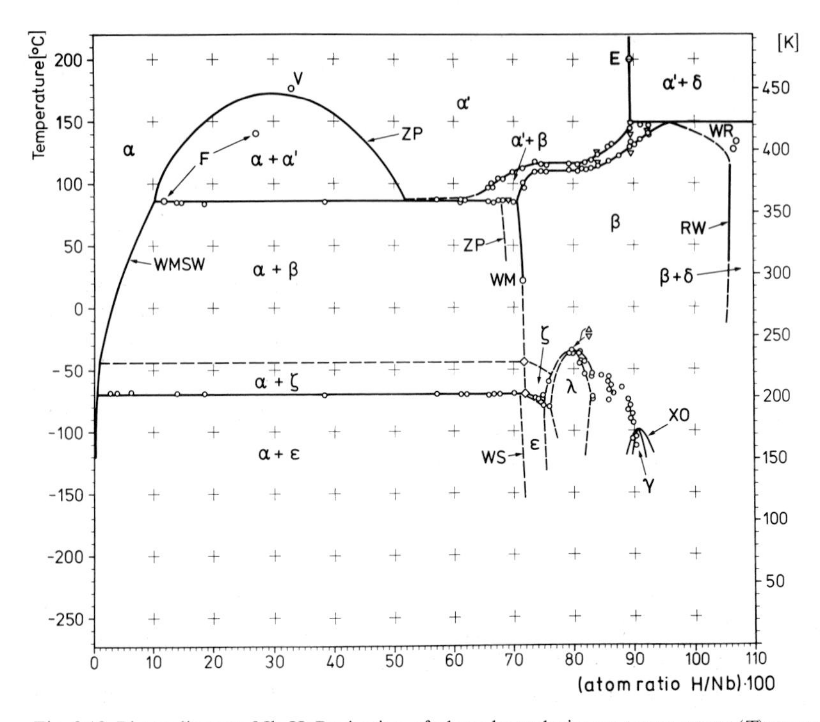

Fig. 2.18. Phase diagram Nb-H. Projection of phase boundaries on temperature (T) concentration (c) coordinates. Corresponding values of molecular hydrogen gas pressure in equilibrium with solid hydride can be found in Fig. 2.43. The Nb vapor pressure is negligible in the temperature region shown. Solid hydrogen does not belong to the family of interstitial alloys under discussion and is neglected in the phase diagram. The solubility limits in the α phase are shown in detail in Fig. 2.19.

Only a selected number of experimental values are indicated on the boundary lines shown, using different symbols to indicate the methods used. ○: differential thermal analysis [2.27, 62]; ▽: electrical resistivity [2.62]; O: optical microscopy [2.67, 68]; ◇: TEM [2.23, 24]; △: n diffraction [2.61]; X: x-ray structure analysis [2.5, 6]; ZP: x-ray lattice parameter measurements [2.80]; RW: [2.81]; V: Gorsky effect [2.70]; WR: electrical resistivity [2.62]; WM: metallography [2.62]; WS: transition from superconducting to nonsuperconducting alloy [2.62]; WMSV: see Fig. 2.19; F: start of incoherent $\alpha-\alpha'$ phase separation on cooling slowly after doping above T_c as measured by γ diffractometry and optical microscopy [2.74, 82]. Further results about coherent phase separation will be published by *Zabel* and *Peisl* [2.80]. The boundary lines as shown are essentially in agreement with various other experiments on restricted regions in the diagram (Mössbauer effect) [2.83, 84], (calorimetry) [2.85, 86], (magnetic susceptibility) [2.87, 74, 88], (NMR) [2.89, 90]

the precipitation process [2.66] $\Delta H_p = 0.12$ eV/H. These results agree with other measurements [2.78, 92–94].

3) The $\alpha-\alpha'$ phase transition is now well understood on the basis of the elastic interaction of disordered hydrogen interstitials via their long-range elastic displacement field ([Ref. 2.3, Chap. 2]). The α and α' phases have bcc structure and differ only in lattice parameter according to the different hydrogen concentration (see Sect. 2.2, Fig. 2.2, and [Ref. 2.3, Chap. 3]).

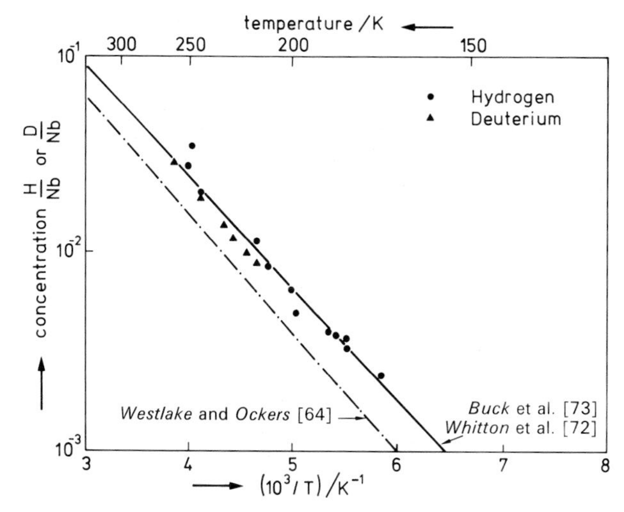

Fig. 2.19. Hydrogen concentration at the solubility limit in the α phase as a function of reciprocal temperature. The method of *Whitton* et al. [2.72] is described in Fig. 2.17 (see critical remarks by *Westlake* [2.91]). Comparison is made with the results of *Westlake* and *Ockers* [2.64] and *Buck* et al. [2.73]

Theoretically, one must distinguish between coherent and incoherent phase separation and take into account that the coherent phase diagram $(\alpha+\alpha')$ depends on sample geometry [see Ref. 2.3, Chap. 2].

Experimentally, mainly incoherent phase diagrams have been determined (see also *Conrad* et al. [2.95]). The term "incoherent" roughly describes a situation in which the coherency stresses have generated dislocations which geometrically reduce the misfit between two phases of different lattice parameter. The total distortion energy is drastically reduced by this process [2.96].

The discrepancies existing between the phase boundaries determined by lattice parameter measurements and by other methods, e.g., optical microscopy, near the triple phase line $\alpha-\alpha'-\beta$ may be due to inherent problems in calculating the hydrogen concentrations from Bragg peaks. These problems may arise due to residual long-range distortions which can exist even in the case of nearly incoherent phase separation. Upon heating, resolution of an incoherent phase generates coherency stresses of reversed sign as compared with cooling. Hysteresis effects result [2.80]. These problems of coherency stresses, incoherent precipitation and related defect structures, and transformation kinetics [2.74] are still under investigation. In addition, the value of the specific lattice parameter change per unit concentration of hydrogen published by different authors varies by as much as 30% [2.5, 6, 80]. Agreement between our phase boundaries $\alpha+\beta$ and the lattice parameter results of *Zabel* and *Peisl* [2.80] can be reached by assuming that $\Delta a/a \approx 5.0 \cdot 10^{-2} c$.

4) *Pryde* and *Titcomb* [2.97] analyzed the phase transitions during isothermal doping with hydrogen (see Fig. 2.43). Their results disagree with the phase boundaries in Fig. 2.18, which again may be due to either coherency stresses or, in addition, to kinetic problems caused by critical slowing down effects near $T_c = 177\,°C$ (*Völkl* [2.70]). For example, *Fenzl* et al. [2.74, 82]

showed by γ diffractometry that electrolytic doping around 100 °C induces reversible or irreversible distortions of the sample. These distortions are probably due to large concentration gradients of hydrogen.

5) The transformation $\alpha' - \beta$ is a disorder-order transformation from the point of view of hydrogen. Only certain tetrahedral sites are occupied by hydrogen in the β phase (see Sect. 2.2). *Pick* and *Bausch* [2.5] analyzed this structure in detail and evaluated H-Nb and H-H interaction parameters. Although direct H-H interaction forces are small compared with H-Nb forces, the direct H-H interaction in the ordered structure is essential for the structural transition from bcc to fco during ordering. Otherwise, a bct structure would result.

6) At about room temperature the β phase NbH_{1-x} has a large homogeneity region in which deviations from stoichiometry have the range $0.3 < x < -0.1$. This indicates relatively small values of the "vacancy" formation energy and large values of the "interstitial" formation energy in the hydrogen superstructure. From the excess temperature dependence of the electrical resistivity (Fig. 2.13, *Welter* [2.62]) the formation energy of a hydrogen "Frenkel defect" is roughly 0.6 eV.

7) Phase transitions of the disorder-order type take place in nonstoichiometric $NbH_{1-x}(\beta)$ at low temperatures due to ordering of the hydrogen vacancies. The disordered structure can decompose into two completely ordered phases or become ordered in one phase according to "branching schemes" proposed by *Somenkov* [2.18] and *Khachaturyan* [2.98]. In Fig. 2.18 two low-temperature phases of this kind are indicated as ε and λ, although only $Nb_4H_3(\varepsilon)$ is firmly established. During cooling in the electron microscope phase transitions were observed in the sequence $\beta \rightarrow \zeta \rightarrow \varepsilon$ for compositions below $NbH_{0.75}$ (*Schober* [2.23, 24]). The concentration of 0.25 disordered vacancies of the β phase $NbH_{1-0.25}$ is partially ordered in the nonstoichiometric ζ phase $NbH_{0.5+0.25}$ and then totally ordered in the ε phase $NbH_{0.75}$. Stoichiometric $NbH_{0.5}(\zeta)$ has not been detected.

8) The γ phase (see Fig. 2.3c) is not member of the family of structures arising from ordering of the hydrogen vacancies in nonstoichiometric β-NbH. Assuming repulsive interaction between nearest and next-nearest neighbors of the hydrogen superstructure, the potential energy of the γ phase is lower than that of the β phase. Accordingly it should be the stable structure for the composition NbH at low temperatures (*Hauck* [2.19, 20]), but for unknown reasons it forms only by cooling below 180 K from the β phase NbH_{1-x} in the region $0.88 < x < 0.92$.

9) Precision structure analysis at low temperatures by neutron diffraction would be necessary to improve this phase diagram and to investigate in more detail disorder-order transitions, which seem so common in this system.

10) The effect of replacing H by D has not been analyzed in detail. The phase boundaries for Nb-D in the region $300 < T < 450$ K and $0 < c < 1$ seem to be nearly the same as in Fig. 2.18 ([2.72, 90]; see also [2.95]). Information about Nb-T is lacking (see [2.99]).

2.3.4 The Ta-H System

The phase diagram cited most frequently in the past is the "DCC" diagram (*Ducastelle* et al. [2.32]) in which there is provision for two ordered phases, β_1 and β_2. This is plausible in view of a number of early studies in which phase transitions were reported in the compound $TaH_{0.5}$ [2.100–102]. An attempt to determine the Ta-D phase diagram[6] was the work by *Slotfeld-Ellingsen* and *Pedersen*, referred to below as the "SEP" diagram [2.103]. Although the SEP diagram contains topological inconsistencies, it gives a fair impression of the difficulties and the complexity of the Ta-D(H) system. Concurrently, two Japanese studies of the Ta-D system (*Asano* et al. [2.36], *Hirabayashi* et al. [2.104] referred to as "AH") brought some clarification. Both the SEP and the AH work are mainly calorimetric studies using powdered Ta-D samples. The inherent difficulties with powder samples are possible inhomogeneities in concentration which may easily amount to several % [2.105]. Also powder samples contain much higher impurity levels and are much more difficult to analyze for hydrogen content than are bulk samples.

At KFA Jülich we decided to make an additional Ta-H DTA study (referred to as the SC study [2.105]) with the following features: 1) the specimens were high-purity single-crystal chips[7], 2) a large number of samples (>150) was used for improved statistics, 3) high and very low DTA scanning rates were employed[8], and 4) the H content was analyzed to better than $\pm 0.2\%$ (see also the last section). Also, the SC study was carried out parallel to the above-mentioned metallographic investigation on Ta-H alloys which allowed one to check optically important features of the SC diagram.

The DTA results of the Ta-H solvus at low concentrations, shown in Fig. 2.20 will be discussed first. Essentially, a straight line was observed below 10 °C[9]. The slope of the solvus (heat of solution) was determined to be 0.148 eV. Figure 2.20 also contains the data by *Lecocq* and *Wert* [2.106].

The reasons for the appreciable discrepancy are still unclear. It may result from the different impurity levels and dislocation densities. *Lecocq* and *Wert* also gave a short review of previous solvus studies of the Ta-H(D) system [2.106]. A more recent Ta-H solvus which is in fair agreement with our work [2.105] was presented by *Rosan* and *Wipf* [2.79].

The TaH phase diagram shown in Fig. 2.21 is based on the DTA-results of the SC-study [2.105] and the magnetic susceptibility data by *Köbler* and *Schober* [2.75]. The data points for $c \leqq 50\%$ are found in [2.105]. The main differences from the previous DCC, SEP, and AH diagrams are that:

1) there exists a high-temperature phase ε,

2) β has a much more limited range in concentration than previously assumed,

---

[6] The Ta-H and Ta-D diagrams are expected to be very similar.

[7] In this way, uniformity in concentration was achieved.

[8] In this way, transitions of widely different transformation enthalpies could be detected and properly analyzed.

[9] The $\beta \rightarrow \varepsilon$ transition reported below caused a discontinuity in the slope at 10 °C.

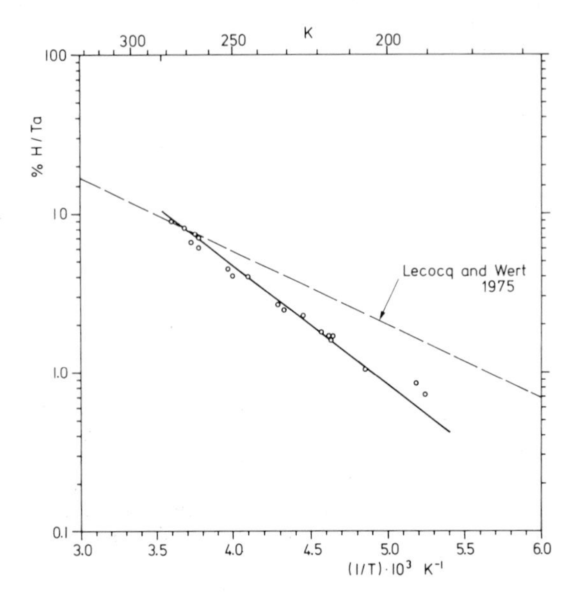

Fig. 2.20. The Ta-H solvus as obtained by DTA [2.105]. The results of *Lecocq* and *Wert* [2.106] are inserted

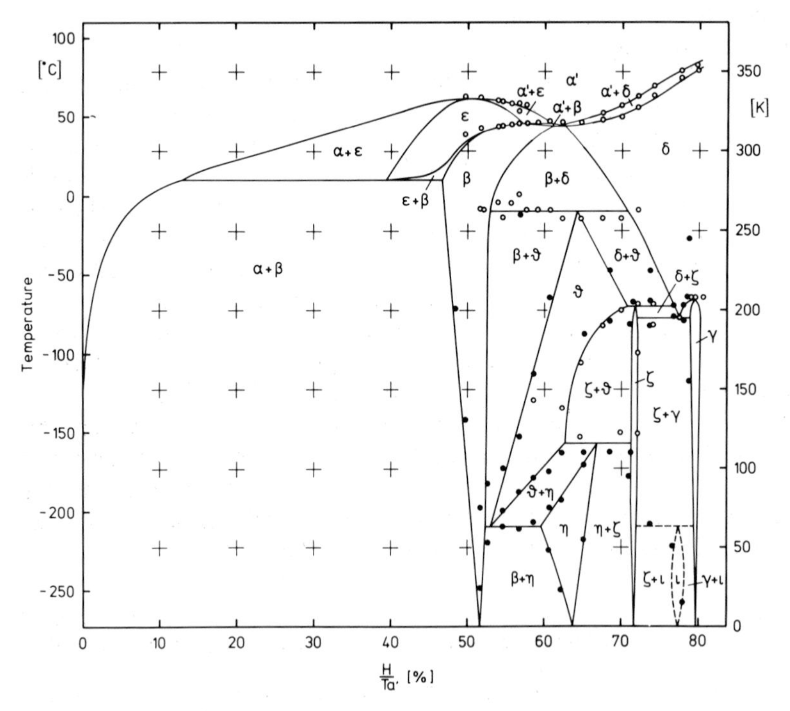

Fig. 2.21. Ta-H phase diagram after *Schober* and *Carl* [2.105] and *Köbler* and *Schober* [2.75]. Open circles: DTA data. Filled circles: magnetic susceptibility data

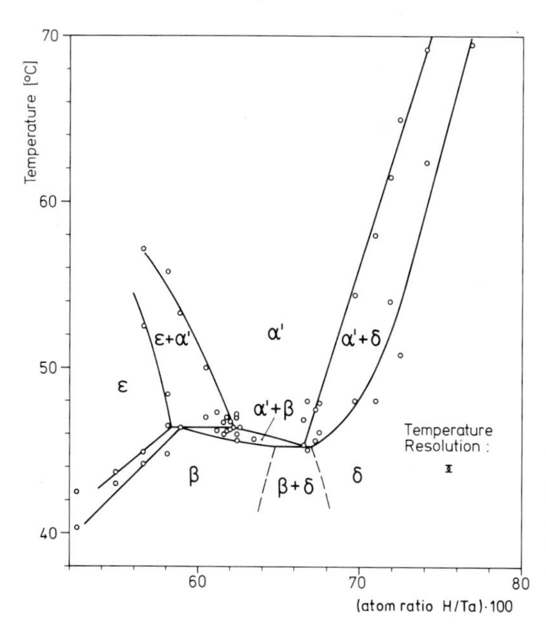

Fig. 2.22. Partial Ta-H phase diagram around 60% [2.105]. This diagram is based on a precise low-scanning rate DTA technique. The four phases ε, β, α', and δ are connected via a peritectoid and a eutectoid transformation

3) α' exhibits eutectoid decomposition into β and δ,

4) there is a monotectoid point at roughly 68% and -20 °C.

In order to clarify the phase diagram further around 63% and at elevated temperatures, a partial phase diagram was measured using a low-speed DTA scanning technique [2.105][10]. The result is shown in Fig. 2.22. The important finding was that the temperature of the peritectoid $\beta \rightarrow \varepsilon + \alpha'$ transformation is 46.3 ± 0.3 °C, whereas the lower temperature of 45.3 ± 0.3 °C was established for the $\alpha' \rightarrow \beta + \delta$ eutectoid transition. Thus, we have the presence of the four phases α', β, δ, ε in the immediate vicinity of about 60% and 47 °C. Figure 2.22 shows how these phases are related via a peritectoid and an eutectoid reaction.

The γ phase has a decomposition temperature of -65 °C and seems to be centered around 80% rather than 75%. Thus, γ is probably not identical with a phase Ta_4H_3 but rather with a phase of composition Ta_5H_4[11,12].

Evidence for the $\varepsilon \rightarrow \alpha + \beta$ eutectoid reaction at 10 °C was obtained by *Zierath* [2.107] in 1969. In that work (see Fig. 2.23), a discontinuity was observed at 10 °C in the slope of the β phase solvus. We clearly attribute this change of slope to the above transformation. The dashed lines and the Greek letters in Fig. 2.23 are taken from the SC work [2.105].

[10] Essentially, this technique consists of using very low scanning rates (of the order of 0.2°/min) at high sensitivities. In this way, the onset, as well as the end of phase transitions, could be determined with an accuracy of about ± 0.2 °C.

[11] A similar situation is encountered in the Nb-H system in which there is the ε phase of composition Nb_4H_3 and a further low-temperature phase λ at about 80% (*Welter* [2.26]).

[12] Phase ζ has a composition near 75%.

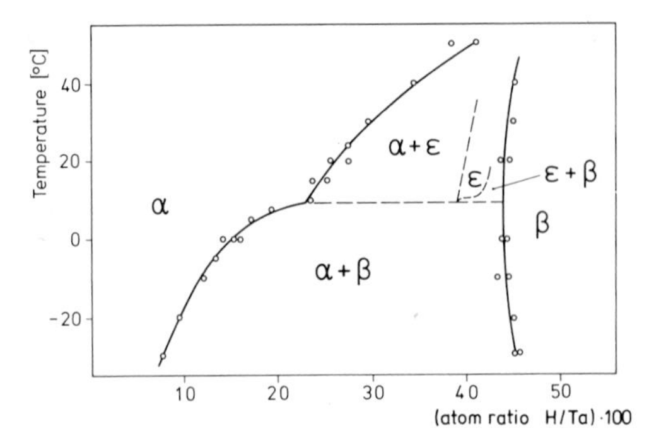

Fig. 2.23. Partial Ta-H diagram (*Zierath* [2.107]). The open circles and the full lines are taken from the original work [2.107]. Dashed lines and Greek letters were inserted by the authors

The diagram in Fig. 2.21 is by no means a final version of the reactions of hydrogen and tantalum. We mention a very recent Ta-H diagram by *Asano* et al. [2.108] which appeared in 1977. It is based on calorimetric results obtained with *powder* samples. This new diagram is essentially in agreement with our work [2.105] up to 55%. Above 55%, however, there are serious discrepancies which may in the view of the present authors be ascribed to inhomogeneities in the powder samples. Notably, the high concentration and low-temperature region needs further study. It is, however, doubtful whether DTA alone can bring more extensive clarification. It is our feeling that neutron diffraction of well-characterized specimens as a function of temperature would be useful in further investigations.

2.3.5 The V-H System

The first frequently cited diagram is the one of *Maeland* [2.41] published in 1964. Quite recently, *Asano* et al. [2.42] presented a calorimetric study and a phase diagram (called below the "AAH" diagram). In parallel, *Fukai* and *Kazama* studied the same system and published a diagram (referred to below as the "FK" work [2.56]). The AAH and FK diagrams are qualitatively in fair agreement. We note, however, that no two-phase regions between disordered and ordered phases are presented in these studies.

The V-H solvus at low concentrations was investigated previously [2.63, 64, 77, 109]. We shall report below results of a DTA study performed at KFA Jülich [2.194, 195]. Techniques similar to those described above in the discussion of the SC study [2.105] were employed. High purity V chips were hydrogen charged from the gas phase or by electrolytic techniques[13].

[13] After electrolytic charging, the samples had to be homogenized at 200 °C to ensure a uniform concentration. Electro-refined vanadium (U.S. Bureau of Mines—Boulder City, Nevada) was used in the study; the interstitial impurity levels were: C<16, N<10, O<70 weight ppm.

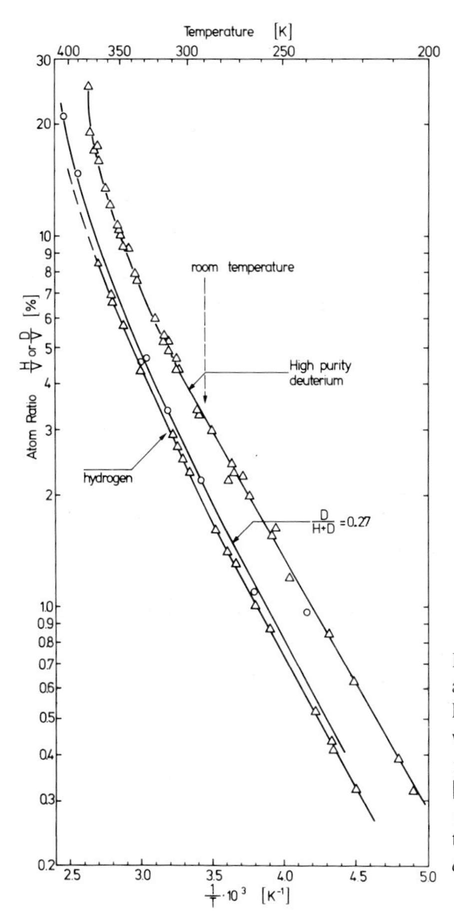

Fig. 2.24. Curve marked hydrogen: VH-solvus as determined with DTA-techniques [2.194]. Expressing the solvus by $c = c_0 \exp(-\Delta H/kT)$ we obtained: $\Delta H = 0.1413 \pm 0.005$ eV and $c_0 = 510 \pm 50\%$. Curve marked deuterium: VD-solvus [2.195]. Here, $\Delta H = 0.135 \pm 0.004$ eV and $c_0 = 700 \pm 70\%$. Curve marked $D/(D+H) = 0.27$: an isotope mixture was used; here $\Delta H = 0.1413$ eV and $c_0 = 600\%$

Fig. 2.24 shows the results of the VH solvus study (curve marked hydrogen). The results are in excellent agreement with data by *Westlake* and *Ockers* [2.64] and in fair agreement with work by *Chang* and *Wert* [2.77].

The results of the DTA study[14] of the V-H system [2.194] are presented in Fig. 2.25. A comparison with the AAH and FK diagrams shows the following differences: 1) In our work, the two-phase regions at high temperatures between α and the ordered phases ε and β were established. 2) Some evidence was obtained for a transformation of phase β to a new phase (η) at -20 °C. 3) The low-temperature transformation found around -50 °C in the AAH and FK study was established to be the peritectoid decomposition of phase δ.

In view of the work by *Boccara* [2.60], further structural studies are needed in order to determine whether continuous transitions are possible between the

[14] The diagram represents the experimental situation as of February 1977.

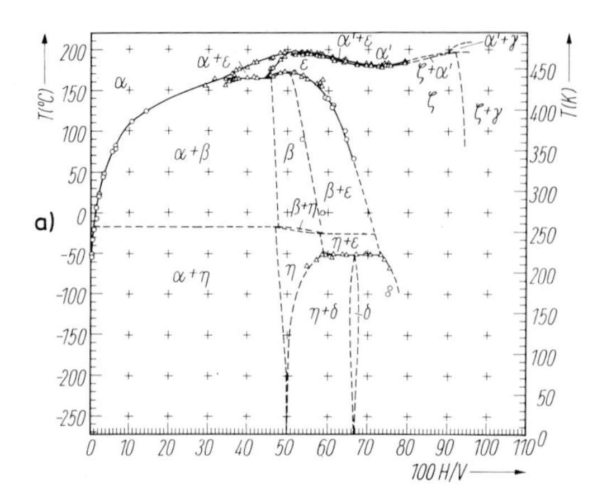

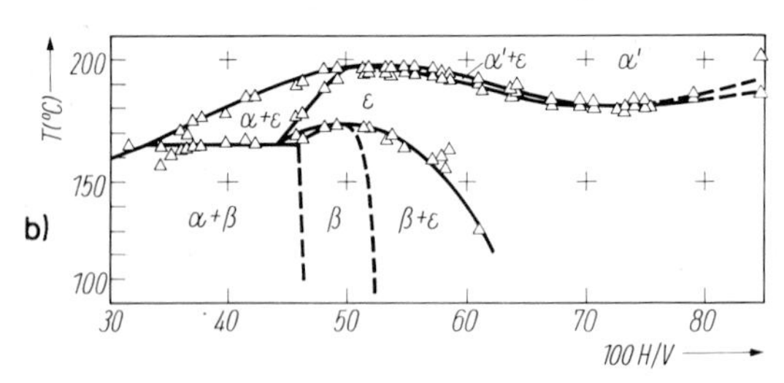

Fig. 2.25a and b. The phase diagram V-H after [2.194] as obtained from DTA results. The triangles denote transitions which were measured with the precise low-scanning rate technique ($\sim 0.5°\ \mathrm{min}^{-1}$). Circles denote transitions seen with the high-rate procedure ($5.0\ \mathrm{min}^{-1}$)

phases β, δ, ε, ζ, and η or if two-phase regions are required. Accordingly, the phase diagram would have to be modified. Two typical DTA curves of alloys with different concentrations are presented in Fig. 2.26.

2.3.6 The V-D System

This system is treated separately here due to the large isotope effect. We first note the work by *Hardcastle* and *Gibb* [2.55] who observed a transformation upon cooling of phase β_D at $-25\ °C$ into phase β'_D. However, the $\delta \rightarrow \gamma$ transition (see below) was not reported in that study. In their calorimetric study of the V-D system, *Asano* and *Hirabayashi* (2.52] found evidence of the $\delta \rightarrow \gamma$ transformation but did not detect the $\beta_D \rightarrow \beta'_D$ transition. Two further proposals of the V-D diagram, which are in qualitative agreement with the above studies, were offered by *Westlake* et al. [2.54] and *Arons* et al. [2.45]. The latter authors found, however, evidence for a new intermediate phase lying between δ and α' in a very narrow temperature range (*Arons* et al. [2.45]). Finally, *Westlake* and *Ockers* [2.64] found that the solubility of deuterium in V is about twice that determined for hydrogen.

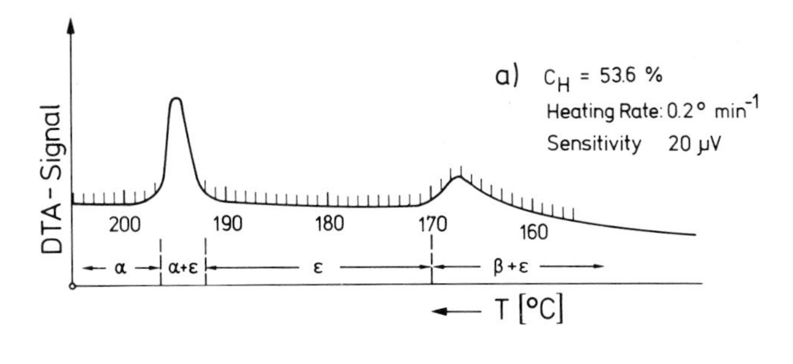

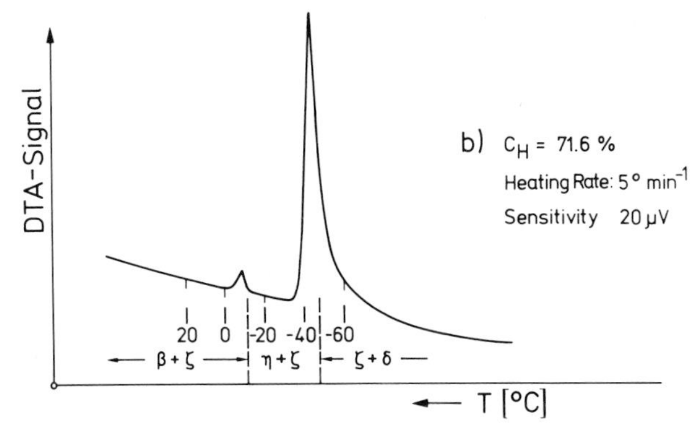

Fig. 2.26a and b. Representative DTA curves of two V-H alloys [2.194]. (a) Low scanning-rate (0.2°/min) curve showing the $\beta \rightarrow \varepsilon \rightarrow \alpha$ transitions. The widths of the one- and two-phase regions are clearly visible. (b) High scanning-rate (5°/min) curve used for detecting low-temperature transitions

Fig. 2.27a shows the results of our DTA-study of the VD-system [2.195, 197]. The diagram is in good qualitative agreement with the work of *Asano* and *Hirabayashi* [2.52]. It contains however a new VD phase (ζ). For a detailed discussion of this VD diagram see [2.195, 197]. A schematic extension of Fig. 2.27a is presented in Fig. 2.27b.

2.4 Phase Morphologies

In the following we shall discuss the morphological aspects of incoherent hydrogen precipitation in Nb, Ta, and V. Coherent precipitation is dealt with in [Ref. 2.3, Chap. 2]. The phase morphologies are important since they govern to a certain extent such properties as mechanical and electrical behavior. Also, the hydride and domain morphologies reflect the crystal structures of the hydrides. Consequently, the appearance or disappearance of morphological features is indicative of phase transitions and may be used for establishing phase boundaries. Emphasis is placed in this work on recent metallographic and electron microscope studies of the morphological aspects of hydride precipitation.

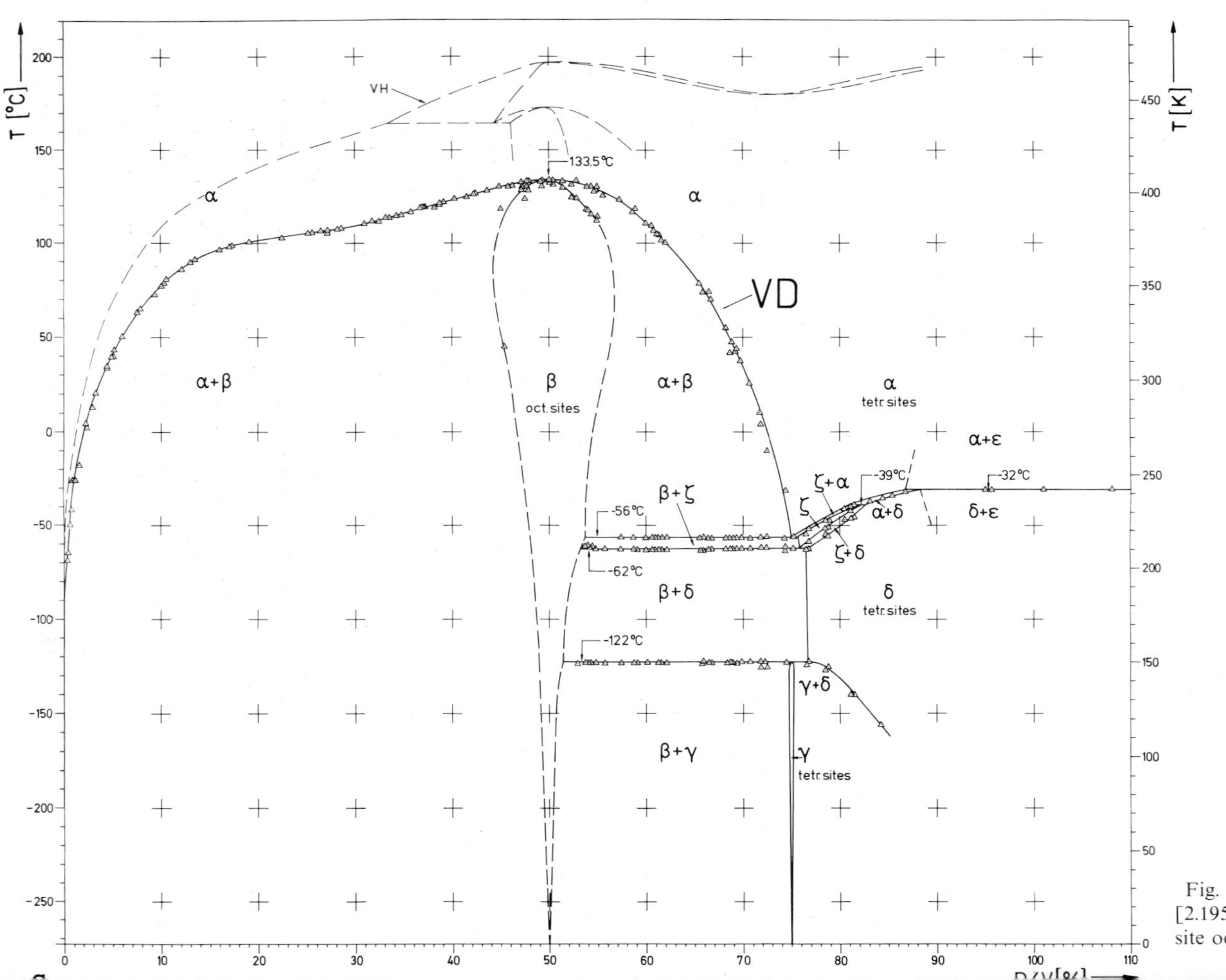

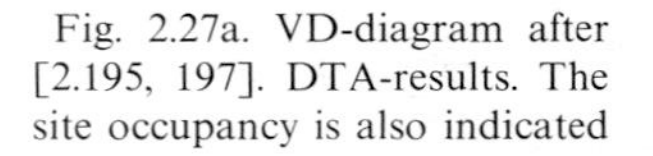
Fig. 2.27a. VD-diagram after [2.195, 197]. DTA-results. The site occupancy is also indicated

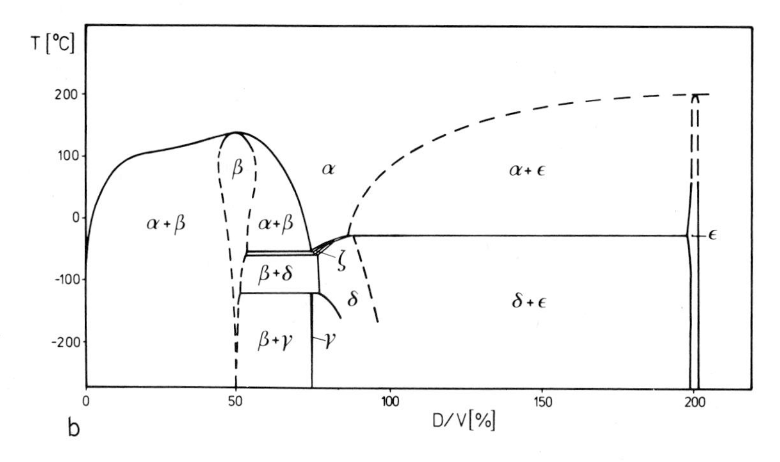

Fig. 2.27b. Schematic VD phase diagram with an extension to 200% [2.195, 197]

2.4.1 Hydride Morphologies in the Nb-H System

There are a number of early studies of the morphology of β phase precipitation in the α matrix [2.5, 6, 110–113]. It was found in these studies that the β phase precipitates in form of plates on $\{100\}_c$ planes. A more extensive study performed at KFA Jülich [2.21–24, 67, 68, 114–116] will be summarized below.

Metallographic Results

In agreement with previous work, a $\{100\}_c$ habit plane was found for β precipitates in α when the samples were slowly cooled from temperatures above 200 °C to room temperature at rates of 1° min^{-1}. (In this way, an approximate "equilibrium morphology" could be obtained.) Three polarized light micrographs of alloys with increasing concentration are shown in Fig. 2.28a–c. It is apparent that the thickness and fractional area of the β phase increase with concentration. We note that the large spacings and plate dimensions are an indication of the extensive diffusion occurring during phase separation in the samples.

Similarly, δ phase (NbH_2) precipitation in β can easily be observed metallographically as in Fig. 2.29a, b. The morphology of β and δ precipitation is schematically presented in Fig. 2.30. Only the application of polarized light observations to the study of Nb-H alloys [2.67, 68, 114–116] has revealed the internal domain structure of β phase areas[15]. In such polarized light micrographs, different domains generally exhibit different shades of grey and are therefore distinguishable[16]. The various domains generally possess contrast

[15] Another very suitable technique is electron microscopy.

[16] We note here that a prerequisite for such domain observations is careful vibratory polishing as discussed in [2.116].

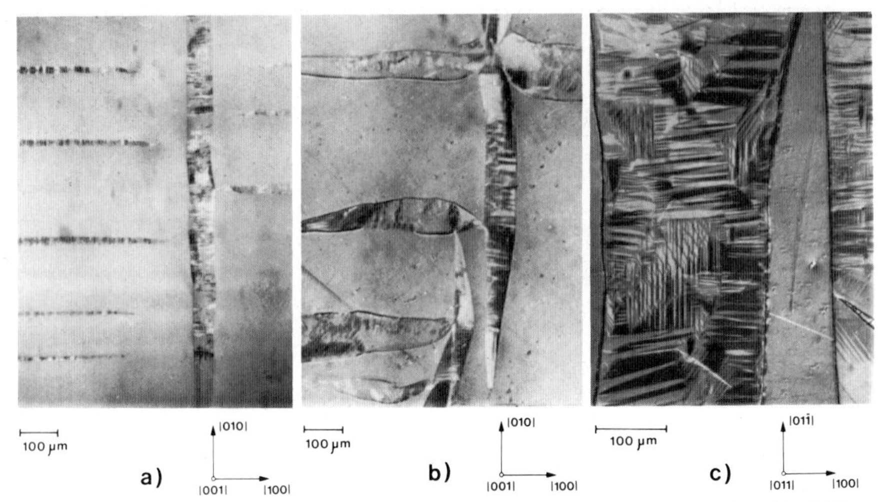

Fig. 2.28a–c. $\alpha-\beta$ phase morphology as observable with polarized light on well-polished Nb-H specimens ([2.67, 68], with kind permission of Elsevier Sequoia Oxford, England). (a) $C_H \approx 10$–15%; $(100)_c$ orientation. β phase is embedded in the matrix in form of thin plates on $\{100\}_c$. (b) $C_H \approx 25$–30%. Thicker plates of β extend through the sample. (c) $C_H \approx 50$–60%. $(110)_c$ orientation. Thin plates of the α phase are embedded in the β matrix

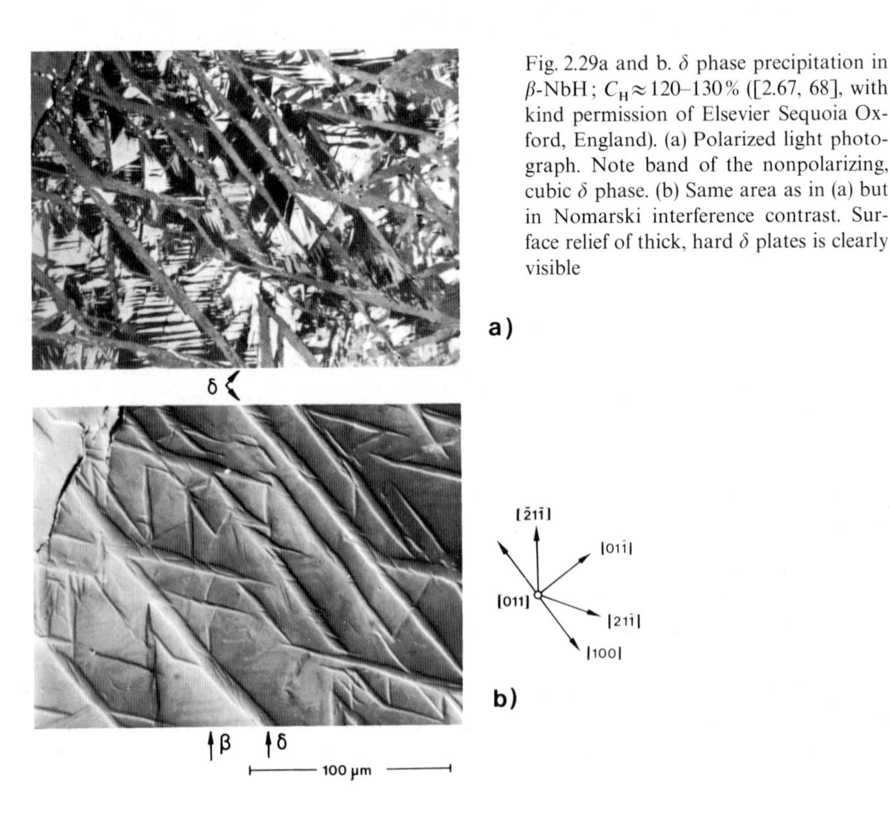

Fig. 2.29a and b. δ phase precipitation in β-NbH; $C_H \approx 120$–130% ([2.67, 68], with kind permission of Elsevier Sequoia Oxford, England). (a) Polarized light photograph. Note band of the nonpolarizing, cubic δ phase. (b) Same area as in (a) but in Nomarski interference contrast. Surface relief of thick, hard δ plates is clearly visible

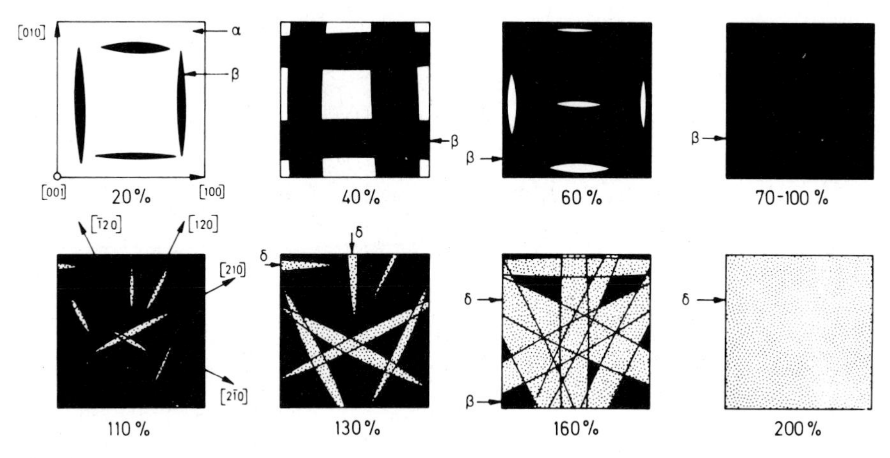

Fig. 2.30. Schematic diagram of α, β, and δ phase morphology as observable on $\{100\}_c$ [2.23, 24]

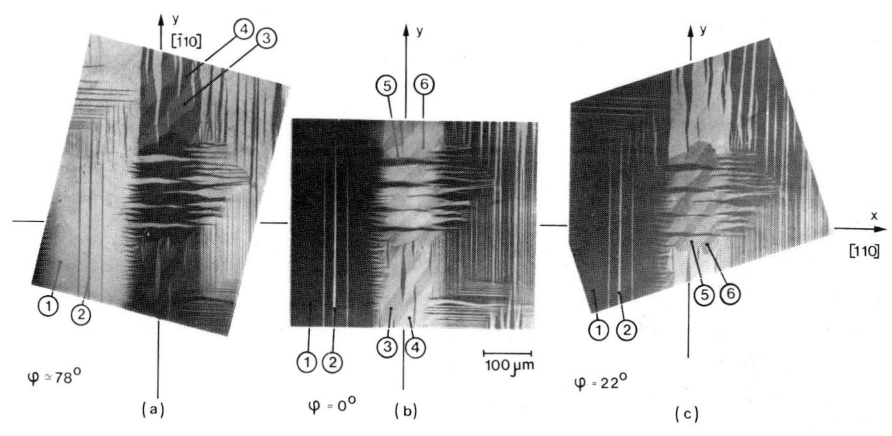

Fig. 2.31a–c. Polarized light photographs of approximately the same area of a β-NbH crystal under different contrast conditions ([2.67, 68], with kind permission of Elsevier Sequoia Oxford, England). (a) Domains 1 and 2 are in strong contrast, 3 and 4 are weak, 5 and 6 are invisible. (b) Contrast of pairs 1–2 and 3–4 is now reversed. (c) Pair 1–2 in strong contrast, 3–4 invisible, 5–6 in fair contrast

shades different from the isotropic cubic matrix (α phase), allowing a clear distinction between β and α phase areas.

It follows from the orthorhombic unit cell in Fig. 2.3a that there are six possible domain variants in the β phase. In Fig. 2.31a–c there are indeed six different shades of grey, or domain configurations for a given area. It is also evident from Fig. 2.3a that $\{100\}_c$ planes are twin planes for the orthorhombic structure and may be used as coherent domain boundary planes as illustrated in Fig. 2.32. Trace analysis of many domain micrographs indeed revealed a frequent occurrence of $\{100\}_c$ domain boundary planes; other low index planes

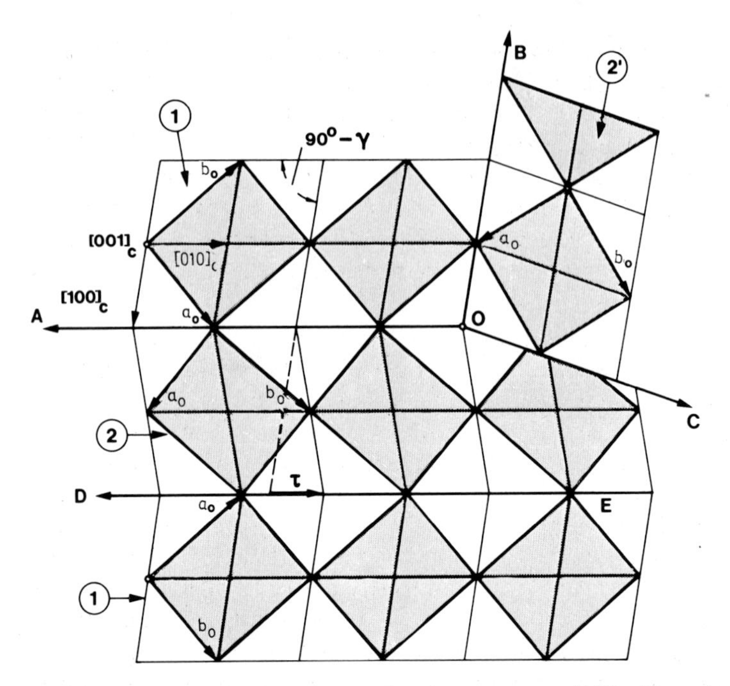

Fig. 2.32. Model for the Nb-H β phase domain geometry on $\{100\}_c$. The orthorhombic unit cells defined by vectors $\boldsymbol{a}_0$, $\boldsymbol{b}_0$, $\boldsymbol{c}_0$ are shaded. Boundaries along OA and OB are coherent, dislocation-free twin boundaries. Boundaries of type OC are incoherent boundaries. (For the sake of clarity, the orthorhombic distortion was exaggerated)

[such as (110)–(210)–(111)–(211)] were also quite common. Typically, the β phase domains were thin plates with thicknesses of a few μm to tens of μm and plate dimensions of a few hundred μm.

The most important condition for the observation of domains by polarized light is the absence of plastically deformed layers on the specimen surface. In view of this point, we readily understand that domain configurations could also be observed on chemically polished faces, on mechanically polished surfaces which were annealed before charging, and, finally, on cleavage faces of the β phase[17].

By using a heating stage, optical microscopy allows one to directly observe the appearance of the disordered α' phase from the domains of the β phase and to determine the temperature width of the $\alpha'-\beta$ two-phase region [2.115][18]. The heating stage work established the existence in the Nb-H system of a

[17] β phase crystals are very brittle and cleave easily on $\{110\}_c$ (and to a much lesser extent on $\{100\}_c$ planes). Cleavage produces so little deformation that β phase domains may immediately be seen on such faces [2.67, 68].

[18] In view of the "Landau rules" mentioned above, two-phase regions must often exist in order-disorder transitions. The reader is also referred to a phenomenological discussion of two-phase regions in ordered alloys [2.117].

distinct ($\alpha'-\beta$) two-phase region over a temperature interval of several degrees C. Furthermore, this work resulted in a morphological description of the disordering of β phase areas when going through the $\beta \rightarrow \beta+\alpha' \rightarrow \alpha'$ phase boundaries (see the similar behavior on the Ta-H system in Fig. 2.35). Analogously, the morphology of β precipitation in α' was determined [2.115]. Also, a partial Nb-H phase diagram for $c_H > 70\%$ obtained in this study was incorporated in Fig. 2.18. Finally, a triple-point temperature of 84.5 ± 0.5 °C was measured with this technique, a value which is 11° above the previously accepted temperature[19].

Regarding the $\beta \rightarrow \zeta \rightarrow \varepsilon$ transformation between 0 and 70%, polished specimens were cooled in a special cooling stage [2.116] to below -65 °C to see whether the domain structure of β would be affected by the $\beta \rightarrow \zeta$ and $\zeta \rightarrow \varepsilon$ transitions occurring at -45 °C and -65 °C, respectively. The results were negative; no changes were observed in the domain patterns. It was concluded that β, ζ, and ε have nearly the same metal atom lattice, i.e., the same orthorhombic cell dimensions. In contrast, cooling experiments with alloys around 82% revealed the sudden transformation of β at ~ -40 °C into a new ordered low-temperature phase[20]. This transition became visible at decreasing temperatures through the dissolution of the β phase domain arrangement.

We briefly mention the influence of elevated or high cooling rates ($>10°\,\mathrm{min}^{-1}$ and $>25°\,\mathrm{min}^{-1}$, respectively) on the morphology for β precipitation in α. Under these conditions of high supersaturation, which are also encountered during electrolytic charging at room temperature, mainly dendritic hydrides were found [2.67] which were, nonetheless, large in size due to the high mobility of the hydrogen.

Finally, in the course of the metallographic study it was established that order→disorder (orthorhombic→cubic) transitions in the Nb-H system may also be seen in Nomarski interference contrast [2.67, 68]. This is easily understood since originally flat surfaces of the β phase develop inclined surface sections after the phase transition which have the same geometry as the domain structure [2.68]. Similarly, previously flat surfaces of the cubic parent phase develop slightly inclined surface facets when the specimen transforms to the noncubic ordered phase. In either case, it is these inclined surface sections which are visible through Nomarski interference contrast as different shades of grey. With this technique, it was possible to confirm the existence of the $\beta \rightarrow \gamma$ (orthorhombic→cubic) transformation which was first observed in the x-ray work of *Pick* [2.5, 6].

The above metallographic study may be compared with similar work by *Birnbaum* et al. [2.118] which is in general agreement with our results. In

[19] Recently, the more accurate value of $T_t = 85 \pm 0.5$ °C was obtained using a very precise low-speed scanning DTA technique.

[20] This new phase (referred to as λ in Fig. 2.18) was found in a recent DTA study of alloys around 80%. It transforms into β at -35 °C upon heating (*Welter* [2.26]).

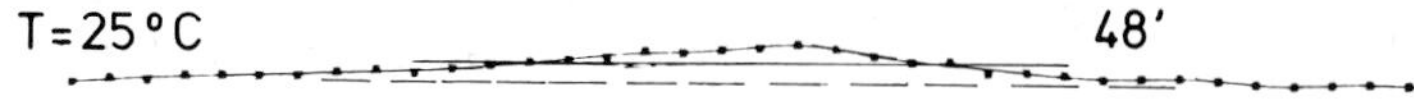

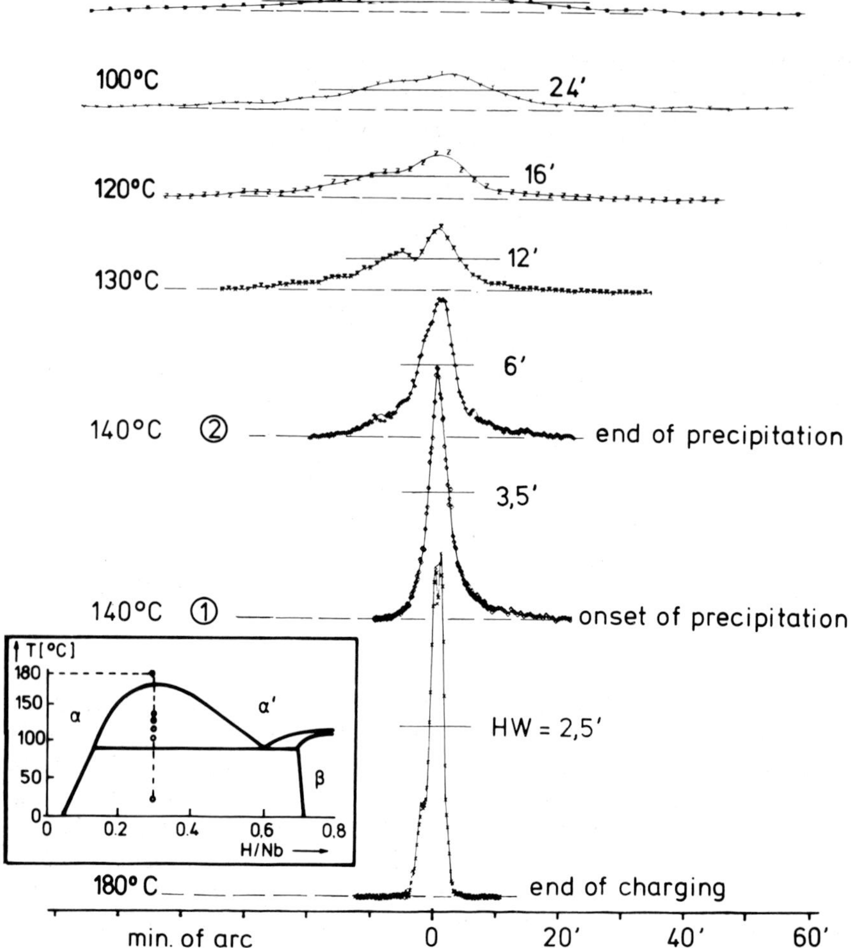

Fig. 2.33. γ-ray rocking curves of a 30%-NbH single crystal after in situ hydrogen charging. The widths increase after the formation of coherent and incoherent α' precipitates

particular, these authors differentiate between a) the equilibrium morphology (obtained at very low cooling rates), b) the "blocky" hydride morphology at rates between 1 and 25 K $\min^{-1}$, and c) the "dendritic" morphology at even higher rates. Additional observations on domain boundary mobility under external (compressive) stress were made. Here, it was found that certain areas were swept free of domain boundaries where in others the boundaries remained immobile. However, the main goal of obtaining a mono-domain crystal by applying a suitable external stress was not achieved[21]. Similar experiments in this laboratory were equally unsuccessful (see below in Sect. 2.4.2).

[21] Such a mono-domain crystal would be very advantageous for a variety of studies. We emphasize the great difficulties, however, in assessing the "single-crystallinity" of such a sample, which presumably could only be done by neutron diffraction.

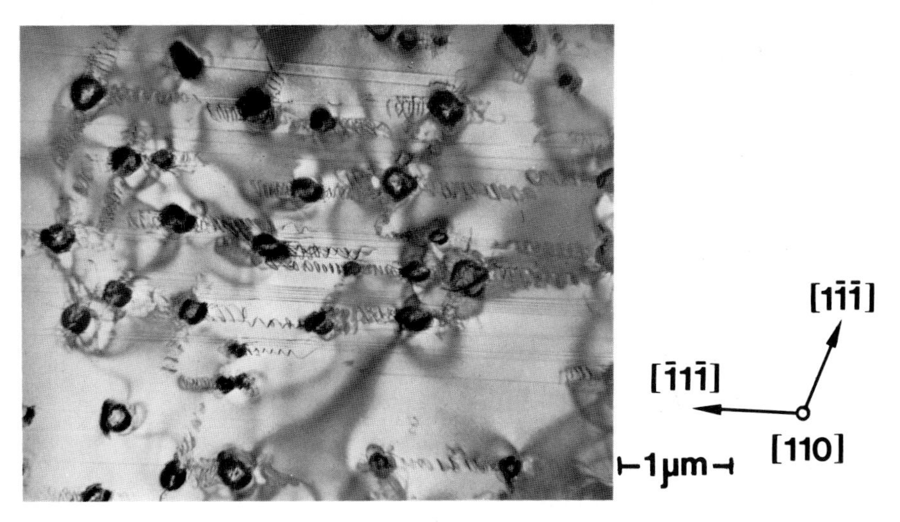

Fig. 2.34. ε phase precipitation in α-NbH at low concentrations and low temperatures. Electron micrograph. The precipitates emit prismatic, interstitial loops with $b=(a/2)\langle 111\rangle$ to relieve the volume constraint ([2.22], with kind permission of Pergamon Press, New York)

We include here the essential results of a recent γ-ray diffraction and metallographic study of the $\alpha-\alpha'$ phase morphology and separation (*Fenzl* et al. [2.74]). Single crystals around 30% were electrolytically charged above the critical point and slowly cooled down into the $\alpha-\alpha'$ two-phase region. Two main effects were observed in situ: 1) With the onset of precipitation, the halfwidth of the γ-ray rocking curve broadened continuously from 3.5′ to 24′ at 100 °C (see Fig. 2.33). This broadening obviously was caused by the precipitates of α' in α. 2) A surface relief was found to develop on the surface parallel to traces of {100} planes arising from the formation of α' plates on {100}. The subsequent formation of β phase plates at 85 °C took place at the location of these α' plates. Thus, the α' plates on {100} are a precursor of the final β phase morphology.

Electron Microscope Results

While a summary of the structural results obtained by electron microscopy (TEM) was given in Section 2.2, we briefly outline here some of the morphological features as discussed in detail in [2.23, 24]. The appearance of β (or ε) phase precipitates at low concentrations and temperatures is illustrated in Fig. 2.34. It is seen that small irregularly shaped hydride particles precipitate inside α. To reduce the total elastic distortion energy, these particles emit prismatic interstitial loops with $b=(a/2\langle 111\rangle$ into the matrix. When these specimens were warmed again to the single-phase α region, dense dislocation tangles or "dislocation skeletons" remained at the sites of the previous hydride particles.

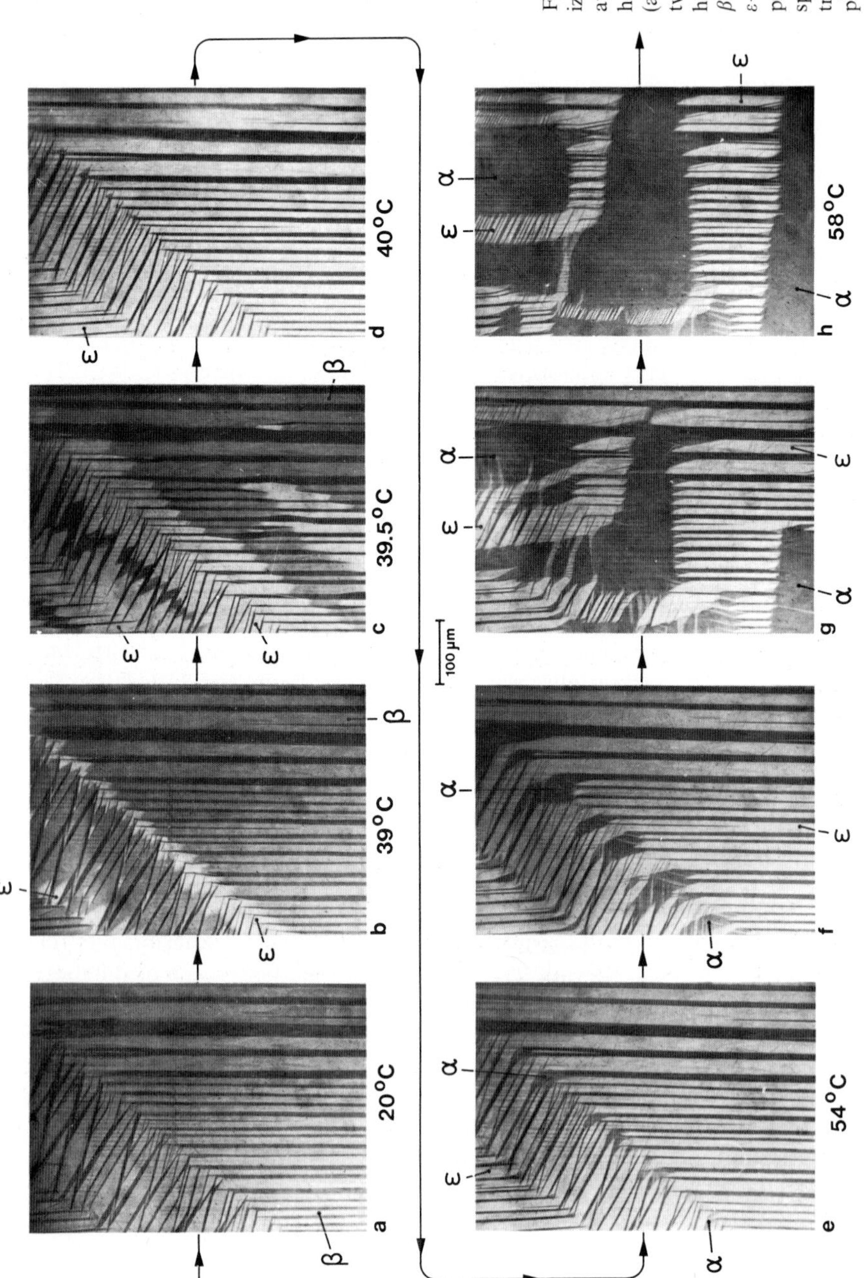

Fig. 2.35a–h. A series of polarized light photographs of a 55 at-% Ta-H specimen which was heated from 20 to 60 °C [2.27]. (a) Pure β phase. (b) and (c) β, ε: two-phase region. The ε phase has much stronger contrast than β. (d) Pure ε phase. (e)–(h) the $\varepsilon \rightarrow \alpha$ transformation starts and progresses. (Above 58 °C, the specimen has the uniform isotropic appearance of the α phase)

Cooling again below the solvus, these dislocation skeletons often provided nucleation sites for the precipitating hydrogen. Morphological studies of the β phase at higher concentrations were not possible due to the limited field of view in TEM. The domain structure of β, however, could easily be observed. Results similar to those reported above were obtained. Finally, further details such as twins, antiphase boundaries, deformation effects, and ε phase precipitation in β were described in the original work [2.23, 24]. We also note other TEM work on Nb-H alloys [2.118, 119] which is in agreement with the above results.

2.4.2 Hydride Morphologies in the Ta-H System

Metallographic Results

Recently, *Westlake* and *Ockers* [2.120] presented a metallographic study of the Ta-H system in which valid criticism of the *Slotfeld-Ellingsen* and *Pedersen* work [2.103] was given. It was found that the observed $\alpha-\beta$ two-phase morphology, as well as the single-phase β structure, was inconsistent with the phase diagram in [2.103]. In the view of our own extensive metallographic study it seems very probable that the phase called β in the above study [2.120] is actually identical with ε.

Summarizing the Ta-H metallography in this laboratory [2.27], we find that primary room temperature precipitation does not occur at low concentration in phase β but in phase ε on $\{100\}_c$ habit planes in form of thin plates. Even at rather high cooling rates ($\approx 10°$ min^{-1}), such a plate morphology could be found. The $\varepsilon \rightarrow \beta$ transformation was clearly visible in polarized light since phase ε had much more contrast than phase β. This phenomenon is illustrated in Fig. 2.35a–h where a series of micrographs at increasing temperatures is shown. The domain configuration remains virtually unchanged upon going through the $\beta \rightarrow \varepsilon$ transformation. We see, however, the onset of the transformation through the appearance of bright areas inside the β phase. Likewise, the final $\varepsilon \rightarrow \alpha$ transition can also easily be seen on the photographs (see figure caption for details).

At concentrations around 59–60%, we observed the peritectoid transition $\beta \rightarrow \varepsilon + \alpha'$. The eutectoid transition to the right, $\alpha' \rightarrow \beta + \delta$, is much more difficult to image; however, the presence of two ordered phases in the vicinity of the eutectoid point could be demonstrated. Returning to the question of two-phase regions in ordered alloys, we also observed in the Ta-H system, in the case of the $\delta \rightarrow \delta + \alpha' \rightarrow \alpha'$ transition, a broad two-phase region extending over several °C. Cursory experiments to see whether the $\delta \rightarrow \beta + \zeta$ and $\gamma \rightarrow \zeta$ transformations could be seen in polarized light showed no variation in the domain pattern present at room temperature when the corresponding phase boundaries were crossed. From this we may tentatively conclude that phases β, δ, and γ have very similar metal atom unit cell dimensions (which in turn govern the domain arrangement).

Finally, attempts were made to produce single-domain crystals by applying elastic stress to the domain arrangement. This was done by elastically bending polished crystal plates in a four-point bending apparatus which simultaneously allowed one to keep the sample at an elevated temperature. Polarized light observations were made before and during the application of the elastic tensile stress. Observations at a few degrees below the order-disorder temperature showed the possibility of moving a few boundaries, but essentially demonstrated that the original domain pattern persisted up to the onset of plastic deformation. Likewise, experiments to grow single domain areas by elastically straining above the order-disorder temperature and then cooling down into the ordered phase were unsuccessful. Again, a multidomain pattern was observed.

Electron Microscopy of Ta-H Alloys

In addition to the electron diffraction study on the β phase [2.37], there is also the work of *Lecocq* [2.121], who investigated precipitation in low concentration alloys ($c \approx 10\%$). The hydrides were found to be surrounded by dense dislocation tangles and loops. A diffraction contrast experiment yielded a loop Burgers vector of $\boldsymbol{b} = (a/2\langle 111 \rangle$. Also, the loops were edge-type in character and lay along a cylinder defined by their Burgers vector. Clearly, the loops were punched out to reduce the distortion energies. While the punched-out loops are of interstitial nature, we may rather formally consider that a corresponding number of "vacancy loops" remain at the site of the precipitate and form the precipitate-matrix dislocation interface.

Summarizing the TEM work done at KFA Jülich, we found that the β phase precipitates at low concentrations and temperatures in the form of thin plates on $\{100\}_c$. The presence of prismatic interstitial loops as previously reported [2.121] was confirmed. With regard to α precipitation in β around 45% at low temperatures, both coherent and incoherent morphologies were observed. In the coherent case, spherical and plate like precipitates of α were found inside the β "matrix"[22]. The incoherent precipitates of α in β were plates on $\{100\}_c$. Domain boundaries were again the distinctive features of β and were located on low index planes of the structure. Diffraction experiments at high concentration and low temperatures revealed the existence of at least two more ordered phases, the structures of which are still under study. Also, a further cubic Ta-H phase was seen at low temperatures, visible through the disappearance of domain boundaries and superstructure reflections at about $-155\,^{\circ}$C. (A similar phase was found in the Nb-H system and has been referred to as γ [2.5, 6].)

[22] We note here in the case of coherent α precipitates in β that the state of coherency is maintained much longer than in the opposite case of β precipitates in α. The reason is that the β "matrix" is a hard and brittle material in which it should be very difficult to move the dislocation loops required to reduce the distortion stresses of the α phase precipitate.

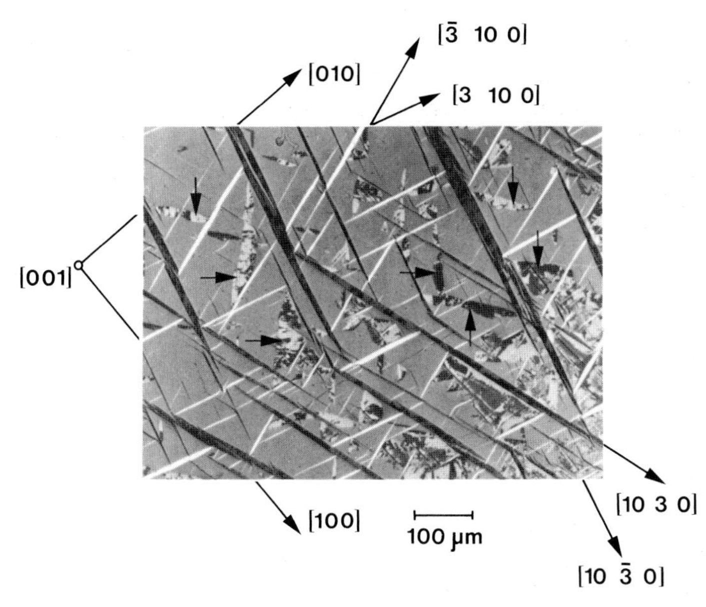

Fig. 2.36. Polarized light micrograph of the $\alpha-\beta$ morphology in the V-H system [2.194]. The thin plates are the original β plates produced by gas-phase charging. The broad hydride patches (see arrows) arose during polishing procedure

2.4.3 Hydride Morphologies in the V-H System

Metallography

Two early studies [2.122, 123] described the habit plane and general characteristics of hydrogen precipitation in V at low concentrations. More extensive work by *Wanagel* et al. [2.124] showed the habit plane of β in α to be (10 3 3) within an accuracy of 1°. The orientation relationship between α and β is such that $[001]_\beta$ is tilted 2.17° along [111] with respect to $[001]_\alpha$. Directions $[\bar{1}10]_\alpha$ and $[\bar{1}10]_\beta$ are identical. Also, the following relationship holds for β and γ[23]:

$$(100)_\beta \parallel (110)_\gamma \quad \text{and} \quad [001]_\beta \parallel [001]_\gamma .$$

Work in this laboratory confirmed the (10 3 3) habit planes for β phase precipitates in α reported in [2.124]. An interesting metallographic feature is the occurrence of polishing-induced surface hydrides. This phenomenon is shown in Fig. 2.36 where the thin plates are the original β phase plates (produced by gas phase charging), whereas the broad hydride patches arose during the polishing procedure[24]. Figure 2.37 shows the $\beta+\varepsilon$ two-phase

---

[23] The notation used here is based on a bct β phase and an fcc γ phase. Strictly speaking, β_H has a monoclinic structure.

[24] Such massive hydrogen pickup is a common feature in the preparation of V alloys, see for example [2.125]. It is our experience that 10 at-% can easily be picked up during the polish. Obviously, this effect would severely affect metallographic studies of V deuterides.

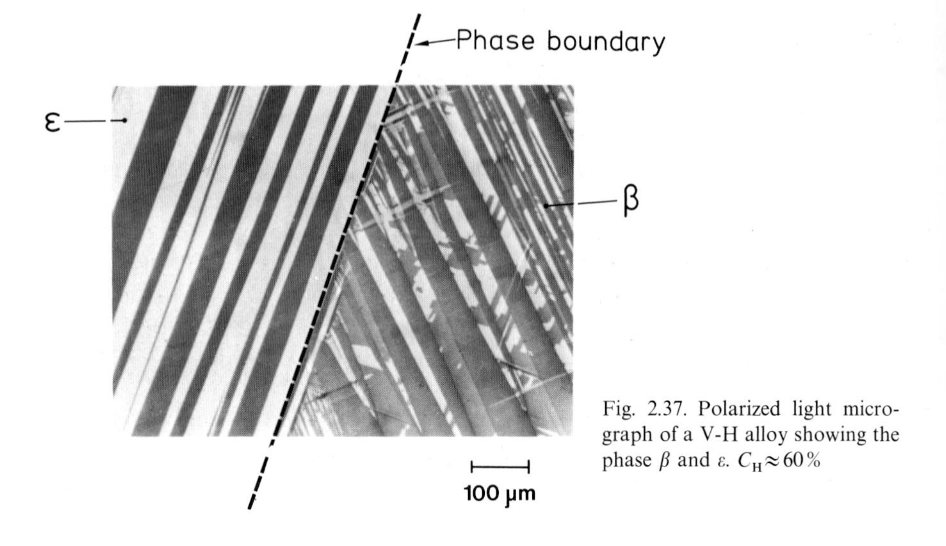

Fig. 2.37. Polarized light micrograph of a V-H alloy showing the phase β and ε. $C_H \approx 60\%$

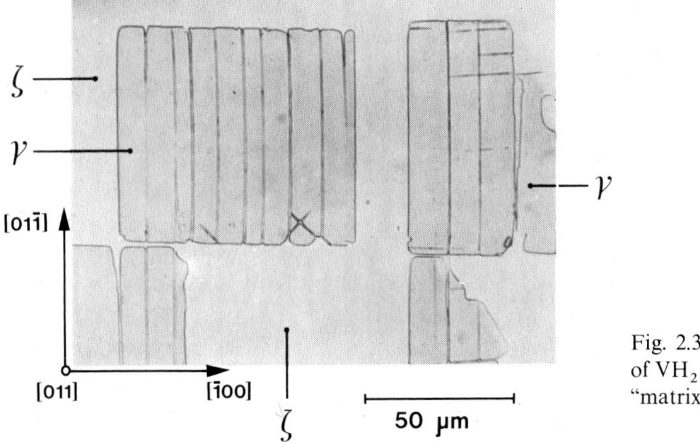

Fig. 2.38. Optical micrograph of VH_2 precipitates in the VH "matrix" [2.27]

domain morphology around 60%, and Fig. 2.38 the appearance of $\gamma(VH_2)$ inside phase ζ(VH). Further work on the above transitions is in progress.

Electron Microscopy of V-H Alloys

The structural electron diffraction results were discussed above [2.46, 47, 49, 124]. There are two morphological studies of the V-H system [2.126, 127], in the first of which the precipitation under external stress is described. In these studies structural aspects, the domain configuration, and crack formation were also discussed. For important new work see [2.198, 199].

2.5 Preparation of Samples

In most cases one starts by preparing crystals of the matrix metal with the proper purity, crystal lattice perfection, and shape. The interstitial sites of these crystals can be filled with a specified concentration of hydrogen by exchange with a surrounding gas phase or a protonic electrolyte as hydrogen source. The large values of the diffusion coefficient usually guarantee fast equilibration in the bulk. The main problem consists in the proper control of the surface penetration rate, which should be high during the doping process and very low afterwards for most experiments, in which sealed samples are required. Below room temperature doping by ion implantation may be useful, although relatively expensive ion accelerators are necessary for this method (Chap. 6). In a recent review [2.69] preparation and characterization procedures were discussed in detail. Valuable information can also be found in a recent monograph [2.128]. In the present work we summarize only a few important facts.

2.5.1 Purification of the Matrix Metal

Three characteristic groups of impurities can be distinguished as shown in Table 2.3 for Nb. The first group consists of substitutional foreign atoms of transition metals, mainly V, Nb, Ta, W, Mo, with concentrations up to 200 at-ppm. They cannot be removed easily because they have physicochemical properties similar to the matrix, but in most cases the quoted concentrations can be tolerated. Substitutional impurities from low melting elements belong to the second group. In general, their properties differ considerably from the properties of the matrix atoms. They are removed to levels far below 10 at-ppm by electron-beam float zone melting.

The last, but most important group consists of interstitial impurities from the light elements like H (which is not listed in Table 2.3), C, N, O. They have a drastic influence on the properties of the metal [2.129–132] and its hydrides [2.12, 64, 78, 133]. Their total concentration must be reduced at least below 50 at-ppm. The commercial raw materials are usually strongly contaminated by these elements. But with standard heating techniques in ultrahigh vacuum they can be removed relatively easily in a nondestructive way to levels below 10 at-ppm, except for V with its low melting point. After such a degassing treatment, various processes for further preparation can follow immediately: a doping process with gaseous elements, e.g., nitrogen, or a coating process [2.134, 135], or hydration [2.97]. If the samples are heated by current flow, electrotransport can be used to improve the final purity [2.136, 137]. The purest vanadium has been made by electrotransport refining [2.138] and zone melting [2.139].

Hydrogen and nitrogen are desorbed as diatomic molecules without interference with other species. The reactions $2[H]_{dissolved} \rightleftharpoons (H_2)_{gas}$ and

Table 2.3. Mass-spectrometric analysis of raw niobium-batches from major procedures (by courtesy of H. Beske, ZCH, Kernforschungsanlage Jülich). Concentrations are given in at-ppm. Limits of detection: 0.1–1 ppm. Wires (∅ = 1.6 mm) made from these materials exhibit a RRR between 1500 and 2000 after being Joule-heated for 3 h at 2600 K in a vacuum of 10^{-9} mbar

Impurity	Electron-beam-remelted niobium	Zone-refined niobium		
	Teledyne, Wah Chang Albany	Metals Research Limited		Materials Research Corporation
	Rod	Wire	Single crystals	Wire
Ta	105	225	92	45
Mo	1.6	11	2.3	0.1
W	3.3	28	66	11
Al	1.4	13		2.2
Cr	4.3	12	7	4.5
Mn	0.6	2.1		0.7
Fe	6.8	16	10	6.8
Ni	0.6	1.4	0.6	0.7
Cu	3.4	8.5	0.2	5.5
C	5	4.7	3.7	2.7
N	150	400	300	40
O	200	1760	487	260

$2[\mathrm{N}]_{\mathrm{dissolved}} \rightleftharpoons (\mathrm{N}_2)_{\mathrm{gas}}$ in the stationary state are determined by Sieverts' law for small concentrations according to $c_\infty = \sqrt{p} \cdot K^0 \exp(-E/RT)$. Values for the preexponential factor K^0 and the activation energy E are listed in Table 2.4.

In the case of oxygen, the partial pressure corresponding to the equilibrium $2[\mathrm{O}]_{\mathrm{dissolved}} \rightleftharpoons (\mathrm{O}_2)_{\mathrm{gas}}$ is so much smaller than the residual oxygen gas pressures present in ultrahigh vacuum systems that one would expect complete oxidation of the sample (*Fromm* [2.145]). But this does not happen. At low temperatures the limitation in diffusion rate restricts oxidation to surface layers only. At elevated temperatures oxide molecules begin to evaporate, which leads to a strong deoxidation of the sample. Mass spectrometric studies [2.146, 147] have revealed that at low pressures ($<10^{-5}$ mbar) of oxygen and water vapor the deoxidizing species is predominantly the molecule NbO. The reaction $[\mathrm{O}]_{\mathrm{dissolved}} + \langle \mathrm{Nb} \rangle_{\mathrm{solid}} \rightarrow (\mathrm{NbO})_{\mathrm{gas}}$ is irreversible, because niobium oxides condense on the cold surface of the vacuum container and cannot return to the sample. At low pressures (see Table 2.4) $c_\infty = p \cdot K^0 \exp(-E/RT)$.

An even more complex problem is the removal of carbon [2.144, 148, 149]. In the presence of oxygen, carbon is volatilized from the surface as carbon monoxide. The simplest way to provide enough oxygen at the sample is heating at moderate temperatures (2000 K) in an oxygen atmosphere (10^{-6} mbar).

Table 2.4. Thermodynamic relationships for niobium-gas systems in equilibrium or stationary condition at low concentration values. c in at-%, p in mbar, T in K

Gas	c-p-T relationship	Remarks			References
		c	p	T	
H_2	$\log c_H = 0.5 \log p_{H_2} - 2.43 + 1840/T$	< 5	$10^{-7} - 1$	400–1200	[2.140]
N_2	$\log c_N = 0.5 \log p_{N_2} - 3.04 + 9300/T$	< 10	$10^{-4} - 3$	1900–2500	[2.141]
O_2	$\log c_O = \log p_{O_2} - 3.23 + 16700/T$	< 2	$10^{-5} - 10^{-3}$	2150–2500	[2.142]
H_2O	$\log c_O = \log p_{H_2O} - 4.95 + 19600/T$	< 0.8	$10^{-5} - 10^{-3}$	2150–2450	[2.143]
CO	$\log c_C \cdot c_O = \log p_{CO} - 6.08 + 14650/T$		$10^{-4} - 10^{-2}$	1800–2150	[2.144]

Under these conditions the oxygen reactions indicated previously will occur simultaneously with the formation and volatilization of carbon monoxide. The reaction energies for hydrogen, nitrogen, oxygen, and carbon are negative. As a consequence, high temperatures just below the melting point are needed to reach low values of c_∞.

In the case of a massive sample, diffusion will be the main rate-determining process. The usual formalism (*Crank* [2.150]) applies with a surface concentration rapidly decreasing towards its final value as boundary condition. Just below the melting point of niobium the H, C, N, O interstitials have approximately the same diffusion coefficient [2.2, 69] and travel a root mean square distance of about 1 cm in 1 h. In the case of a foil the rate will be determined by surface reactions, whereas the concentration differences within the sample are small (e.g., *Grünwald* et al. [2.151]). As hydrogen is one of the most important residual gases in an ultrahigh vacuum apparatus, the sample must be cooled rapidly and its surface must be passivated with the help of oxygen gas deliberately injected into the ultrahigh vacuum system as soon as the sample has cooled to a few hundred degrees to avoid uncontrolled doping with H (*Faber* and *Schultz* [2.94]).

For the purification of vanadium, additional methods have been used. Oxygen can be removed efficiently by electrotransport (*Carlson* [2.138]) or by heating in a molten getter material like Ba (*Peterson* et al. [2.152]).

2.5.2 Analysis of Impurity Contents

The most universal multielement method for the chemical analysis of all elements down to the sub-ppm is mass spectrometry. Gaseous elements can be determined by other methods with varying success [2.153, 154]. For nitrogen the Kjeldahl method seems to be the most reliable. Extraction by vacuum-fusion methods at very high temperatures ($T > 2800$ K) can also be used (*Friedrich* et al. [2.155]). The surface layers can be characterized by Auger electron spectroscopy or secondary ion mass spectrometry [2.156–158].

The average purity of samples can be analyzed very accurately by measuring the residual electrical resistivity at liquid helium temperatures (*Wenzl* and *Welter* [2.69]). A popular specification for the purity of a sample is the residual resistance $R(295\,\mathrm{K})/R(4.2\,\mathrm{K})$ which is essentially equal to the ratio of specific resistivities. High RRR means high purity. Samples with less than 50 at-ppm impurities have RRR >500. It is not difficult to obtain samples of Nb and Ta on a large scale (one sample/day) with RRR up to 3000. For V a value of above 1000 can be reached by electrotransport refining or zone melting [2.138, 139]. Otherwise the RRR stays below 1000. Such samples are pure enough for most experiments on hydrides. By using the whole machinery of purification methods, higher purities of Nb have been reached with RRR between 10,000 and 30,000 [2.137, 159, 160].

2.5.3 Growth, Shaping, and Characterization of Single Crystals

Single crystals are mostly grown with an electron beam floating zone technique [2.161, 162]. Standard commercial crystals have a maximum diameter of 13 mm and lengths up to 250 mm. These crystals usually contain a high density of dislocations and many subgrain boundaries but also small regions of quite good crystal lattice perfection. As the crystals are usually not grown in ultrahigh vacuum conditions, they must be degassed after growth. It appears that during this heat treatment the regions of good quality stabilize, whereas those of low quality deteriorate (*Fenzl* [2.163]). Crystals with a very low dislocation density can be obtained with special techniques, like growth from the melt by the "pedestal" method (*Naramoto* and *Kamada* [2.164, 165]) or growth in the solid state by the "strain-anneal" method [2.166, 167].

The most flexible way to shape crystals is by spark machining. But the generation of dislocations at the erosion front cannot be avoided; the depth of the damage zone varies between 0.1 and 1 mm depending on the cutting speed (*Guberman* [2.168]). The damaged surface layers must be etched away, e.g., with a mixture of HF and HNO_3. To prevent hydrogen pickup during the etching, one must use hydrogen-free electrolytes [2.28, 169]. Monocrystalline foils 10 μm thick with large areas of several cm^2 have been made by jet polishing (*Wombacher* [2.170]).

The easiest method to check the density of dislocations is to generate etch pits and to analyze them in the optical microscope [2.171, 172]. More information on the size of the mosaic blocks can be gained from conventional x-ray topography and double-crystal rocking curves (see, e.g., *Achter* et al. [2.173]). Figure 2.39 shows a Berg-Barrett x-ray topograph of a typical commercial crystal. The spatial distribution of mosaic blocks separated by small angle grain boundaries is clearly revealed by this technique. With x-rays the registered depth below the surface is limited to some μm. Only in nearly perfect crystals can this depth be extended to 1 mm [2.174, 175]. Larger volumes (up to some cm^3) can be analyzed by using γ-rays with shorter

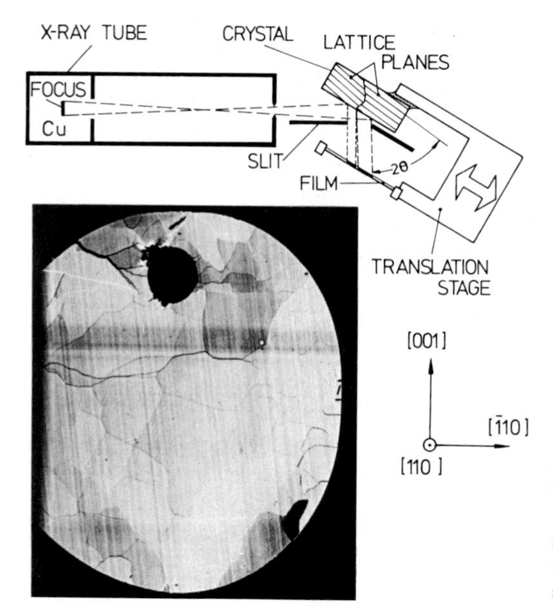

Fig. 2.39. Berg-Barrett topography of the surface of a slice cut from a niobium crystal. The hole was drilled by spark machining. Various subgrains of slightly different orientation are seen

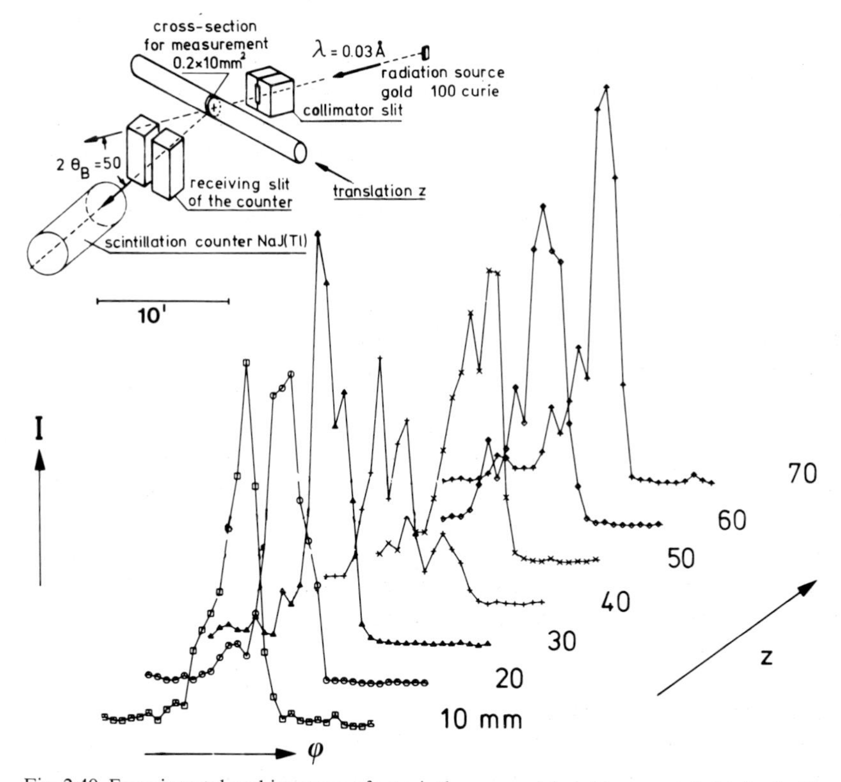

Fig. 2.40. Experimental rocking curve of a typical commercial niobium crystal obtained with γ-rays (412 keV). The cylindrical crystal (φ: 13 mm) was analyzed at different positions along its length (parameter z) with a beam of 0.2×10 mm cross section (by courtesy of *H. J. Fenzl*)

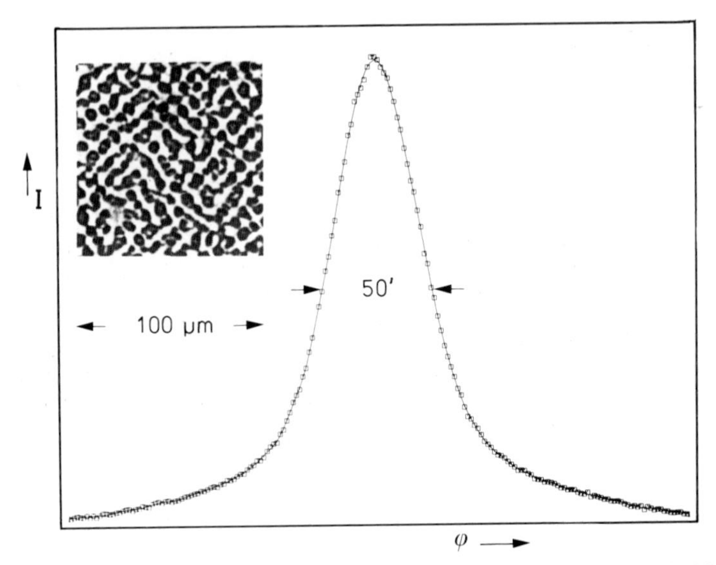

Fig. 2.41. Rocking curve obtained with γ-rays for NbH with $c \approx 0.45$ quenched from 400 °C to room temperature within less than 1 s (by courtesy of *H. J. Fenzl*). Such a crystal can be used as neutron monochromator [2.178]. The inset shows an optical micrograph of a section through such a crystal. Quite regular arrangements of small α and β phase regions can be seen (black and white contrast)

wavelengths and larger absorption and extinction lengths than x-rays [2.74, 176, 177]. The larger extinction length leads to a better resolution of the mosaic structure.

Figure 2.40 shows an example of such an analysis of the structural quality of a commercial Nb crystal similar to the one analyzed in Fig. 2.39 using 412 keV γ-rays emitted from activated gold. Perfect crystals would show one sharp peak with a line width of 10″ of arc, which corresponds essentially to the divergence of the beam. Thus the γ diffractometer permits the selection of regions with sufficient perfection which can be cut out for further sample preparation. The γ-ray rocking curve of a hydrided niobium crystal is shown in Fig. 2.41. The peak is smooth and wide due to internal deformation of the sample as a result of incoherent separation of different hydride phases. At higher concentrations, samples may even disintegrate during cooling if two-phase regions are crossed too slowly. Crystals with smooth peaks such as shown in Fig. 2.41 can be used to generate a neutron beam with a smooth spectrum of wavelengths with a width adjusted optimally for neutron scattering experiments [2.178].

2.5.4 Synthesis of Hydrides

The hydrides of V, Nb, and Ta are usually synthesized by doping properly prepared crystals with hydrogen from a reservoir of hydrogen gas or liquid

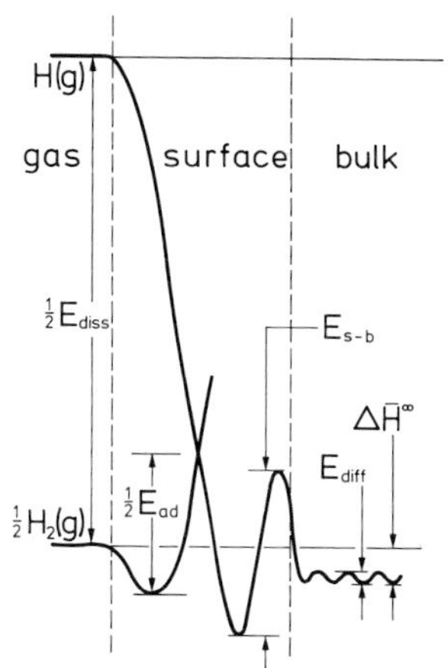

Fig. 2.42. Schematic potential energy diagram of a molecule H_2 and an atom H between gas phase and bulk of niobium. E_{diss} is the dissociation energy of the molecule (2.2 eV/atom H) in the gas phase; E_{ad}, the activation energy of dissociation at the surface; E_{s-b}, the activation energy of transfer of atom from surface into bulk; $\Delta\bar{H}^{\infty}$, the enthalpy of exothermic dissolution (−0.3 eV/half molecule for Nb); E_{diff} is the classical diffusion activation energy (0.1 eV/atom for Nb)

electrolyte containing protons. The basic reactions are $H_2 \rightleftharpoons 2H$ and $H^+ + e^- \rightleftharpoons H$. The doping rate is determined by energy barriers on the surface as schematically indicated in Fig. 2.42 for gas doping and by diffusion in the bulk. The surface of the samples is usually poisoned by impurities which modify and enhance the energy barriers (Fig. 2.42). An important part of the doping process consists in cleaning the surfaces or modifying them in a way to increase or decrease the surface penetration rate. Special techniques have been developed for experiments where the dissolution of hydrogen in niobium is studied at low temperatures ($T < 600$ K). Then hydrogenation must take place in the *UHV* vessel immediately after the degassing of the sample (*Pryde* and *Titcomb* [2.97]) or the surface must be coated with nonpoisonable materials like palladium (*Boes* and *Züchner* [2.179]). In general, the surface layers become permeable at more elevated temperatures (see below). An advantage of the surface barriers is undoubtedly the fact that many experiments can be performed in air up to temperatures of 500 K without any loss of hydrogen. Special sealing techniques have been developed for higher temperatures (*Klatt* et al. [2.180]).

Gas doping is based on the *p-c-T* diagrams shown in Figs. 2.43, 2.44, and 2.45. The useful temperature range lies between 700 and 850 K. At lower temperatures the permeation rate of hydrogen through the surface decreases drastically. Figure 2.46 shows a modern gas doping apparatus for hydrogen pressures below 3 bar (*Klatt* et al. [2.185]). Pure hydrogen gas can be obtained by permeation through a Pd-Ag cell or by decomposition of a hydride like Fe-Ti (*Klatt* et al. [2.11]). The major residual contamination of the gas results from its reaction with the walls of the tubing and of the vessel (e.g., water vapor may form by reaction with oxygen adsorbed on the walls.)

Figure 2.47 represents an electrochemical doping cell with the sample inserted as the cathode in the electrolyte. Concentrated orthophosphoric acid (H_3PO_4) or sulfuric acid (H_2SO_4) can be used as an electrolyte up to

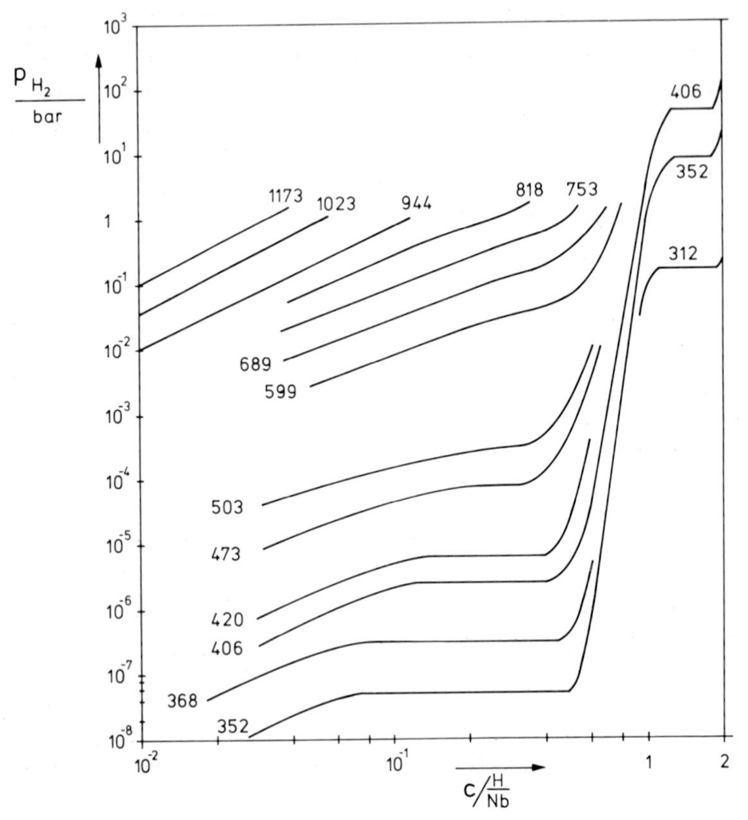

Fig. 2.43. Survey of experimental solubility curves for niobium, plotted as logarithm of hydrogen gas pressure vs concentration (ratio H/Nb) with temperature (in K) as parameter [2.97, 140, 181–184]

temperatures of at least 500 K. Below room temperature, alcoholic solutions of HCl and H_2SO_4 can be used (*Boes* and *Züchner* [2.179] and Chap. 6). Compared with doping from the gas phase, the electrochemical method has various advantages—simple experimental setup, easy extension to higher concentrations and lower temperatures, and less contamination. In principle, the adjustment of the hydrogen concentration can be accomplished by fixing the electrochemical potential of the sample in relation to the electrolyte. In practice, side reactions occur and the basic phenomena involved in the process are not yet well understood. For instance, the reaction $2H_{ad} \rightarrow H_2$(gas) uses up most of the electric current flowing in the cell. Therefore, the efficiency of doping is often limited to a few percent and depends on sample shape and current density. The normally used cell voltages of 100 mV to 1 V are much higher than the equilibrium voltages for the hydride. In this case concentrated hydrides are formed rapidly near the surface. They act as diffusion barriers and

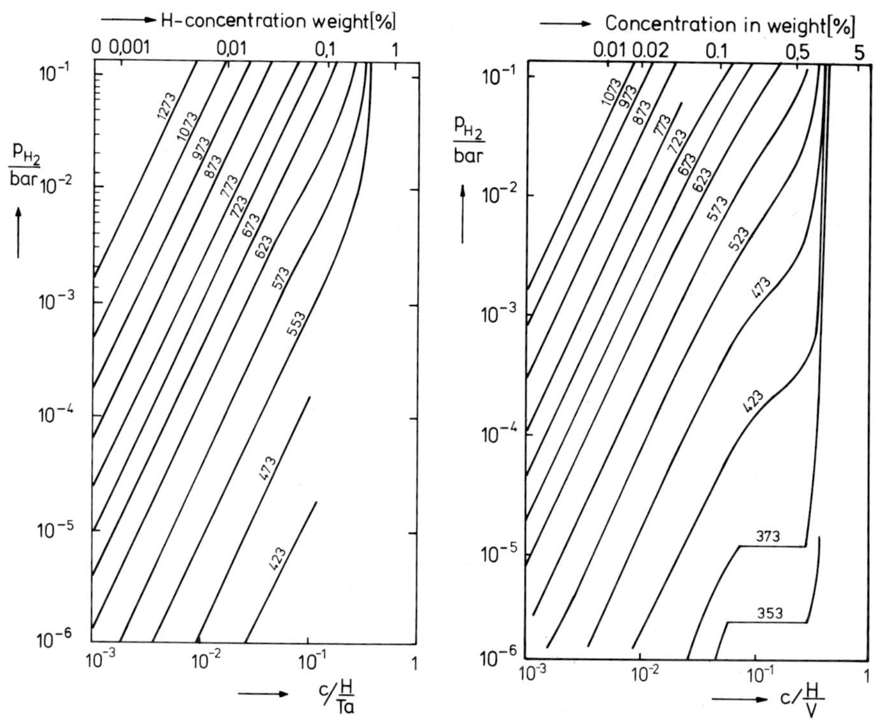

Fig. 2.44. Solubility curves for tantalum (collected by *Fromm* and *Gebhardt* [2.128] and *Völkl* and *Alefeld* [2.2])

Fig. 2.45. Solubility curves for vanadium (collected by *Fromm* and *Gebhardt* [2.128] and *Völkl* and *Alefeld* [2.2])

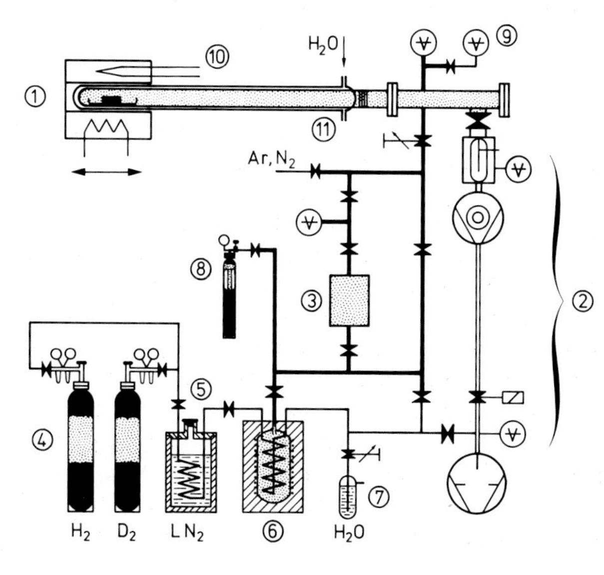

Fig. 2.46. Apparatus for hydriding with hydrogen gas: *1* removable furnace positioned over the sample, *2* pump unit with cold baffle: turbomolecular pump and backing pump, *3* calibrated volume, *4* containers for pressurized gas, *5* cold trap to remove H_2O, *6* palladium-silver permeation cell to purify hydrogen from pressure containers, *7* hydrogen exhaust, *8* container with metal hydride from which very pure hydrogen is released, *9* precision pressure gauges, *10* thermocouple, *11* jacket for rapid cooling with water

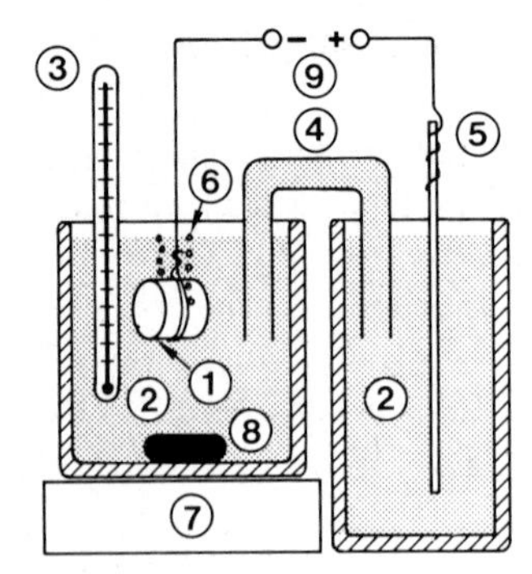

Fig. 2.47. Schematic drawing of arrangement for electrochemical doping of samples with hydrogen: *1* sample polarized as cathode, *2* electrolyte, *3* thermometer, *4* current bridge between hot cathodic (left) and cold anodic (right) compartment, *5* cathode, *6* hydrogen gas bubbles formed by H-H recombination at anode surface during current flow (unwanted but unavoidable side reaction), *7* heater and motor for magnetic stirrer *8*, *9* electric connection to power supply, e.g., potentiostat

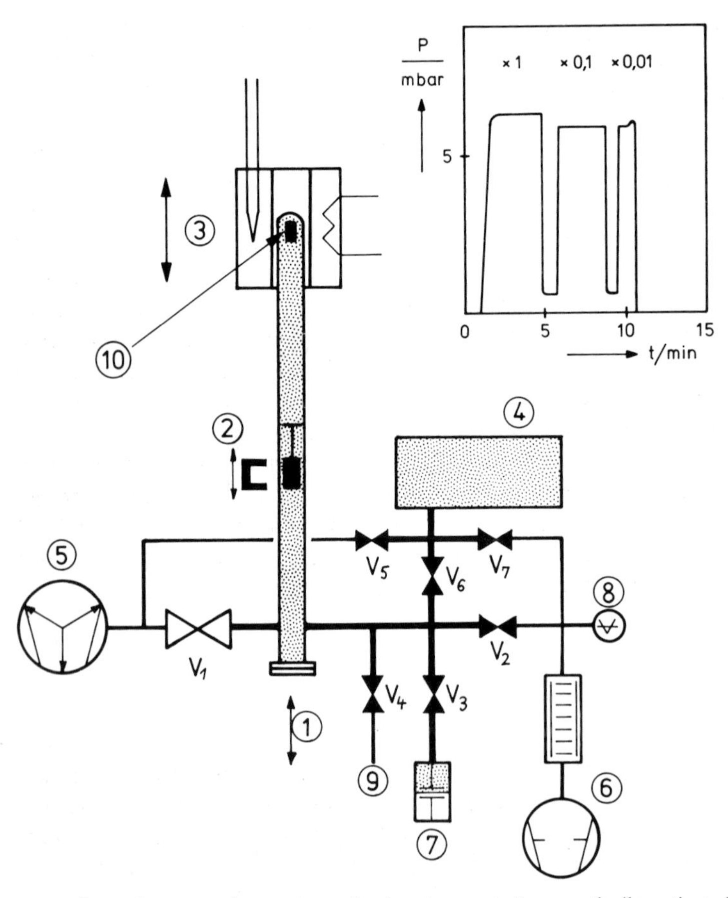

Fig. 2.48. Hydrogen analyzer: *1* sample charging port; *2* magnetically activated sample pusher; *3* movable furnace; *4* calibrated volume for gas sampling; *5* sputter-ion pump; *6* mechanical pump with sorbent trap; *7* capacitance manometer, range 10^{-4} to 10 mbar; *8* backing pressure gauge; *9* vent valve; *10* sample. The inset shows an extraction cycle; when the hydrogen pressure reaches a steady value in the apparatus, valve V_6 is closed and the sampling volume is evacuated down to 10^{-7} mbar. Then valve V_6 is again opened until new steady-state conditions are established. This process is repeated until essentially all the hydrogen is removed from the sample and measured. Detection limit: 5 ng hydrogen

induce excessive elastic stresses which may lead to plastic deformation and surface cracks. Furthermore electrochemical doping is accompanied by chemical dissolution of the metal, especially when sulfuric acid is used.

With energetic proton beams, dilute and concentrated solid solutions can be prepared at the temperature of liquid helium (*Stritzker* [2.186] and Chap. 6). Various problems are associated with this ion implantation technique which concern location of the hydrogen, lattice defects, as well as the morphology and structure of metastable phases.

NbH_2 has been prepared at room temperature by mixing niobium powder and zinc in hydrochlorid acid (*Brauer* and *Müller* [2.28]). Atomic hydrogen is produced by the reaction $Zn+2HCl \rightarrow ZnCl_2+2H$. Another way to produce H-atoms is the thermal dissociation of hydrogen molecules on hot tungsten filaments (*Oates* and *Flanagan* [2.187]). Further preparation methods are discussed in Chapter 6.

In general, it is difficult to get reliable information regarding the hydrogen content from the operating parameters of the doping process. Measuring the change in mass is an easy and fast method of checking the concentration. The best and most common method for the determination of the amount of hydrogen dissolved in niobium is the vacuum-extraction technique. The sample is degassed in a calibrated volume at approximately 1200 K; the change of pressure is used to determine the hydrogen content. Figure 2.48 shows such a hydrogen analyzer (*Lengeler* et al. [2.188]). If the analyzer is designed according to *UHV* technology and a starting pressure of less than 10^{-7} mbar is used, the whole pressure increase can be attributed to hydrogen.

Secondary ion mass spectrometry is directly sensitive to hydrogen. Furthermore, it permits one to determine the depth distribution of hydrogen although it does this destructively by sputtering. Nuclear reactions like $^{1}H(^{19}F, \alpha\gamma)^{16}O$ (*Leich* and *Tombrello* [2.189]) allow one to analyze depth distributions nondestructively, but without good lateral resolution.

2.6 Conclusion

We summarize the results of this chapter by pointing out some common features of the different systems utilizing lattice gas models (see, for instance, [2.4, 190, 191]). The application of these models is based on several properties:

1) With the notable exception of some V-H(D) phases, hydrogen occupies the same type of interstitial site in V, Nb, and Ta: the tetrahedral site.

2) Six tetrahedral sites are available per matrix atom. These sites can be filled by exothermic transfer of hydrogen from the molecular gas phase (see Sect. 2.5.1 and Fig. 2.42), although blocking of more than one site by an interstitial atom has prevented so far the theoretical possibility of exceeding the maximum concentration of two interstitials per metal atom reached experimentally in V and Nb.

3) The interaction between H and the surrounding matrix ions is relatively weak in comparison with other interstitials as revealed by the small increase of the average atomic volume (Fig. 2.2). Therefore, the unit cells of the matrix atoms of even the concentrated hydrides MH or MH_2 can be generated from the initial bcc unit cell by small relative distortions not exceeding 10%.

4) Migration of hydrogen through the lattice is characterized by a large diffusion coefficient which has only a weak temperature dependence. In spite of this high mobility, the occupation of tetrahedral sites is normally well defined as was shown by *Sizmann* and his collaborators [2.14, 33].

Phase transitions occur due to long-range and short-range interactions between the hydrogen lattice gas atoms. The experimental phase diagrams were presented in the form of projections of pressure-concentration-temperature surfaces onto the temperature-concentration plane (Figs. 2.1, 18, 21, 25, 27) and onto the pressure-concentration plane (Figs. 2.43–45). These phase diagrams have several features in common:

1) Slightly above room temperature the lattice gas or α phase with disordered hydrogen extends over a vast temperature and concentration range until the composition MH is reached. At higher concentrations, one type of the various "interstitial crystal" phases appears with composition MH_2 (called δ phase in Nb, γ phase in V).

2) Decreasing the temperature below this region results in disorder-order transitions of the lattice gas accompanied by structural transitions of the matrix. Therefore, these transitions are not continuous and are of first order. The ordered phase exists in a smaller concentration range than the disordered gas phase, roughly between $MH_{0.5}$ and MH. In terms of a "hydrogen interstitial crystal" a disordered arrangement of vacancies or "vacancy gas" is present in this region of the phase diagrams if the composition deviates from MH_x with $x<1$. The β phase in V, Nb, Ta; the ε phase in V, Ta; the δ phase in Ta; and the ζ phase in V belong to this type of ordered interstitial crystal with disordered vacancies.

3) Further cooling induces disorder-order transitions of the "vacancy gas" according to certain branching schemes [2.18]. In principle, only perfectly ordered structures exist at 0 K. In reality, the ordering process cannot always proceed to equilibrium because the mobility is no longer sufficient below a temperature of about 70 K. Therefore, only a few phases with different types of vacancy ordering were detected (ε and λ phases in Nb, phases ζ and γ in Ta, and phases η and δ in V).

4) The gas-liquid phase transition $\alpha-\alpha'$ with a critical point was observed explicitly in NbPd and Pd alloys. In V and Ta, T_c lies below the highest temperatures of disorder-order transitions (T_c is about 310 K in V and 210 K in Ta [2.71, 192]). Therefore, the region called α' in the V and Ta phase diagrams still corresponds to the interstitial gas phase α. Unfortunately, a theoretical model which explains both types of transitions, gas-liquid and disorder-order (gas-liquid-crystal), is not available. Interaction effects via elastic distortions

alone are not sufficient for understanding the complexity of the experimental phase diagrams as already indicated by *Pick* for the case of the β phase in NbH [2.5, 6].

Acknowledgment. Helpful discussions are acknowledged with Drs. Welter, Pick, and Hauck and with Mr. Fenzl. Special thanks are due to Drs. D. Braski and J. Roberto for a revision of the English text. The technical assistance of B. Bischof, A. Brusch, H. J. Bierfeld, F. Birmans, A. Carl, K. H. Klatt, U. Linke, C. Mambor, and Dr. V. Sorajić was appreciated. Much of the photographic work is due to W. Fisseler. The expert typing was done by Mrs. Klein.

References

2.1 R.B. McLellan, C.G. Harkins: Mater. Sci. Eng. **18**, 5–35 (1975)
2.2 J. Völkl, G. Alefeld: "Hydrogen Diffusion in Metals"; in *Diffusion in Solids, Recent Developments*, ed. by A.S. Nowik and J.J. Burton (Academic Press, New York 1974)
2.3 G. Alefeld, J. Völkl (eds.): *Hydrogen in Metals I. Basic Properties*, Topics in Applied Physics, Vol. 28 (Springer, Berlin, Heidelberg, New York 1978)
2.4 G. Alefeld: Phys. Stat. Sol. **32**, 67 (1969)
2.5 M.A. Pick, R. Bausch: J. Phys. F: Metal Phys. **6**, 1751 (1976)
2.6 M.A. Pick: Ph.D. Thesis, University of Aachen (1973); "Strukturelle Phasenübergänge im NbH-System", Techn. Rpt. Jül-951-FF, KFA Jülich (1973)
2.7 W.B. Pearson: *Lattice Spacings and Structures of Metals and Alloys* (Pergamon Press, Oxford, London, Edinburgh, New York, Paris, Frankfurt 1964)
2.8 H. Hillmann: Z. Metallk. **66**, 69–73 (1975)
2.9 D.M. Gruen (ed.): *The Chemistry of Fusion Technology* (Plenum Press, New York 1972)
2.10 G. Alefeld: Wärme **81**, 89 (1975)
2.11 K.H. Klatt, A. Carl, M.A. Pick, H. Wenzl: DPA P2607156.7 (1976); M.A. Pick, H. Wenzl: Int. J. Hydrogen Energy **1**, 413–420 (1977)
2.12 R.J. Walter, W.T. Chandler: Trans. Met. Soc. AIME **233**, 762 (1965)
2.13 V.A. Somenkov, A.V. Gurskaya, M.G. Zemlyanov, M.E. Kost, N.A. Chernoplekov, A.A. Chertkov: Sov. Phys.—Solid State **10**, 1076 (1968)
2.14 H.D. Carstanjen, R. Sizmann: Ber. Bunsenges. Phys. Chem. **76**, 1223 (1972)
2.15 C. Wainwright, A.J. Cook, B.E. Hopkins: J. Less-Common Metals **6**, 362 (1964)
2.16 M.S. Rashid, T.E. Scott: J. Less-Common Metals **30**, 399 (1973)
2.17 V.A. Somenkov, V.F. Petrunin, S.Sh. Shil'stein, A.A. Chertkov: Sov. Phys.—Cryst. **14**, 522 (1970)
2.18 V.A. Somenkov: Ber. Bunsenges. Phys. Chem. **76**, 733 (1972)
2.19 J. Hauck: Acta Cryst. A **33**, 208 (1977)
2.20 J. Hauck, H.J. Schenk: J. Less-Common Metals **51**, 251 (1977)
2.21 T. Schober, M.A. Pick, H. Wenzl: Phys. Stat. Sol. (a) **18**, 175 (1973)
2.22 T. Schober: Scripta Met. **7**, 1119 (1973)
2.23 T. Schober: Phys. Stat. Sol. (a) **29**, 395 (1975)
2.24 T. Schober: Phys. Stat. Sol. (a) **30**, 107 (1975)
2.25 W. Pesch: To be published
2.26 J.-M. Welter: To be published
2.27 T. Schober: Proc. 2nd Int. Congr. Hydrogen in Metals, Paris, 1977, Section 1D2. Proc. 9th Int. Congr. Electron Microscopy, Toronto, 1978, p. 644
2.28 G. Brauer, H. Müller: Angew. Chem. **70**, 53 (1958)
2.29 J.J. Reilly, R.H. Wiswall: Inorg. Chem. **9**, 1678 (1970)
2.30 V.A. Somenkov, A.V. Gurskaya, M.G. Zemlyanov, M.E. Kost, N.A. Chernoplekov, A.A. Chertkov: Sov. Phys.—Solid State **10**, 2123 (1969)
2.31 V.F. Petrunin, V.A. Somenkov, S.Sh. Shil'stein, A.A. Chertkov: Sov. Phys.—Crystallography **15**, 137 (1970)

2.32 F. Ducastelle, R. Caudron, P. Costa: J. Phys. Chem. Sol. **31**, 1247 (1970)
2.33 M. Antonini, H. D. Carstanjen: Phys. Stat. Sol. (a) **34**, K 153 (1976)
2.34 B. Stalinski: Bull. Acad. Polon. Sci. **2**, 245 (1954)
2.35 A. G. Khachaturyan, G. A. Shalatov: Acta Met. **23**, 1089 (1975)
2.36 H. Asano, Y. Ishino, R. Yamada, M. Hirabayashi: J. Solid State Chem. **15**, 45 (1975)
2.37 J. Wanagel, S. L. Sass, B. W. Batterman: Phys. Stat. Sol. (a) **11**, K 97 (1972)
2.38 T. Schober, H. Wenzl: Scripta Met. **10**, 819 (1976)
2.39 H. Asano: Private communication
2.40 V. A. Somenkov, A. Y. Chervyakov, S. Sh. Shil'stein, A. A. Chertkov: Sov. Phys.—Crystallography **17**, 274 (1972)
2.41 A. J. Maeland: J. Phys. Chem. **68**, 2197 (1964)
2.42 H. Asano, Y. Abe, M. Hirabayashi: Acta Met. **24**, 95 (1976)
2.43 H. Asano, Y. Abe, M. Hirabayashi: J. Phys. Soc. Japan **41**, 974 (1976)
2.44 V. A. Somenkov, I. R. Entin, A. Y. Chervyakov, S. Sh. Shil'stein, A. A. Chertkov: Sov. Phys.—Solid State **13**, 2178 (1972)
2.45 R. R. Arons, H. G. Bohn, H. Lütgemeier: Proc. Int. Meeting Hydrogen in Metals, Jülich 1972, Jül-Conf-6, Vol. I, p. 272; J. Phys. Chem. Sol. **35**, 207 (1974)
2.46 J. Wanagel, S. L. Sass, B. W. Batterman: Phys. Stat. Sol. (a) **10**, 49 (1972)
2.47 M. Cambini, R. Serneels, R. Gevers: Phys. Stat. Sol. (a) **21**, K 57 (1974)
2.48 S. Takano, T. Suzuki: Acta Met. **22**, 265 (1974)
2.49 J. Wanagel, S. L. Sass, B. W. Batterman: Phys. Stat. Sol. (a) **11**, 767 (1972)
2.50 D. G. Westlake, S. T. Ockers, W. R. Gray: Met. Trans. **1**, 1361 (1970)
2.51 H. Asano, M. Hirabayashi: Phys. Stat. Sol. (a) **16**, 69 (1973)
2.52 H. Asano, M. Hirabayashi: Phys. Stat. Sol. (a) **15**, 267 (1973)
2.53 A. Y. Chervyakov, I. R. Entin, V. A. Somenkov, S. Sh. Shil'stein, A. A. Chertkov: Sov. Phys.—Solid State **13**, 2172 (1972)
2.54 D. G. Westlake, M. H. Müller, H. W. Knott: J. Appl. Cryst. **6**, 206 (1973)
2.55 K. I. Hardcastle, T. R. P. Gibb: J. Phys. Chem. **76**, 927 (1972)
2.56 Y. Fukai, S. Kazama: Scripta Met. **9**, 1073 (1975)
2.57 D. G. Westlake, S. T. Ockers, M. H. Müller, K. D. Anderson: Met. Trans. **3**, 1709 (1972)
2.58 I. R. Entin, V. A. Somenkov, S. Sh. Shil'stein: Sov. Phys.—Solid State **16**, 1569 (1975)
2.59 H. Horner: Private communication
2.60 N. Boccara: Ann. Phys. **47**, 40–64 (1968)
2.61 B. Alefeld et al.: To be published
2.62 J.-M. Welter: To be published
2.63 R. Heller, H. Wipf: Phys. Stat. Sol. (a) **33**, 525 (1976)
2.64 D. G. Westlake, S. T. Ockers: Met. Trans. **4**, 1355 (1973); **6** A, 399 (1975)
2.65 J. A. Pryde, C. G. Titcomb: Trans. Faraday Soc. **65**, 2758 (1969)
2.66 R. A. Swalin: *Thermodynamics of Solids* (Wiley and Sons, New York, London, Sydney, Toronto 1972)
2.67 T. Schober, U. Linke: J. Less-Common Metals **44**, 63 (1976)
2.68 T. Schober, U. Linke: J. Less-Common Metals **44**, 77 (1976)
2.69 H. Wenzl, J.-M. Welter: "Properties and Preparation of NbH Interstitial Alloys", in *Current Topics in Materials Science*, ed. by E. Kaldis, Vol. 1 (North-Holland, Amsterdam 1978)
2.70 J. Völkl: Ber. Bunsenges. Phys. Chemie **76**, 797–805 (1972)
2.71 J. Tretkowski: „Einfluß der Probenform auf thermodynamische Eigenschaften von Wasserstoff in Metallen"; Techn. Rpt. Jül-1049-FF, KFA Jülich (1974)
2.72 J. L. Whitton, J. B. Mitchell, T. Schober, H. Wenzl: Acta Met. **24**, 483–490 (1976)
2.73 O. Buck, D. O. Thompson, C. A. Wert: J. Phys. Chem. Sol. **32**, 2331–2344 (1971)
2.74 H. J. Fenzl, M. A. Pick, H. Wenzl: Scripta Met. **11**, 271 (1977)
2.75 U. Köbler, T. Schober: J. Less-Common Metals **60**, 101 (1978)
2.76 G. Hörz: Metall **30**, 728–736 (1976)
2.77 H. Y. Chang, C. A. Wert: Acta Met. **21**, 1233 (1973)
2.78 G. Pfeiffer, H. Wipf: J. Phys. F: Metal Phys. **6**, 167–179 (1976)
2.79 K. Rosan, H. Wipf: Phys. Stat. Sol. (a) **38**, 611 (1976)

2.80 H. Zabel, H. Peisl: Phys. Stat. Sol. (a) **37**, K 67–K 70 (1976)
2.81 J.J. Reilly, R.H. Wiswall: Inorg. Chem. **9**, 1678–1682 (1970)
2.82 H.J. Fenzl: To be published
2.83 M. Ableiter, U. Gonser: Z. Metallk. **66**, 86–92 (1975)
2.84 M. Amano, Y. Sasaki: J. Japan Inst. Metals **38**, 969 (1974)
2.85 R. Heibel, H. Wollenberger: Scripta Met. **10**, 945–947 (1976)
2.86 S. Aronson, J.J. Reilly, R.H. Wiswall: J. Less-Common Metals **21**, 439–442 (1970)
2.87 J.-M. Welter et al.: To be published
2.88 D. Zamir, R.M. Cotts: Phys. Rev. **134** A, 666–675 (1964)
2.89 H. Lütgemeier, R.R. Arons, H.G. Bohn: J. Magn. Resonance **8**, 74 (1972)
2.90 K. Nakamura: Trans. Nat. Res. Inst. Met. **17**, 305–322 (1975)
2.91 D.G. Westlake: Scripta Met. **10**, 75 (1976)
2.92 D.G. Westlake: Trans. AIME **245**, 287–292 (1969)
2.93 G. Schaumann, J. Völkl, G. Alefeld: Phys. Stat. sol. **42**, 401–413 (1970)
2.94 K. Faber, H. Schultz: Scripta Met. **6**, 1065–1070 (1972)
2.95 H. Conrad, G. Bauer, G. Alefeld, T. Springer, W. Schmatz: Z. Physik **266**, 239–244 (1974)
2.96 H. Trinkhaus et al.: Private communication
2.97 J.A. Pryde, C.G. Titcomb: Trans. Faraday Soc. **65**, 2758 (1969)
2.98 A.G. Khachaturyan: "Static Concentration waves in the Theory of Order-Disorder Phenomena in Substitutional and Interstitial Solid Solutions", in *Order-Disorder Transformations in Alloys*, ed. by H. Warlimont (Springer, Berlin, Heidelberg, New York 1974)
2.99 D. Chandra, T.S. Ellemann, K. Verghese: J. Nucl. Mater. **59**, 263–279 (1976)
2.100 W.G. Saba, W.E. Wallace, H. Sandmo, R.S. Craig: J. Chem. Phys. **35**, 2148 (1961)
2.101 P. Kofstad, R. Butera: J. Appl. Phys. **34**, 1517 (1963)
2.102 T.G. Berlincourt, P.W. Bickel: Phys. Rev. B **2**, 4838 (1970)
2.103 D. Slotfeld-Ellingsen, B. Pedersen: Phys. Stat. Sol. (a) **25**, 115 (1974)
2.104 M. Hirabayashi, S. Yamaguchi, H. Asano, K. Hiraga: In *Order-Disorder Transformations in Alloys*, ed. by H. Warlimont (Springer, Berlin, Heidelberg, New York 1974) p. 266
2.105 T. Schober, A. Carl: Scripta Met. **11**, 397 (1977)
2.106 P. Lecocq, C. Wert: Thin Solid Films **25**, 77 (1975)
2.107 J. Zierath: Ph.D. Thesis, University of Münster (1969)
2.108 H. Asano, R. Yamada, M. Hirabayashi: Trans. Jap. Inst. Met. **18**, 155 (1977)
2.109 D.G. Westlake: Trans. TSM-AIME **239**, 1341 (1967)
2.110 G.C. Rauch, R.M. Rose, J. Wulff: J. Less-Common Metals **8**, 99 (1965)
2.111 Y. Sasaki, M. Amano: Trans. Jap. Inst. Met. **8**, 276 (1967); **13**, 296 (1972)
2.112 D. Hardie, P. McIntyre: Met. Sci. J. **6**, 40 (1972)
2.113 M. Amano, Y. Sasaki: Phys. Stat. Sol. (a) **19**, 405 (1973)
2.114 T. Schober, U. Linke, H. Wenzl: Scripta Met. **8**, 805 (1974)
2.115 T. Schober, H. Wenzl: Phys. Stat. Sol. (a) **33**, 673 (1976)
2.116 T. Schober, U. Linke: Metallography **9**, 309 (1976)
2.117 G.V. Raynor: In *Physical Metallurgy*, ed. by R.W. Cahn (North-Holland, Amsterdam 1970) p. 374
2.118 H.K. Birnbaum, M.L. Grossbeck, M. Amano: J. Less-Common Metals **49**, 357 (1976)
2.119 M. Amano, Y. Sasaki: J. Jap. Inst. Met. **38**, 969 (1974)
2.120 D.G. Westlake, S.T. Ockers: J. Less-Common Metals **42**, 255 (1975)
2.121 P. Lecocq: Ph.D. Thesis, University of Illinois at Urbana Champaign (1974)
2.122 D.G. Westlake: J. Less-Common Metals **23**, 89 (1971)
2.123 M.S. Rashid, T.E. Scott: J. Less-Common Metals **31**, 377 (1973)
2.124 J. Wanagel, S.L. Sass, B.W. Batterman: Met. Trans. **5**, 105 (1974)
2.125 E. Lang, J. Bressers: Z. Metallk. **67**, 66 (1976)
2.126 S. Takano, T. Suzuki: Acta Met. **22**, 265 (1974)
2.127 T. Chiba, S. Takano: J. Phys. Soc. Japan **31**, 1113 (1971)
2.128 E. Fromm, E. Gebhardt (eds.): *Gase und Kohlenstoff in Metallen* (Springer, Berlin, Heidelberg, New York 1976)
2.129 G. Elssner, G. Hörz: J. Less-Common Metals **21**, 451–455 (1970)

2.130 S. Miura, J. Takamura, M. Yamashita: Review of Doshisha University **14**, 13–39 (1973)
2.131 W. DeSorbo: Phys. Rev. **132**, 107–121 (1963)
2.132 C.C. Koch, J.O. Scarbrough, D.M. Kröger: Phys. Rev. B **9**, 888–897 (1974)
2.133 D. Richter, J. Töpler, T. Springer: J. Phys. F: Metal Phys. **6**, L93–L97 (1976)
2.134 K.H. Klatt, J.-M. Welter, H. Wenzl: Z. Metallk. **67**, 567–572
2.135 N. Boes, H. Züchner: Z. Naturforsch. **31**a, 754–759 (1976)
2.136 R. Kirchheim, E. Fromm: High Temp.—High Pressures **6**, 329–339 (1974)
2.137 K. Schulze, E. Fromm, W. Grünwald, F. Haeßner, S. Hofmann, R. Kirchheim, K.-D. Rasch, H. Schultz, M. Winterkorn: Reinstoffe aus Wissenschaft und Technik (to be published)
2.138 O.N. Carlson, F.A. Schmidt, D.G. Alexander: Met. Trans. **3**, 1249 (1972)
2.139 J. Bressers, R. Creten, G. van Holsbeke: J. Less-Common Metals **39**, 7 (1975)
2.140 E. Veleckis, R.K. Edwards: J. Phys. Chem. **73**, 683 (1969)
2.141 E. Gebhardt, E. Fromm, D. Jakob: Z. Metallk. **55**, 423 (1964)
2.142 E. Fromm, H. Jehn: Z. Metallk. **58**, 61 (1967)
2.143 E. Fromm, H. Jehn: Z. Metallk. **58**, 120 (1967)
2.144 E. Fromm, G. Spaeth: Z. Metallk. **59**, 65–68 (1968)
2.145 E. Fromm: J. Less-Common Metals **14**, 113–125 (1967)
2.146 L.H. Rovner, A. Drowart, F. Degreve, J. Drowart: Contract AF **61** (052)–700 (1967)
2.147 J. Jupille, J.M. Michel: J. Less-Common Metals **39**, 17–34 (1975)
2.148 G. Melchior: Vakuum-Technik **18**, 1–7 (1969)
2.149 G. Hörz, K. Lindemaier: Z. Metallk. **63**, 240–247 (1972)
2.150 J. Crank: *The Mathematics of Diffusion* (Clarendon Press, Oxford 1975)
2.151 W. Grünwald, F. Haeßner, W. Hemminger, H.L. Lukas: Z. Metallk. **65**, 184–190 (1974)
2.152 D.T. Peterson, R.G. Clark, W.A. Stensland: J. Less-Common Metals **30**, 169 (1973)
E. Fromm, H. Kirchheim: J. Less-Common Metals **26**, 403 (1972)
2.153 L.M. Melnick, L.L. Lewis, B.D. Hold (eds.): *Determination of Gaseous Elements in Metals* (Wiley and Sons, New York 1974)
2.154 Ph. Albert: Informationsheft des Büro Eurisotop 90, Serie: Monographien **34** (1974)
2.155 K. Friedrich, E. Lassner, T. Kraus, G. Paesold, F. Schlät: Hochtemperatur-Werkstoffe (6. Plansee-Seminar, Juni 1968) p. 1016–1027
2.156 H.H. Farell, H.S. Isaacs, M. Strongin: Surface Sci. **38**, 31–52 (1973)
2.157 M. Romand, J.S. Solomon, W.T. Barn: Mat. Res. Bull. **11**, 517 (1976)
2.158 S. Hofmann, G. Blank, H. Schultz: Z. Metallk. **67**, 189 (1976)
2.159 R.W. Meyerhoff: J. Electrochem. Soc.: Solid State Sci. **118**, 997–1001 (1971)
2.160 J. Barthel, K.H. Berthel, K. Fischer, R. Gebel, G. Güntzler, M. Jurisch, W. Neumann, J. Kunze, P. Müll, H. Oppermann, R. Petri, G. Sobe, G. Weise, W. Wisner: Fiz. Metal. Metalloved. **35**, 921 (1973)
2.161 R.E. Reed, H.D. Guberman, T.O. Baldwin: J. Phys. Chem. Solids, Suppl. **1**, 829 (1967)
2.162 J. Barthel, R. Scharfenberg: In *Crystal Growth*, ed. by H.S. Peiser (Pergamon Press, New York 1968) p. 133–139
2.163 H.-J. Fenzl: Private communication
2.164 H. Naramoto, K. Kamada: J. Cryst. Growth **24/25**, 531–536 (1974)
2.165 H. Naramoto, K. Kamada: J. Cryst. Growth **30**, 145–150 (1975)
2.166 T.G. Digges, Jr., M.R. Achter: Transactions of the Metallurgical Society of AIME, June 12 (1964)
2.167 H.J. Fenzl, M. Beyß: In „Kristall-Labor-Führer und Bericht 1969–1973", ed. by H. Wenzl; Jül-944-FF, KFA Jülich (1973) p. 3, 4
2.168 H.D. Guberman: J. Appl. Phys. Techn. Rpt. **39**, 2975 (1968)
2.169 V. Sorajić: Angew. Elektrochemie **27**, 80–83 (1973)
2.170 P. Wombacher: J. Phys. F: Scientific Instr. **5**, 243–245 (1972)
2.171 R.G. Vardiman, M.R. Achter: Transactions of the Metallurgical Society of AIME **242**, 196–205 (1968)
2.172 H.D. Guberman: ORNL-3878, Radiation Metallurgy Section, Solid State Division, Progress Report for Period Ending August 1965

2.173 M.R. Achter, C.L. Vold, T.G. Digges, Jr.: Transactions of the Metallurgical Society of AIME **236**, 1597–1605 (1966)
2.174 R.E. Reed, H.D. Guberman, T.O. Baldwin: ORNL-3949, Radiation Metallurgy Section, Solid State Division, Progress Report for Period Ending, February 1966
2.175 H. Wenzl: In „Kristall-Labor-Führer und Bericht 1969–1973", ed. by H. Wenzl; Techn. Rpt. Jül-944-FF, KFA Jülich (1973) p. 7.2
2.176 J. Schneider: J. Appl. Cryst. **7**, 541 (1974)
2.177 A. Freund, J.R. Schneider: In „Dreiländer-Jahrestagung über Kristallwachstum und Kristallzüchtung", Techn. Rpt. Jül-Conf-18, KFA Jülich (1976)
2.178 J.R. Schneider, N. Stump: Nucl. Instr. Methods **125**, 605–608 (1975)
2.179 N. Boes, H. Züchner: Z. Naturforsch. **31**a, 760–768 (1976)
2.180 K.-H. Klatt, J.-M. Welter, H. Wenzl: Z. Metallk. **67**, 567–572 (1976)
2.181 J.J. Reilly, R.H. Wiswall: Inorg. Chem. **9**, 1678–1682 (1970)
2.182 E. Veleckis: Ph.D. Thesis, Illinois Institute of Technology (Chicago, 1960)
2.183 R. Burch, N.B. Francis: Acta Met. **23**, 1129 (1975)
2.184 O.J. Kleppa, P. Dantzer, M.E. Melniach: J. Chem. Phys. **61**, 4048 (1974)
2.185 K.-H. Klatt et al.: To be published
2.186 B. Stritzker: Z. Physik **268**, 261–264 (1974)
2.187 W.A. Oates, T.B. Flanagan: Can. J. Chem. **53**, 694–701 (1975)
2.188 B. Lengeler et al.: To be published
2.189 D.A. Leich, T.A. Tombrello: Nucl. Instr. Methods **108**, 67–71 (1973)
2.190 J.R. Lacher: Proc. Roy. Soc. London A **161**, 525 (1937)
2.191 F.D. Manchester: *Hydrogen in Metals*, ed. by I.R. Harris, J.P.G. Farr (Elsevier Sequoia S.A., Lausanne 1976)
2.192 H.C. Bauer, J. Tretkowski, J. Völkl, G. Alefeld: Z. Physik **B29**, 17 (1978)
2.193 H. Lütgemeier, H.G. Bohn, R.R. Arons: J. Magn. Resonance **8**, 80 (1972)
2.194 T. Schober, A. Carl: Phys. Stat. Sol. (a) **43**, 443 (1977)
2.195 T. Schober: Scripta Met. **12**, 549 (1978)
2.196 G. Mair: To be published
2.197 T. Schober, A. Carl: Proc. 2nd World Hydrogen Energy Conf., Zürich, Vol. 3 (1978), p. 1575
2.198 J.S. Bowles, B.C. Muddle, C.M. Wayman: Acta Met. **25**, 513 (1977)
2.199 M.P. Cassidy, B.C. Muddle, T.E. Scott, C.M. Wayman, J.S. Bowles: Acta Met. **25**, 829 (1977)

3. Hydrogen in Palladium and Palladium Alloys

E. Wicke and H. Brodowsky, with cooperation by H. Züchner

With 35 Figures

Dedicated to Prof. Dr. Dr. h. c. *Carl Wagner* in honor of his pioneer work on the palladium/hydrogen system

3.1 Overview

Of all metal-hydrogen systems the Pd/H_2 system is the first, historically, that attracted research activities (*Graham* [3.1]) and, as *Lewis* [3.2] pointed out, has been most extensively investigated since then. A condensed review of the progress in knowledge about this system was given by *Flanagan* [3.3] at the occasion of the Centennial of Graham's discovery of the absorption of hydrogen by palladium and palladium-silver alloys. The broad and continual interest is primarily due to the high solubility and mobility of the hydrogen in the fcc Pd lattice, where the hydrogen atoms occupy octahedral sites, as has been shown by neutron diffraction on β phase Pd(H) and Pd(D) [3.4], and recently by applying the channeling technique with ^{3}He on low-content α phase Pd(D) [3.5]. Palladium hydride represents one of the most transparent and instructive models for a metal-hydrogen system from structural, thermodynamic, and kinetic points of view. Much fundamental knowledge and experimental technique have been developed in investigations on this system, and applied later to other more complicated cases. Nevertheless, there are quite a number of details in the mechanism of hydrogen diffusion as well as in the behavior of electronic states in this system not yet fully understood. Surprises, therefore, are still possible, like the discovery of superconductivity in hydrogen-rich β phase Pd(H) by *Skośkiewicz* [3.6] in 1972.

A basic problem still in discussion is which type of model will be most adequate for handling the attractive hydrogen-hydrogen interactions induced by the elastic strains of the host lattice. In one type of model, particular emphasis is laid upon the long-range part of these interactions, and it is suggested that short-range attractions do not occur, i.e., that the short-range interactions are purely repulsive [Ref. 3.100, Chap. 2].

Macroscopic elasticity theory is involved in this model, and mean-field approximations are used; the statistics are rather cumbersome if the simple Bragg-Williams approximation is not applied. The other type of model is based upon short-range attractive interactions, acting predominantly between nearest neighbors, such that the concept of pair interaction and the statistical method of the quasichemical approximation can be applied. It is this model that will be preferred in the following, for two reasons: first, because the detailed elaboration of the quasichemical method and its atomistic approach enable straightforward evaluation of experimental results, and second, because quite a number of those results indicate the efficacy and hence the existence of short-range attractive interactions. The real condition will be somewhere intermediate, as often in nature, between the two simplifying concepts of short-range and of long-range interactions, but the full treatment of the general case is still lacking.

Besides this problem, the situation in the palladium-hydrogen field at present can be characterized by the impression that quite a number of different lines of research now begin to converge into what may be called a composite and generally valid conception that enables a better understanding of the backgrounds and of the functional connections of the different effects.

Such converging lines are, for instance, the determination of the frequencies of the local modes of H and D in the Pd host lattice by neutron scattering methods, and the measurement of the equilibrium isotope effects by thermodynamic means. Particularly important is the extension of our knowledge from the bulk to the hydrogen chemisorption states at the surface. The results achieved by surface spectroscopy and diffraction methods are well in line with the measurements of isotope separation effects (inverted in comparison to the bulk), and with the mechanism of surface kinetics, developed in investigations of the rate of hydrogen entering the lattice from the gas phase or leaving it. Another example is the convergence of the results on the electronic band structures of Pd, Pd alloys, and their hydrides, obtained by measurements of electronic properties, by thermodynamic investigations, by Mössbauer spectroscopy, and by APW band calculations.

It is the aim of this review to expose such convergent lines and to promote the advancement of adequate concepts. Thus, by taking into account the screening effects of solute ions, a "shifting band model" is presented, that comprises the results on electronic effects mentioned above, and, besides this, can be correlated to the change of activation energies of hydrogen diffusion in Pd alloys with composition.

These diffusion measurements have been performed with time-lag methods. It is the concern of the last section of this review to point up the reliability and efficiency of these methods, and to represent the convergence of their results with those of Gorsky effect measurements and of recent neutron diffraction studies.

3.2 Thermodynamics

3.2.1 Basic Relationships

The chemical potential of hydrogen dissolved in Pd can be set equal to

$$\mu_H = \mu_H^0 + RT \cdot \ln \frac{n}{1-n} + \Delta\mu_H \tag{3.1}$$

according to *Wagner* [3.7][1]. Here μ_H^0 is the standard potential, the second term on the right-hand side is the configurational term for ideal statistic distribution of the hydrogen atoms among the octahedral sites (n = H/Pd atomic ratio), and $\Delta\mu_H$ stands for the deviations from ideal solution behavior. *Wagner* himself took the excess potential $\Delta\mu_H$ as a contribution of the hydrogen electrons $\Delta\mu_e$; later *Brodowsky* [3.9] added a protonic contribution

$$\Delta\mu_H = \Delta\mu_{H^+} + \Delta\mu_e\,. \tag{3.1a}$$

This does not mean, however, that the H-atom dissociates into proton and electron when entering the Pd lattice. Actually, the electronic states of the H-atoms are broadened by the field of the surrounding Pd atoms and the electrons delocalized by exchange interaction with the high density d-band states at the Fermi level. So the nature of the H-atoms in the octahedral sites can be imagined as electronically screened protons, the screening charge of which occupies part of the broadened states in the $4d$ band and of the delocalized states in the $5s$ band of the host lattice. The corresponding ascent of the Fermi level gives rise to the contribution $\Delta\mu_e$, while $\Delta\mu_{H^+}$ originates in the expansive strain which the vibrating protons with their screening charges exert on the octahedron of adjacent Pd atoms. In view of this mechanical nature $\Delta\mu_{H^+}$ can be denoted as "elastic" contribution to the excess potential.

In the case of solution equilibrium the chemical potential of the dissolved hydrogen atoms equals the one of hydrogen gas

$$\mu_H = \tfrac{1}{2}\cdot\mu_{H_2} = \tfrac{1}{2}\mu_{H_2}^0 + \tfrac{1}{2}RT\cdot\ln p_{H_2}\,. \tag{3.2}$$

Introducing Sieverts' constant

$$K = \exp[(\mu_H^0 - \tfrac{1}{2}\mu_{H_2}^0)/RT]\,, \tag{3.3}$$

one obtains with (3.1)

$$\ln\sqrt{p_{H_2}} = \ln\left(K\cdot\frac{n}{1-n}\right) + \frac{1}{RT}(\Delta\mu_{H^+} + \Delta\mu_e)\,. \tag{3.4}$$

When the last term of the right-hand side is omitted, (3.4) represents the behavior in the ideal solution state, i.e., Sieverts' law. The temperature

[1] In an earlier statistical treatment *Lacher* [3.8] supposed two different types of interstitial sites—one type (60%) accessible for H-atoms, the other (40%) not accessible—in order to account for the steep increase of μ_H at $n > 0.6$. *Wagner* [3.7] was the first to relate this increase with the electronic contribution to the chemical potential of the dissolved hydrogen. Accordingly—and as it turned out later to be correct—he based the statistics on the mutual equivalence of all interstitial (octahedral) sites.

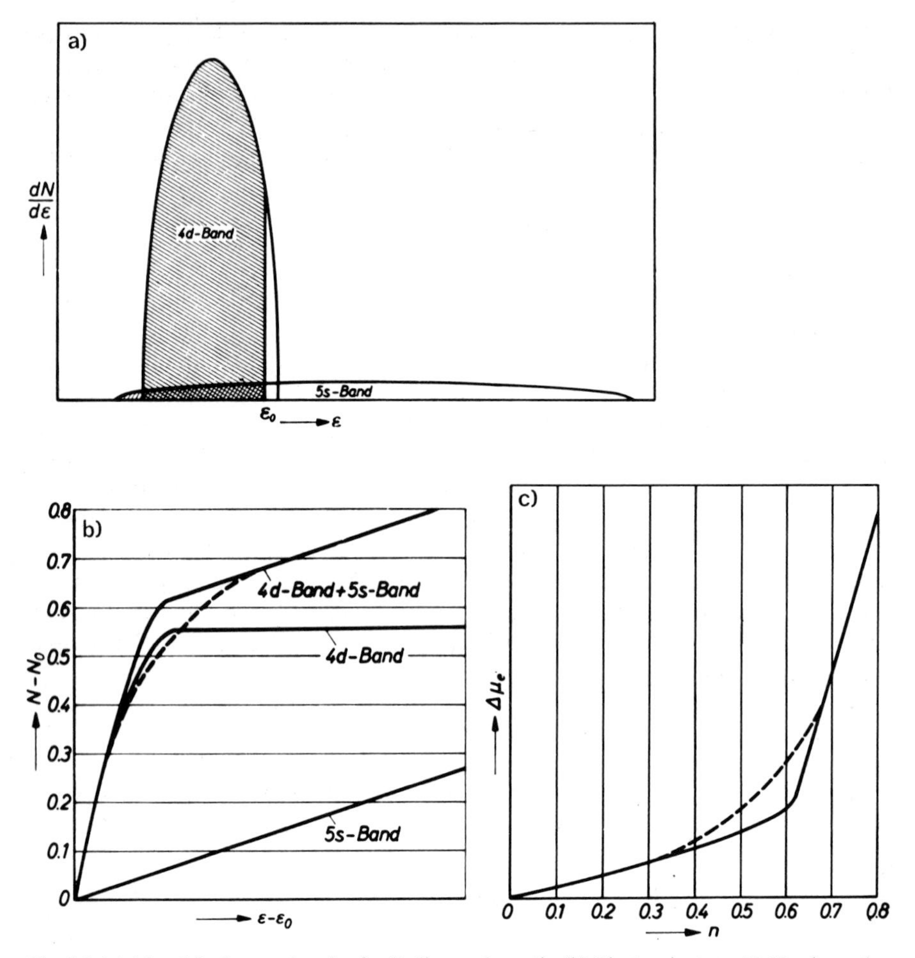

Fig. 3.1. (a) $4d$ and $5s$ electron bands of palladium, schematic. (b) Electronic states $N-N_0$ above the Fermi level; broken line: at finite temperature $T>0$ K. (c) Elrctronic contribution $\Delta\mu_e$ to the excess potential (*Brodowsky* [3.9])

dependence of Sieverts' constant is determined, according to

$$\ln K = -\frac{\Delta H^0}{2RT} + \frac{\Delta S^0}{2R} \tag{3.5}$$

by the molar enthalpy of desorption, ΔH^0, and the standard molar entropy of desorption without configurational contributions, ΔS^0, both in the limit $n \to 0$.

The *elastic contribution* $\Delta\mu_{H^+}$ to the nonideal behavior, originating in the dilatation of the octahedral "cages" by the occupying hydrogen atoms, leads to attractive interaction between the H-atoms in the lattice as was first pointed out in 1964/65 [3.9, 10]. This attraction causes $\Delta\mu_{H^+}$ to take negative values, and brings about cluster formation and disintegration into two phases at appropriate values of hydrogen concentration and temperature. At small hydrogen contents $\Delta\mu_{H^+}$ is proportional to n: $\Delta\mu_{H^+}=E_H(T)\cdot n$, where $E_H(T)<0$ means an interaction parameter, dependent on temperature. Introduction into (3.4) yields

$$\ln\sqrt{p_{H_2}}=\ln\left(K\cdot\frac{n}{1-n}\right)+\frac{E_H(T)\cdot n}{RT} \tag{3.6}$$

as a good approximation for the $p(n)$ isotherm in the range $0<n\lesssim 0.02$ [3.11, 12].

The *electronic contribution* $\Delta\mu_e$ to nonideality represents the ascent of the Fermi level by the electrons of the dissolved hydrogen. Figure 3.1 illustrates in the approximation of the rigid band model, that at first the 4*d* band with its high density of states will be filled up, causing a slight and gradual increase of $\Delta\mu_e$ in the range $0<n\lesssim 0.6$. Above this zone only the low-density states of the 5*s* band are available; $\Delta\mu_e$ accordingly bends off to a steep further increase with constant slope. At these high hydrogen concentrations the strong and linear increase of $\Delta\mu_e$ with n predominates in the chemical potential of the dissolved hydrogen. Therefore, the $p(n)$ isotherms in this region are of the simple logarithmic form

$$\ln p=-A(T)+B(T)\cdot n\,, \tag{3.7}$$

with temperature-dependent coefficients A and B [3.11] as was recognized empirically as early as 1930 [3.13].

3.2.2 Phase Diagram, Methods, and Results

The *lattice structure* of palladium hydride represents an isotropically expanded form of the fcc host lattice with the hydrogen atoms occupying part of the octahedral sites. Below about 300 °C the homogeneous solid solution disintegrates into an α phase with low hydrogen content and an expanded, hydrogen-rich β phase. The lattice of the β phase, when built up by discontinuous expansion of the α phase, is highly distorted, which made precise x-ray determinations of the phase limits difficult (see the discussion by *Lewis* [Ref. 3.2, p. 138ff.]). At room temperature the best values of the lattice constants for pure Pd and for the coexisting α and β phases ($n=\alpha_{max}=0.008$, and $n=\beta_{min}=0.607$) are, respectively,

$$a(0)=3.890\,\text{Å}\,,\qquad a(\alpha_{max})=3.894\,\text{Å}\,,\qquad a(\beta_{min})=4.025\,\text{Å}\,.$$

From these data follows an increase of volume of the host lattice in the $\alpha\rightarrow\beta$ transition of: $\Delta V=1.57$ cm^3/g-atom of H.

On the other hand the lattice of the homogeneous β phase increases gradually with further increase of hydrogen content above β_{min}. *Schirber* and *Morosin* [3.14] recently studied the increase of the lattice constant in the range $0.8 < n < 0.98$ at 77 K by x-ray determinations and obtained the result

$$a(n) = a(\beta_{min}) \cdot [1 + 0.044(n - \beta_{min})] . \tag{3.8}$$

With this relationship the partial molar volume of the atomic hydrogen in the hydrogen-rich β phase region studied can be calculated as $\bar{V}_H = 1.30$ cm^3/g-atom of H. Comparison with the value $\Delta V = 1.57$ cm^3/g-atom, derived above from the jump of volume in the $\alpha \rightarrow \beta$ transition, shows that the lattice expansion by additional uptake of hydrogen decreases if the hydrogen content is already high[2]—as is to be expected.

The method of primary importance for settling the phase limits in the Pd/H_2 system is the measurement of *pressure-composition isotherms*. The hydrogen content of the solid under the actual hydrogen pressure used is to be determined by means of gas volumetry, gravimetric methods, or by applying electric resistivity relationships. The details of these methods have been dealt with extensively in literature and need not be considered here. Special techniques for measurements under high-pressure conditions will be found in Chapter 4.

Using bulk Pd with bright surface, the gas-solid equilibrium is difficult to obtain below about 120 °C. This is due to impurity layers at the surface inhibiting the dissociation of impinging H_2 molecules into atoms. The early investigations at lower temperatures were therefore carried out with Pd black, the rough surface of which is an active dissociation catalyst. When the surface of bulk Pd is brought into contact with dissociation catalysts, the hydrogen atoms from the catalyst surface spill over to the Pd surface and there enter the lattice. This is the action of "hydrogen transfer catalysts" [3.16, 17], the application of which extended the temperature range of measurements with bulk Pd to well below 0 °C. Approved transfer catalysts are finely dispersed uranium hydride (UH_3) or copper powder [3.11], or finally a thin coating of Pd or Pt black on the bulk Pd surface.

A valuable tool for research on the Pd/H_2 system is the *electrochemical* charging technique for solubility determinations, originating also from a discovery of *Graham* [3.18]. The hydrogen uptake or loss by the Pd electrode, polarized as cathode or anode, respectively, is measured by coulometry—or, with wire shaped electrodes, by electric resistance—while the pressure is calculated from the hydrogen potential. The applicability of the method is based on the nature of Pd as a noble metal—the surface can easily be activated

[2] This agrees qualitatively with the results of *Baranowski* et al. [3.15] on the volume increase of fcc metals and alloys due to interstitial hydrogen in single phase solution. Up to $n \approx 0.75$ the authors found a linear increase, yielding $\bar{V}_H = 1.75$ cm^3/g-atom; then the curve bends off and approaches $n = 1$ with $\bar{V}_H \approx 0.4$ cm^3/g-atom.

for quick attainment of electrochemical equilibrium—and on the metallike consistency of the hydride. There are some difficulties inherent in the method, and controversy has arisen in literature concerning its reliability; these problems, however, have been solved, as discussed extensively by *Lewis* [Ref. 3.2, p. 31 ff.]. As a matter of fact, the technique, when applied with due precautions, yields results in good agreement with p-n-T measurements.

The electrochemical methods can be made applicable for less noble metals, like the Vb metals Ta, Nb, V, and others, by purifying their surfaces in ultrahigh vacuum and simultaneously coating them with an evaporated Pd film, as *Boes* and *Züchner* [3.19] recently showed. The favorable properties of the Pd surface mentioned above are then effective for hydrogen transfer into the lattice of these metals and for the establishment of electrochemical equilibrium at the surface.

The development of *UHV techniques* with their procedures to maintain high purity conditions made it possible to measure the gas-solid equilibria with bulk Pd also at lower temperatures without any transfer catalyst. Applying these methods, *Clewley* et al. [3.20] in a very careful investigation measured $p(n)$ isotherms and $p(T)$ isochores in the low concentration range $0.0005 < n < 0.01$ between -80 and $+350\,°C$, using a Pd wire and a rod as samples. In the temperature range 0 to 100 °C the results of the isochore measurements agree within the limits of error with earlier values of *Nernst* [3.11],

$$\Delta H^0 = 4620 \pm 150 \text{ cal/mol of } H_2\,; \qquad \Delta S^0 = 25.5 \pm 0.5 \text{ cal/K·mol of } H_2\,.$$

These values were obtained by extrapolation towards $n \to 0$ of results from $p(n)$ isotherms on Pd foils that had been measured by gas volumetry with reduced copper powder as hydrogen transfer catalyst, the adsorption capacity of which is negligible. The enthalpy value $\Delta H^0 = 4620$ cal/mol of H_2 in the limit $n \to 0$ had been confirmed earlier by measurements of *Simons* and *Flanagan* [3.21] with Pd wires applying electrochemical techniques. The recent investigation of *Clewley* et al. [3.20], covering a rather broad temperature range, indicates a small but significant decrease of ΔH^0 with temperature.

The most complete determination of the phase diagram of Pd/H_2 till now seems to be the work of *Frieske* [3.22], based upon measurements of *magnetic susceptibility isotherms* between 20 and 300 °C and at hydrogen pressures between 0.01 and 140 atm. Simultaneously, pressure-composition points were taken gravimetrically while the magnetic field was switched off.

Figure 3.2 demonstrates $\chi(n)$ and $p(n)$ isotherms at 120 °C, obtained with bulk Pd. The two branches of the $p(n)$ isotherm, absorption and desorption runs, represent the well-known hysteresis between their plateau regions, which indicate the phase transition range. The branches of the $\chi(n)$ isotherm also diverge in this range, running linearly (lever rule) across the middle of the field.

As was shown earlier [3.11] the desorption branches can be taken as the equilibrium curves within the heterogeneous and the β phase regions; on the

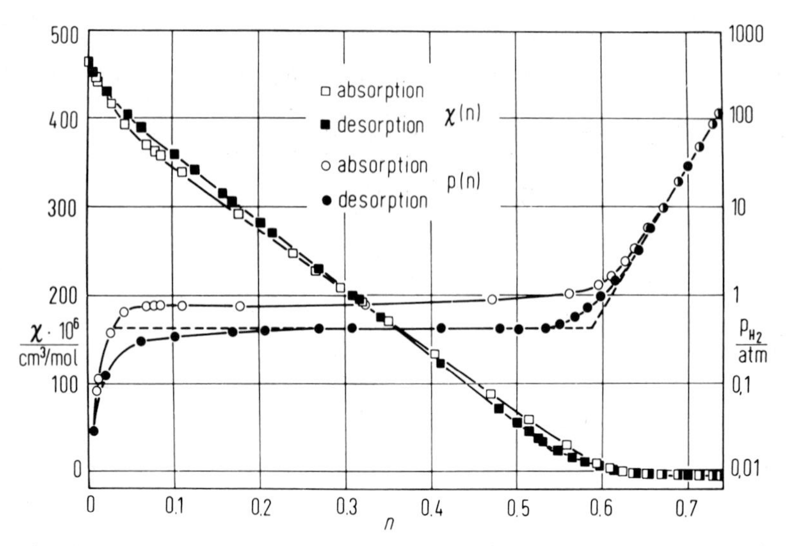

Fig. 3.2. $\chi(n)$ and $p(n)$ isotherms of Pd(H) with bulk palladium at 120 °C (*Frieske* [3.22])

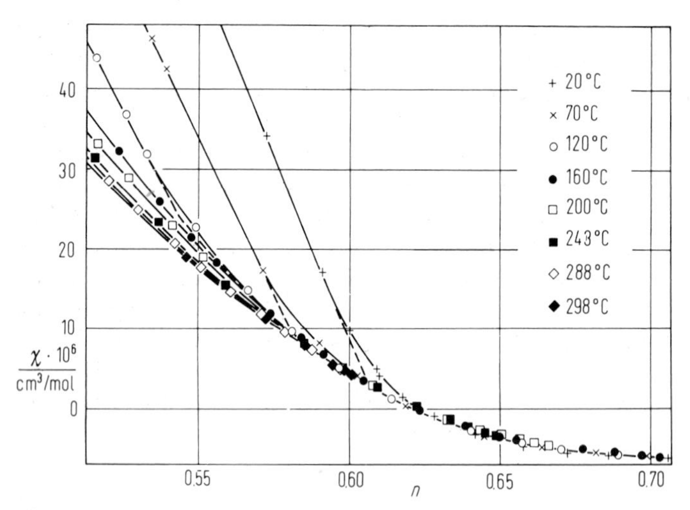

Fig. 3.3. $\chi(n)$ isotherms near the limit to the β phase [3.22]

other hand, if the two branches diverge in the α phase region the absorption curve is nearer to equilibrium. According to this statement, the limits of the *two-phase region* may be determined from the $p(n)$ isotherms after *Nernst* [3.11] by extrapolating the desorption plateau as indicated by the broken line in Fig. 3.2. The intersection on the left with the measured absorption branch in the α phase yields the value of α_{max}. In a similar way the linear part of the $\chi(n)$ desorption branch can be extended to intersect the α phase absorption curve. The values of α_{max} determined by the two methods agree within experimental

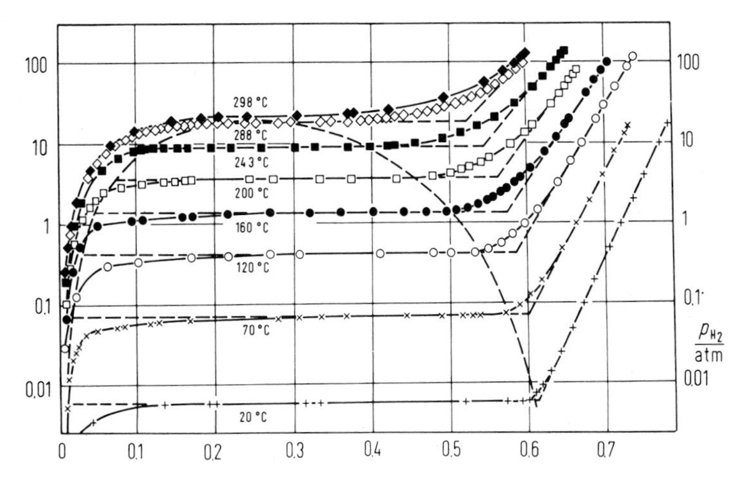

Fig. 3.4. $p(n)$ isotherms of Pd(H) with bulk palladium at different temperatures, including the critical region [3.22]

error. The values of β_{min} can be determined most reliably by means of the $\chi(n)$ isotherms. Figure 3.3 illustrates how all the $\chi(n)$ isotherms between 20 and 298 °C converge into a common envelope which is independent of temperature and crosses the $\chi=0$ line at $n=0.62$ (it then levels out to the value $\chi=-6.1\cdot10^{-6}$ cm^3/mol which was found independent of n in the range $n=0.7$ up to 0.8). The linear parts of the $\chi(n)$ desorption branches in the two-phase region have been extended to this envelope, and the intersections taken as β_{min}.

The boundaries of the two-phase region thus determined from the $\chi(n)$ isotherms yielded the dashed curve in Fig. 3.4. Numerical data are summarized in Table 3.1, together with values from *Nernst* [3.11] for lower temperatures. The position of the critical point is obtained easily by graphical interpolation. For comparison with earlier data see *Mueller* et al. [Ref. 3.23, p. 635f].

The *plateau pressures* in the two-phase region of Fig. 3.4—decomposition pressures—increase with temperature according to

$$\ln p = -\frac{\Delta H}{RT} + \frac{\Delta S^0}{R} \tag{3.9}$$

with $\Delta H = 9800 \pm 100$ cal/mol of H_2; $\Delta S^0 = 23.3 \pm 0.2$ cal/K·mol of H_2 in the temperature range 20 to 300 °C. This is to be compared with Nernst's values

$$\Delta H = 9325 \pm 100 \text{ cal/mol}; \qquad \Delta S^0 = 21.8 \pm 0.2 \text{ cal/K}\cdot\text{mol}$$

in the range −80 to 50 °C. *Clewley* et al. in their above-mentioned investigation under UHV conditions [3.20] reported a value of $\Delta H = 8460$ cal/mol from measurements with bulk Pd in the range −80 to −10 °C. Calorimetric

Table 3.1. Limits of the two-phase region of PdH_n

$T/°C$	α_{max}	β_{min}	
−78.5		0.635 ± 0.003	from
−30		0.629 ± 0.003	$p(n)$ isotherms
0	0.0055 ± 0.001	0.615 ± 0.003	[3, 11]
20	0.008 ± 0.002	0.607 ± 0.002	
70	0.017 ± 0.002	0.575 ± 0.002	from
120	0.030 ± 0.002	0.540 ± 0.003	$\chi(n)$ isotherms
160	0.046 ± 0.002	0.504 ± 0.003	
200	0.075 ± 0.002	0.459 ± 0.004	[3.22]
243	0.117 ± 0.002	0.399 ± 0.005	
288	0.21 ± 0.01	0.29 ± 0.01	

Critical point: $T_c = 292 \pm 2$ °C; $n_c = 0.250 \pm 0.005$
(corresponding pressure: $p_c = 19.7 \pm 0.2$ atm)

determinations by *Nace* and *Aston* [3.24] of the heat of absorption of hydrogen by Pd black were reviewed by *Aston* [3.25] who reported

$$\Delta H_f = -9440 \pm 45 \text{ cal/mol of } H_2$$

as enthalpy of formation from the elements of $PdH_{0.5}$ at 30 °C. At this temperature, $PdH_{0.5}$ is a two-phase mixture the composition of which follows by application of the lever rule with the data of α_{max} and β_{min} of Table 3.1 to be

$$PdH_{0.50} = 1/6\,\alpha\text{-}PdH_{0.01} + 5/6\,\beta\text{-}PdH_{0.60}\,.$$

Hence, only 3‰ of the total hydrogen absorbed formed α phase Pd(H), which means that the calorimetric value obtained represents the enthalpy of formation of β phase Pd(H) from the elements.

The ΔH values obtained by evaluation of the temperature dependence of plateau decomposition pressures by means of (3.9) represent *$\beta \to \alpha$ transition enthalpies* per mol of H_2, i.e., ΔH and ΔS^0 in (3.9) are related to the reaction

$$\frac{2}{\beta-\alpha} PdH_\beta \to \frac{2}{\beta-\alpha} PdH_\alpha + H_2$$

where α and β stand for α_{max} and β_{min}, respectively. There has been some controversy in the literature about this point [3.26]; a short thermodynamic consideration may, therefore, be inserted here.

Imagine a three-phase system PdH_α, PdH_β, and hydrogen gas in equilibrium. An amount of Δn moles of H_2 may be desorbed from PdH_β, thereby transforming a corresponding amount of PdH_β to PdH_α. By this conversion in continual equilibrium the entropy of the system increases by

$$\Delta s = \left[S_{H_2} + \frac{2}{\beta-\alpha} \cdot (S_\alpha - S_\beta) \right] \cdot \Delta n\,,$$

where S_{H_2} is the molar entropy of H_2 at the equilibrium pressure and temperature in question, and S_α, S_β are the molar entropies of the hydrides PdH_α and PdH_β at the phase limits α_{max} and β_{min}, respectively. The heat supply necessary for the isothermic, isobaric conversion is

$$\Delta h = \left[H_{H_2} + \frac{2}{\beta-\alpha}(H_\alpha - H_\beta) \right] \cdot \Delta n = \Delta H_{\beta\to\alpha} \cdot \Delta n \,,$$

where H_{H_2}, H_α, H_β mean the molar enthalpies of the species mentioned. Application of the equilibrium condition $\Delta s = \Delta h/T$, and of the expression $S_{H_2}(T,p) = S^0_{H_2}(T) - R \cdot \ln p$ yields

$$\ln p = -\frac{\Delta H_{\beta\to\alpha}}{RT} + \frac{\Delta S^0_{\beta\to\alpha}}{R}\,, \tag{3.10}$$

equivalent to (3.9), with $\Delta S^0_{\beta\to\alpha} = S^0_{H_2} + [2/(\beta-\alpha)]\,(S_\alpha - S_\beta)$.

In view of the importance of the subject, the problem may be illustrated in another way, too. As pointed out by *Wagner* [3.7], the complete thermodynamic equilibrium between PdH_α and PdH_β is based on the conditions

$$\mu^\alpha_H = \mu^\beta_H \qquad\qquad \mu^\alpha_{Pd} = \mu^\beta_{Pd}$$

$$d\mu^\alpha_H/dT = d\mu^\beta_H/dT \qquad d\mu^\alpha_{Pd}/dT = d\mu^\beta_{Pd}/dT\,,$$

where the μ^i_H, μ^i_{Pd} mean the chemical potentials of H and Pd in PdH_i at the phase limits ($i = \alpha_{max}$, β_{min}), respectively, and the temperature coefficients have to be taken along these limits.

Accordingly, $S^\alpha_H = S^\beta_H$, $H^\alpha_H = H^\beta_H$ along the phase limits, and analogously for Pd in PdH_i[3]. The H_2 equilibrium pressure increases along the phase limits according to

$$\frac{d\ln\sqrt{p}}{dT} = \frac{H_{H_2}/2 - H^\alpha_H}{RT^2}\,; \qquad \frac{d\ln\sqrt{p}}{dT} = \frac{H_{H_2}/2 - H^\beta_H}{RT^2}\,.$$

Multiplying the equations by α and β, respectively, and subtracting yields

$$\frac{\beta-\alpha}{2} \cdot d\ln p/dT = \left(\frac{\beta-\alpha}{2} H_{H_2} + \alpha H^\alpha_H - \beta H^\beta_H \right) / RT^2\,.$$

The difference $\alpha H^\alpha_H - \beta H^\beta_H$ can be set equal to

$$\alpha H^\alpha_H + H^\alpha_{Pd} - (\beta H^\beta_H + H^\beta_{Pd}) = H_\alpha - H_\beta\,,$$

where H_α, H_β are the enthalpies of PdH_α, PdH_β at the phase limits, respectively, as introduced before. Hence it follows

$$d\ln p/dT = \left[H_{H_2} + \frac{2}{\beta-\alpha}(H_\alpha - H_\beta) \right] / RT^2 = \Delta H_{\beta\to\alpha}/RT^2\,, \tag{3.11}$$

comparable to (3.10).

---

[3] The meaning of the term "along the phase limits" shows up from the relation

$$\frac{d\mu^\alpha_H}{dT} \equiv -S^\alpha_H = \left(\frac{\partial\mu_H}{\partial T}\right)_n + \left(\frac{\partial\mu_H}{\partial n}\right)_T \cdot \frac{d\alpha_{max}}{dT}$$

where $-(\partial\mu_H/\partial T)_n = \bar{S}_H$ is the molar partial entropy of the hydrogen in α-Pd(H) at the limit $n \to \alpha_{max}$.

If the *enthalpies of formation* of PdH_α and PdH_β from the elements, taken per mol of H_2, are denoted by ΔH_α and ΔH_β, the following relation with $\Delta H_{\beta\to\alpha}$ holds

$$\alpha \cdot \Delta H_\alpha - \beta \cdot \Delta H_\beta = (\beta - \alpha) \cdot \Delta H_{\beta\to\alpha} .$$

As long as $\alpha_{max} \ll \beta_{min}$, which is the case at normal and lower temperatures (see Table 3.1), the equality $\Delta H_{\beta\to\alpha} = -\Delta H_\beta$ is a good approximation. The calorimetric value of ΔH_β determined by *Nace* and *Aston* [3.24] at 30 °C, as mentioned above, can therefore be taken as the enthalpy of the $\alpha \to \beta$ transition (see also the discussion by *Mueller* et al. (Ref. [3.23], p. 641f).

The experimental values of ΔH in (3.9), presented above, seem to indicate a slight increase with increasing temperature. The question why these values are so little dependent on temperature, while α_{max} and β_{min} change appreciably, was discussed by *Flanagan* and *Lynch* [3.26].

3.2.3 Bulk and Surface Behavior

The technique of measuring $\chi(n)$ and $p(n)$ isotherms simultaneously was applied by *Frieske* [3.22] also to *palladium black* over the same range of temperatures and hydrogen pressures as with bulk Pd. The decomposition pressures in the two-phase region as well as the critical parameters T_c, p_c, n_c, agreed with the values of bulk Pd within the limits of error. The systematic study revealed, however, the following characteristic differences in behavior of Pd black compared to bulk Pd, as illustrated in Fig. 3.5:

1) smaller hysteresis loops,
2) shifts of the $p(n)$ and $\chi(n)$ isotherms towards higher n values,
3) smaller susceptibility at $n \to 0$, i.e., $470 \cdot 10^{-6}$ cm^3/mol at 20 °C compared with $562 \cdot 10^{-6}$ cm^3/mol for bulk Pd.

These differences can be attributed to the special features of the dispersed state of Pd black with high surface area and strong lattice distortions.

1) The narrower loops result from the fact that the nucleation of the β phase is facilitated in the highly defective lattice of the black. Moreover, less mechanical stress is accumulated in the small crystallites of the black—compared to Pd foils or wires—when the phase transition proceeds from the surface towards the bulk (for discussion in detail see [3.11]).

2) The shift of the $p(n)$ isotherms to higher n values results in the α phase from the chemisorption of the hydrogen at the surface of the sample. The chemisorption is almost complete at low hydrogen pressures before the hydrogen enters the lattice. The shift amounts to $\Delta n \approx 0.02$ up to 0.05, dependent on surface area, and corresponds to a monolayer of chemisorbed H atoms (one H per surface Pd atom). The χ values in this region decrease less steeply than in the case of bulk Pd (see Fig. 3.5), indicating that adsorbed hydrogen diminishes the susceptibility less than dissolved hydrogen. In the β phase region the shift of the $p(n)$ isotherm is smaller, because here the deficit of dissolved hydrogen due to the large number of Pd surface atoms compensates part of the adsorption shift.

3) The fact that the susceptibility at $n \to 0$ decreases with increasing dispersity, compared to bulk Pd, indicates the presence of hydrogen irreversibly trapped at the numerous lattice defects of

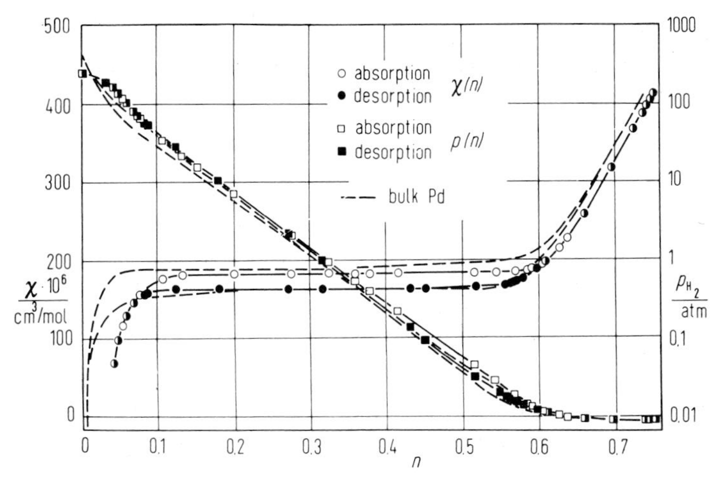

Fig. 3.5. $\chi(n)$ and $p(n)$ isotherms of Pd(H) at 120 °C with Pd black and with bulk Pd for comparison [3.22]

Pd black. This hydrogen amounts to the same order of magnitude, $\Delta n \approx 0.02$ up to 0.05, as the chemisorbed hydrogen, and can be removed only at higher temperatures, 500 up to 1000 °C [3.27, 28], together with sintering of the black. This amount, however, is sufficient only for explaining part of the observed decrease of the susceptibility. The other part may be due, as *Kubicka* [3.27] suggested, to Pd atoms and small clusters that have only loose connection with the bulk, and the electronic states of which are therefore more localized than the normal electron band states; but this question is still being discussed (see also *Selwood* [3.29]).

Considering the above mentioned effects—chemisorption, deficiency of Pd atoms in the bulk, and trapping—it becomes clear why it is not possible to give reliable values of α_{max} and β_{min} for Pd black, although these effects in part compensate mutually in the β phase region.

The particular properties of Pd black lead to the field of *chemisorption*. The chemisorption of hydrogen at Pd surfaces has been studied several times in recent years making use of UHV techniques with metal surfaces of high purity. *Aldag* and *Schmidt* [3.30], applying flash filament desorption to polycrystalline Pd wires between 100 and 1000 K, observed four different states of chemisorbed hydrogen: three states β_1, β_2, β_3, desorbing at temperatures above 350 K with activation energies of about 22, 25, and 35 kcal/mol of H_2, respectively, and another state α, which occurred when the filament temperature was decreased to 100 K, and desorbed at about 250 K with an activation energy of 13–14 kcal/mol. This α state has a small sticking coefficient ($\approx 10^{-3}$), but the amount of hydrogen taken up largely exceeded that of the β states. The authors therefore assumed that this state represented the β phase[4] of their hydrogenated Pd wire which principally can be formed at 100 K with the applied

[4] Unfortunately the desorption spectroscopists use the same notation $\alpha, \beta, \ldots$ for the chemisorption states as is customary for the solid hydride phases.

hydrogen pressure of $5 \cdot 10^{-9}$ Torr. *Conrad* et al. [3.31] studied the chemisorption at single crystal faces of Pd by means of LEED, thermal desorption spectroscopy, and contact potential measurements. The electron work function φ was found to increase with hydrogen coverage, indicating a net electron transfer from the metal to the chemisorbed H-atoms. In analogy to experiences with the system Ni/H_2, the increase $\Delta\varphi$ was assumed to be proportional to the coverage. With this assumption, chemisorption "isotherms" could be measured in the range 35 to 125 °C and hydrogen pressures from 10^{-9} to 10^{-5} Torr, making use of the steady-state condition of the continuous but slow changeover of the H-atoms into the lattice. The shape of the isotherms at low coverage, $\Delta\varphi \sim \sqrt{p_{H_2}}$, confirmed the atomic type of chemisorption.

From the temperature dependence of the isotherms at Pd (110) and (111) surfaces isosteric enthalpies of desorption, 24.4 ± 1 and 21.0 ± 1 kcal/mol of H_2, respectively, could be obtained. Thermal desorption from the Pd (110) surface yielded an activation energy of 23 kcal/mol of H_2; the close agreement with the enthalpy value indicates that the dissociative chemisorption proceeds with practically zero activation energy. Obviously the results of *Conrad* et al. are related to chemisorbed hydrogen in the β_1, β_2 states observed by *Aldag* and *Schmidt* [3.30], although no one-to-one coordination seems possible at present. *Conrad* et al. assumed Pd (111) faces—as the most densely packed planes—predominant in polycrystalline surfaces such as used by *Aldag* and *Schmidt.* The LEED analysis of hydrogen chemisorbed at Pd (111) planes yields a 1×1 structure; this leads to an arrangement where the hydrogen atoms are situated above triangles of Pd atoms (see Fig. 3.6a) and constitute an octahedral plane of their own. This arrangement agrees with the result of *model calculations* for hydrogen chemisorption on Pt (111) planes which *Weinberg* and *Merrill* [3.32] performed, making use of the concept of crystal field surface orbitals of fcc transition metals, and applying linear correlations between bond energy and bond order of Pt–Pt and Pt–H bonds. The most remarkable result of these calculations on an empirical base is the occurrence of a potential well for hydrogen molecules on the Pt (111) plane as represented in Fig. 3.6b, position 1. In this position with a nuclear distance of 0.962 Å (compared to 0.742 Å in the undistorted H_2 molecule), the molecular orbitals overlap with two t_{2g} d orbitals of the surface Pt atoms. The dissociation of the chemisorbed molecule then proceeds along a minimum energy path towards position 2 in Fig. 3.6b, representing the transition state where each hydrogen atom overlaps with one e_g d orbital, and terminates in position 3, identical to the arrangement in Fig. 3.6a, where overlapping with three e_g d orbitals and with p orbitals of the Pt atoms occurs. For the molecular potential well, position 1, the chemisorption energy was calculated as about half the value of the atomic potential wells, position 3 (17 and 31 kcal/mol of H_2, respectively) whereas the transition state, position 2, represents a rather high levelled saddle point with a chemisorption energy of only 4 kcal/mol of H_2. *Weinberg* and *Merrill* took the molecular state as a precursor to atomic adsorption, and reported upon a LEED pattern for molecular adsorption of deuterium on Pt (111) planes that could be obtained as

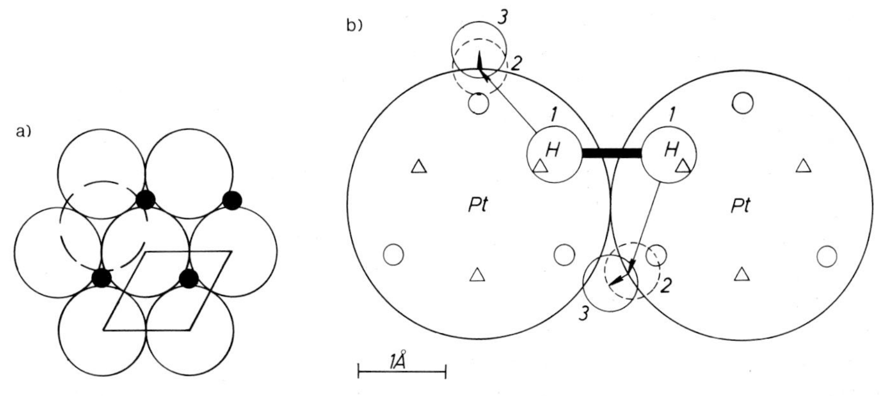

Fig. 3.6. (a) Arrangement of chemisorbed hydrogen atoms on Pd (111) (1 × 1 structure) (*Conrad* et al. [3.31]). Occupied sites identical to positions 3 in (b), free sites presumably passages from surface into bulk. (b) Top view of a Pt (111) plane with the trajectory of two hydrogen atoms from position 1 (molecular well) via position 2 (saddle point) to position 3 (atomic well) (*Weinberg* and *Merrill* [3.32])

○: e_g d orbitals ; △: t_{2g} d orbitals

a transient 2 × 3 structure during adsorption as well as during subsequent slow desorption of the deuterium.

Conrad et al. [3.31] failed to show the existence of such a 2 × 3 structure in the case of hydrogen on Pd (111) in spite of detailed efforts. Nevertheless, there are a number of arguments which support the application of *Weinberg* and *Merrill*'s results (Fig. 3.6b) also to hydrogen on Pd (111), at least qualitatively. These arguments may be summarized as follows:

1) The α desorption peak observed by *Aldag* and *Schmidt* [3.30] after adsorption at low temperatures (100 K) seems not to represent dissolved hydrogen, as the authors believed. If this were true, the peak would be expected to shift to higher temperatures and to broaden appreciably with increasing amounts of adsorbed hydrogen, as demonstrated in the thermal desorption spectra of *Conrad* et al. [3.31]. Instead, the α peaks remain remarkably sharp and keep their position at 250 K, irrespective of the amount of hydrogen accumulated. The α state, therefore, represents presumably a *molecular state* of chemisorbed hydrogen, corresponding to the position 1 in Fig. 3.6b.

2) The small sticking coefficient for the α state adsorption reported by *Aldag* and *Schmidt* may be attributed to the low rotation probability of hydrogen molecules at 100 K, resulting in a small steric factor of orientation of the molecular axis into position 1. At higher temperatures this restraint vanishes, and the atomic chemisorption in position 3 may proceed via the molecular state with high sticking coefficients of the order 0.1 as observed. Moreover, the passage through the molecular potential well does not imply an effective activation energy for the dissociative chemisorption, because the transition state in position 2 (Fig. 3.6b) lies well below the energy level of hydrogen in the gas phase.

3) With the molecular adsorption state, position 1, the model offers a precursor also for the penetration of the hydrogen into the lattice. This process cannot proceed, however, via the states of strong chemisorption, position 3, because these states lie too deep below the hydrogen levels in the lattice. Actually, the hydrogen will occupy first these chemisorption sites before entering the lattice, as was demonstrated already by the shifts of the $p(n)$ isotherms in Fig. 3.5. When the chemisorption

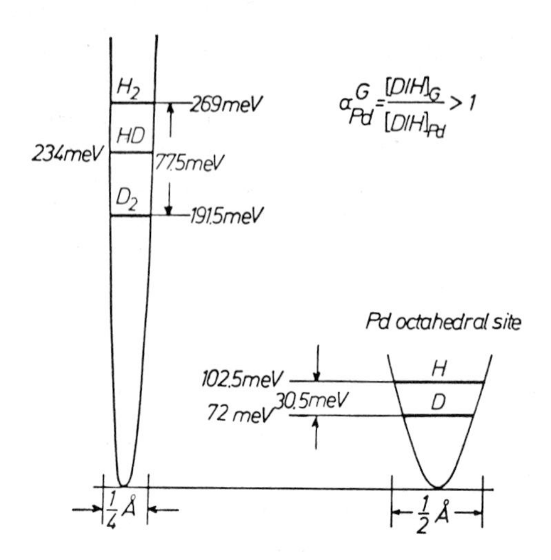

Fig. 3.7. Potential wells and zero-point energies of the gas molecules (left) and the atoms in the octahedral sites (right) (*Sicking* [3.33]). The zero-point energies on the right are scaled up per atom for three-fold degeneracy

layer is complete, as represented by Fig. 3.6a, there remain still empty tetrahedral holes, symmetrical to the occupied octahedral surface sites, but less favored by Pd d orbitals (Fig. 3.6b); these holes suggest themselves as the *passages for* hydrogen atoms *entering the lattice.*

4) Recently *Lynch* and *Flanagan* [3.12] studied the weak, reversible adsorption of hydrogen at Pd black after completion of the strong chemisorption that had been analyzed before by titration methods as a H/Pd = 1/1 layer. $p(n)$ isotherms of the weak adsorption were obtained by subtracting from the total amount of hydrogen taken up the fraction dissolved in the lattice (10 to 20% only, because the sample of Pd black used had a rather large specific surface area of 40 m^2/g). Measurements between 0 and 37.5 °C yielded isosteric enthalpies of desorption ranging from 11 to 8.5 kcal/mol of H_2 with increasing coverage. Deuterium showed up higher equilibrium pressures—at comparable coverage—than hydrogen. This inverse isotope effect indicates atomic adsorption in rather flat potential wells, similar to those sketched in Fig. 3.7. Considering the (111) lattice plane as the most frequent one in polycrystalline Pd black, the adsorption sites are most probably the open tetrahedral surface sites of Fig. 3.6a, i.e., the passages into the interior of the lattice. Presumably, part of the hydrogen taken as "weakly adsorbed" in this study will be located below the uppermost lattice plane, because here less dilatation work is necessary for penetration (by "lifting" the upper plane) than in the interior of the lattice. This lifting effect was verified recently by careful adsorption and absorption measurements on a series of samples of differently sized Pd microcrystallites by W. Minnerup: Dissertation, Münster (1977).

In this way the chemisorption concept developed by *Weinberg* and *Merrill* [3.32] gains particular importance in connection with the transfer mechanism of hydrogen from the molecular state in the gas phase to the atomic state in the lattice, as will be shown in detail in Section 3.5.1.

3.2.4 Equilibrium Isotope Effects

The most remarkable feature of palladium hydrides with the heavier hydrogen isotopes is the higher equilibrium pressure of deuterium or of tritium, compared with normal hydrogen, at the same temperature and composition.

This "inverse" isotope effect (normally one would expect the heavier isotope to prefer the condensed phase) is due to the fact that the potential wells for the vibrating hydrogen atoms in the interstices of the metal lattices are much broader and flatter than in the diatomic gas molecules, and therefore the zero-point energies smaller. Figure 3.7 demonstrates the situation for the systems Pd/H_2 and Pd/D_2. The *zero-point energies* given for the octahedral site are based on studies of inelastic neutron scattering in low concentration α-Pd(H) and α-Pd(D) by *Drexel* et al. [3.34] who observed the first harmonics of the local modes of H and D at

$$\hbar\omega_H = 68.5 \pm 2\ \text{meV}\ ; \qquad \hbar\omega_D = 48 \pm 4\ \text{meV}\ ,$$

respectively. As Fig. 3.7 shows, the zero-point level of deuterium is deeper below that of hydrogen in the gas phase than in the metal, although it must be considered that the energies in the octahedral site have to be counted twice (for two atoms) in order to be comparable with the molecular state. Deuterium is therefore favored, compared to hydrogen, in the gas, hydrogen in the metal. The effect is even stronger for tritium. On the other hand, if the frequency of the local mode, $\hbar\omega_H$, exceeds about 95 meV, the difference of the zero-point energies changes sign, and at $\hbar\omega_H \gtrsim 124$ meV the isotope effect accordingly is reversed to "normal" behavior [see (3.17) below]. This is true in the case of the Vb group hydrides where the local frequencies are $\hbar\omega_H = 145$; 148.8; 165 meV for H in Ta, Nb, V, respectively [3.35], and the deuterides indeed have the smaller equilibrium pressures [3.36, 37].

The second reason for equilibrium isotope effects is the smaller *amplitude of vibration* of the D-atoms in the interstitial sites compared to the H-atoms. (The root-mean-square displacement has been determined from neutron scattering [3.4] in the β phases to be 0.23 Å for H and 0.20 Å for D). This results in smaller volume requirements and thereby in smaller D–D attractive interaction

$$|\Delta\mu_{D^+}| < |\Delta\mu_{H^+}|\ ;$$

for details see [3.9, 38].

The differences in the position of the phase limits of Pd(H) and Pd(D) are small and do not exceed the range of error of the lattice constants or of the α_{max} and β_{min} values. However, the lower critical temperature and higher critical pressure of the Pd/D_2 system [3.39]

$$T_c \approx 276\ ^\circ\text{C}\ (292)\ ; \qquad p_c \approx 35\ \text{atm}\ (19.7)$$

compared to Pd/H_2 (values in brackets) point conclusively to the smaller D–D attraction. The order of magnitude of the differences becomes clear from the energies of pair formation of H- or D-atoms in the host lattice,

$$W_{HH} = -685\ \text{cal/mol of H–H pairs}\ ; \qquad W_{DD} = -665\ \text{cal/mol of D–D pairs}$$

at 25 °C as calculated by *Brodowsky* [3.9] from isotherm measurements (details in the next section).

Recently, the question came up whether there are also *electronic contributions* to the isotope effects. This question has two roots. The first one is the phenomenon of a shift of *s–p* band states from above the Fermi level in pure Pd to a position 5 to 6 eV below E_F in Pd(H), brought about by strong interaction with the 1*s* electron states of the hydrogen atoms in the lattice. This shift was demonstrated by *Switendick* [3.40] in APW energy band calculations of Pd and PdH; the resulting low-lying electron states have been observed directly in photoemission studies on pure Pd and β-Pd(H) by *Eastman* et al. [3.41]. The second reason for the idea of an electronic contribution is the reverse isotope effect in the superconducting behavior of Pd(H) and Pd(D), first observed by *Stritzker* and *Buckel* [3.42]. According to results with metals, the critical transition temperature decreases with increasing isotope mass ("normal" isotope effect). Contrary to this, careful measurements by *Skośkiewicz* et al. [3.43] with Pd(H) and Pd(D) of high H and D content, respectively (the samples were charged under equilibrium conditions at high pressures), yielded

$$T_c(\text{PdH}) = 9.5\ \text{K}\ ; \quad T_c(\text{PdD}) = 11.7\ \text{K}\ .$$

The values are extrapolated to $n \to 1$ from $T_c - n$ curves measured up to $n = 0.98$ to 0.99. *Miller* and *Satterthwaite* [3.44] called attention to the fact that a shift of the $T_c - n$ curve of the hydride by $\Delta n = -0.05$ brings the curve in near coincidence with the one of the deuteride. From this empirical perception the authors concluded that in the hydride about 0.05 electrons (of each H-atom) fewer than in the deuteride are effective in lifting the Fermi level above the *d*-band holes. These electrons are engaged, as the authors assumed, additionally in the low-lying states discovered by *Switendick*; the difference relative to deuterium is attributed to the stronger overlapping with the Pd electronic states due to the larger vibrational amplitudes of the hydrogen atoms, i.e., to their smaller distance from the surrounding Pd atoms in the time average.

If this conception were true, the other electronic properties of palladium hydride would be expected to exhibit corresponding isotope effects. The magnetic susceptibility, for instance, should arrive at its limiting diamagnetic value (see Fig. 3.3) in the deuteride with the composition $\text{PdD}_{n-0.05}$ when this limit is reached in the hydride with the composition PdH_n. A look at Fig. 3.22, p. 128, as well as at the careful early measurements of *Sieverts* and *Danz* [3.45] shows, however, that there are no systematic differences in the $\chi(n)$ values of Pd(H) and Pd(D) from $n = 0$ up to the diamagnetic limit. With regard to the narrow range of errors in these investigations, it is obvious that a shift of the order $\Delta n = 0.05$ would have been shown up as a rather large effect.

Accordingly, the idea of electronic contributions to the isotope effects in palladium hydride can be set aside at present. An explanation to the reverse isotope effect of the superconducting transition temperatures of Pd(H) and Pd(D) must be looked for in other ways, taking into account the electron-phonon interactions in connection with the particular frequency distributions generated by the hydrogen or deuterium components in the superconducting lattice (see for instance *Chowdhury* [3.46]).

The most important quantities for the equilibrium isotope effects in the ternary system Pd(H, D) are the ratios of the equilibrium pressures $p^0_{D_2}/p^0_{H_2}$ of the binary systems Pd(H) and Pd(D) at the same isotope content n. Applying (3.4) to each of the binary equilibrium pressures at the same n one obtains

$$\ln(p^0_{D_2}/p^0_{H_2}) = 2 \cdot \ln(K_D/K_H) + 2 \cdot (\Delta\mu_{D^+} - \Delta\mu_{H^+})/RT. \quad (3.12)$$

The electronic term $\Delta\mu_e$ in (3.4) has been set equal for the two isotopes according to the foregoing discussion; it therefore vanishes in this equation. For small concentrations in the α phase, only the first term on the right-hand side has to be considered; at higher concentrations and in the β phase region both terms must be taken into account.

The ratio of the equilibrium pressures is narrowly connected with the *separation factor* [3.9]

$$\alpha^G_{Pd} = \frac{(D/H)_G}{(D/H)_{Pd}} = \frac{(2p_{D_2} + p_{HD})/(2p_{H_2} + p_{HD})}{n_D/n_H}. \quad (3.13)$$

Replacing p_{HD} by means of the exchange equilibrium

$$\tfrac{1}{2}H_2 + \tfrac{1}{2}D_2 \rightleftarrows HD : K_{HD} = p_{HD}/\sqrt{p_{H_2} \cdot p_{D_2}}$$

and taking approximately $K_{HD} \approx 2$ (true values: 1.7 to 1.87 in the range 200 to 400 K [3.47]) one obtains,

$$\alpha^G_{Pd} = \sqrt{p_{D_2}/p_{H_2}} \cdot n_H/n_D = \sqrt{p^0_{D_2}/p^0_{H_2}}. \quad (3.14)$$

In the last step of this calculation Raoult's law has been taken as valid for the atomic partial pressures $p_{H(D)}$ dependent on the atomic fractions $n_{H(D)}/(n_H + n_D)$. Accordingly, $p^0_{D_2}$ in (3.14) means the equilibrium pressure of PdD_n with $n = n_H + n_D$, and $p^0_{H_2}$ the equilibrium pressure of PdH_n with the same total n.

For the *low concentration range* in the α phase follows with (3.12)

$$\alpha^G_{Pd} = K_D/K_H \equiv \alpha^0. \quad (3.15)$$

By insertion of K_H and K_D from the measurements of *Nernst* [3.11],

$$\ln K_H = -\frac{1163\ \mathrm{K}}{T} + 6.45 \quad (3.16) \qquad \ln K_D = -\frac{949\ \mathrm{K}}{T} + 6.40 \quad (3.16a^*)$$

* For tritium in Pd was obtained $\ln K_T = 832/T + 6.25$ between 25 and 70 °C. [S. Schmidt: Diplomarbeit, Münster (1978)]

one obtains

$$\ln \alpha^0 = \frac{214\ \text{K}}{T} - 0.05\ ; \qquad \alpha^0_{298} = 1.95\ .$$

Clewley et al. [3.20] measured K_H and K_D values by means of UHV techniques, using a 3 mm Pd rod and a 1 mm Pd wire as samples with different pretreatment. The two sets of data, inserted into (3.15), give as results

$$\text{rod:}\ \ln \alpha^0 = \frac{189\ \text{K}}{T} + 0.15\ ; \qquad \alpha^0_{298} = 2.19$$

$$\text{wire:}\ \ln \alpha^0 = \frac{166\ \text{K}}{T} + 0.20\ ; \qquad \alpha^0_{298} = 2.13\ .$$

There seem to exist in literature no direct measurements of the separation factor characteristic for the α phase region. Measurements by *Botter* [3.48] that covered this region were done with highly dispersed Pd, precipitated on porous sintered corundum, i.e., with samples of large specific surface area. Under the conditions applied in this investigation the characteristics of the α phase were lost within the overlapping effects of β phase Pd(H, D) and of chemisorbed hydrogen (see below).

A rather reliable approximation of the separation factor in the low concentration range of the α phase can be obtained when use is made of the local mode frequencies of H and D as determined by *Drexel* et al. [3.34]; $\hbar\omega_H = 68.5 \pm 2$ meV; $\hbar\omega_D = 48 \pm 4$ meV. The calculation is based upon the combination of (3.3) and (3.15),

$$RT \cdot \ln \alpha^0 = \mu^0_D - \mu^0_H - \tfrac{1}{2} \cdot (\mu^0_{D_2} - \mu^0_{H_2})\ . \tag{3.17}$$

The standard potentials of the molecules in the gas phase can be approximated around room temperature by $\mu^0(T) = H^0_{298} - T \cdot S^0_{298}$, the standard potentials of the atoms in the interstitial sites accordingly by making use of the enthalpy and entropy at 298 K of Einstein oscillators with the above quoted frequencies and threefold degeneracy. The zero-point energies, as demonstrated in Fig. 3.7, have been included in the enthalpy values. The difference of the strain energy part of μ^0_H and μ^0_D has been neglected, as well as the influence of the anharmonicities of the H and D vibrations in the interstitial sites[5]. The calculation yields

$$\ln \alpha^0 = \frac{167\ \text{K}}{T} + 0.168\ ; \qquad \alpha^0_{298} = 2.07\ ,$$

[5] Due to the larger vibrational amplitudes the strain energy, part of the standard chemical potential, is higher for H than for D; the anharmonicity, on the other hand, lowers the zero-point energy. Hence the two effects that have been neglected cancel partially.

in fair agreement with the values listed above from $p-n-T$ measurements. Smaller frequencies of the H- and D-atoms in the lattice result in higher separation factors; higher frequencies diminish the isotope effect, as is to be expected from Fig. 3.7. With $\hbar\omega = 124$ meV for H and $124/\sqrt{2} = 87.7$ meV for D, the right-hand side of (3.17) vanishes; the enrichment of the isotopes in the two phases then changes sign.

For application at *higher concentrations* and particularly in the β phase region the term of nonideal behavior in (3.12) must also be considered. Denoting the separation factor in this case by α'', the combination of (3.12–15) yields

$$\ln \alpha''/\alpha^0 = (\Delta\mu_{D^+} - \Delta\mu_{H^+})/RT. \tag{3.18}$$

From measurements of the separation factors as a function of total hydrogen content, n, in the lattice the difference of the elastic contributions of D and H to nonideality can thus be determined. In the β phase of Pd(H, D), separation factors have been measured by *Nernst* [3.11] in the range -78.5 to $+50\,°C$ at 1 atm total pressure and with D/H = 1/1 in the total hydrogen amount used. The measurements were done on a Pd black sample with small specific surface area, so the effects of chemisorption could be neglected. The results are plotted in Fig. 3.8 and can be described by

$$\ln \alpha'' = \frac{245\ \text{K}}{T} + 0.055\,; \qquad \alpha''_{298} = 2.41\,.$$

The mean value of n in Nernst's measurements was $n = 0.67$.

For evaluation of (3.18) the difference of the chemical potential terms can be expressed in good approximation by the difference of the pair interaction energies [3.38]

$$\ln \alpha''/\alpha^0 = 6n(W_{DD} - W_{HH})/RT. \tag{3.18a}$$

With $\alpha''_{298} = 2.41$, and from the α phase measurements, $\alpha^0_{298} = 1.95$, one obtains

$$W_{DD} - W_{HH} = 32\ \text{cal/mol of pairs}\,,$$

in good agreement with the difference of the interaction energies calculated from $p(n)$ isotherm measurements [3.9]. The increase of the separation factor from α to β phase is thus due to the higher pairing tendency of the H-atoms compared to the D-atoms, brought about by the higher elastic strain to the lattice in consequence of the larger vibrational amplitudes.

Another effect of the high concentrations in the β phase, corresponding to lattice expansion and nearest neighbor interaction, is the decrease of the local

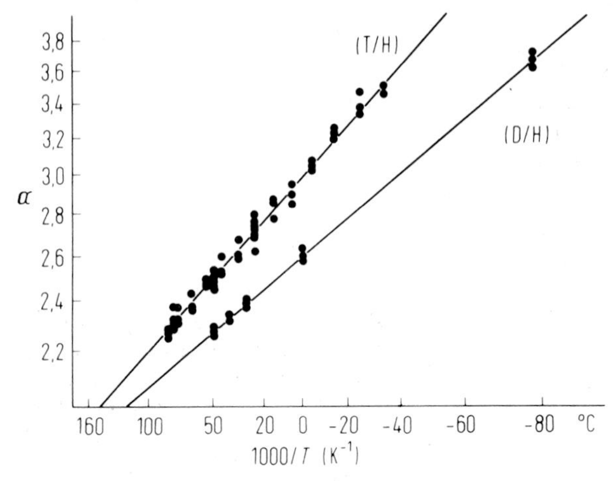

Fig. 3.8. Equilibrium separation factors $\alpha_{Pd}^{G}(D, H)$ (*Nernst* [3.11]) and $\alpha_{Pd}^{G}(T, H)$ (*Sicking* [3.33, 49]) measured in the β phase region

mode frequencies compared to the α phase. Neutron scattering experiments of several research groups with β-Pd(H) [3.50, 51] and β-Pd(D) [3.52] resulted in

$$\hbar\omega_H = 57 \pm 2 \text{ meV} ; \quad \hbar\omega_D = 40 \pm 2 \text{ meV} .$$

When these frequencies are used directly for evaluation of (3.17), one obtains

$$\ln \alpha'' = \frac{229 \text{ K}}{T} + 0.075 ; \quad \alpha''_{298} = 2.33 ,$$

in fair agreement with the experimental results of *Nernst*.

Measurements of the separation factor between *hydrogen and tritium* in the β phase of Pd(H, T) have been performed by *Sicking* [3.49] in the range -34 to $+83$ °C at 1 atm hydrogen pressure and with about 10^{-4} at.% of tritium. The sample was a Pd sheet with a thin coating of black; the tritium analysis was done with an ionization chamber in the gas cycle. The results are plotted also in Fig. 3.8; the averaging straight line yields

$$\ln \alpha''(\text{H, T}) = \frac{320 \text{ K}}{T} - 0.075 ; \quad \alpha''_{298} = 2.71 .$$

Sicking used these results together with the known data of the exchange equilibrium [3.53],

$$\tfrac{1}{2}H_2 + \tfrac{1}{2}T_2 \rightleftarrows HT ; \quad K_{HT} = p_{HT} / \sqrt{p_{H_2} \cdot p_{T_2}} ;$$

$$\ln K_{HT} = -\frac{83.5 \text{ K}}{T} + 0.753$$

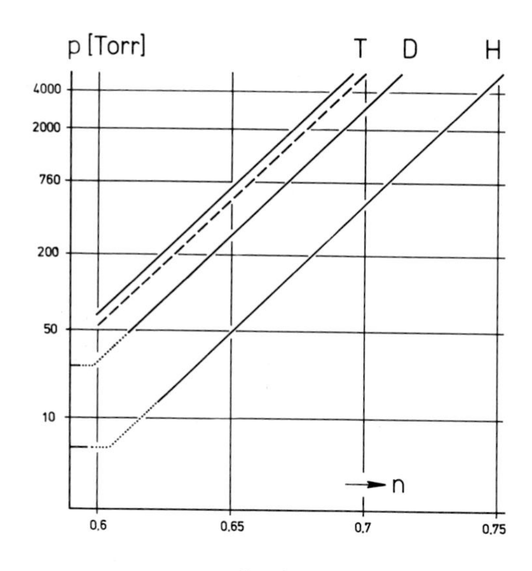

Fig. 3.9. β phase isotherm of Pd(T) at 25 °C, calculated from separation factor measurements by *Sicking* [3.33, 49]. Dashed line: without considering the elastic part $\Delta\mu_{T^+}$ of the excess potential. Pd(H) and Pd(D) isotherms after *Nernst* [3.11]

to calculate values of the ratio $p^0_{T_2}/p^0_{H_2}$ by means of equations analogous to (3.13) and (3.14).

Therefrom he obtained approximate $p(n)$ isotherms of PdT_n in the β phase range $0.650 \leqq n \leqq 0.775$. Subsequently, he estimated the pair interaction energy of tritium by means of the relation $W_{TT} - W_{DD} = \frac{1}{2} \cdot (W_{DD} - W_{HH})$, and used this value ($W_{TT} = -655$ cal/mol of T–T pairs at 298 K) in order to account for the elastic term of nonideality, $\Delta\mu_{T^+}$, in the $p(n)$ isotherms.

His results for 298 K are plotted in Fig. 3.9, where the dashed line represents the approximate isotherm, the solid line the corrected one. For comparison, the isotherms of PdH_n and PdD_n, interpolated for this temperature from the measurements of *Nernst*, have been included in the figure. The complete $p-n-T$ relationship takes the form

$$\ln p = -\frac{\Delta H(n)}{RT} + 2 \cdot \ln \frac{n}{1-n} + \frac{\Delta S^0}{R} \tag{3.19}$$

with the following specifications:

	$\Delta H(n)$ cal/mol	ΔS^0 cal/K·mol
H_2	$23970 - 21500\,n$	25.6
D_2	$22820 - 21500\,n$	25.4
T_2	$22050 - 21500\,n$	24.6 .

Plots of these relationships yield logarithmic straight lines like those in Fig. 3.9, corresponding to the form of (3.7).

The plateau pressures in the two-phase regions of the one-isotope systems Pd/H_2, Pd/D_2, and Pd/T_2 have no particular importance for separation effects. In mixed systems like Pd(H, D), the plateau pressure after equilibration with the whole amounts of α and β phase present consists of p_{H_2}, p_{HD}, and p_{D_2} corresponding to a separation factor that represents a weighted average of those of the α and β phases.

For application in separation processes, the isotope enrichment effects between liquid *water and* Pd *hydride* are even more important than the enrichment between hydrogen gas and Pd hydride. The corresponding isotope separation factors are interrelated by

$$\alpha_{Pd}^{L} = \alpha_{G}^{L} \cdot \alpha_{Pd}^{G} \tag{3.20}$$

wherein the liquid gas separation factor in case of the system H/D is defined by

$$\alpha_{G}^{L} = \frac{(D/H)_L}{(D/H)_G} . \tag{3.21}$$

At small deuterium contents the approximation holds,

$$\alpha_{G}^{L} = K_{HDO} \cdot \frac{p_{H_2O}^0}{p_{HDO}^0} , \tag{3.21a}$$

where K_{HDO} is the constant of the exchange equilibrium

$HD + H_2O \rightleftharpoons H_2 + HDO$ in the gas phase, and the p_i^0 are the vapor pressures of the isotopic water molecules extrapolated to the activity 1 of each species, respectively. At 298 K the three quantities determining the separation factor, α_{Pd}^{L}, between liquid H_2O/HDO and β-Pd(H, D) take the following values:

$$\alpha_{Pd}^{L} = K_{HDO} \cdot \frac{p_{H_2O}^0}{p_{HDO}^0} \cdot \alpha_{Pd}^{G} = 3.47 \cdot 1.07 \cdot 2.41 = 9.0 . \tag{3.22}$$

This rather high separation factor was confirmed by *Sicking* [3.33] in measurements of the rate of H/D distribution between water and Pd black that had been initially charged with D_2 to β-PdD_n. This high value of α_{Pd}^{L} is one of the reasons of the large electrolytic H/D separation factors (9 to 16) observed by *Gibmeyer* [3.54] with cathodic deposition of hydrogen at Pd/Ag diffusion electrodes in the temperature range +60 to −30 °C.

For the system *tritium/hydrogen* the analogy to (3.22) can be set up,

$$\alpha_{Pd}^{L} = K_{HTO} \cdot \frac{p_{H_2O}^0}{p_{HTO}^0} \cdot \alpha_{Pd}^{G} \approx 4.05 \cdot 1.095 \cdot 2.71 = 12.0 \text{ at } 298 \text{ K} . \tag{3.23}$$

The value $K_{HTO}^{298} \approx 4.05$ has been extrapolated from measurements of the exchange equilibrium: $HT + H_2O \rightleftharpoons H_2 + HTO$ in the gas phase between 90 and 140 °C by *Gans* [3.55] (see also *Sicking* [3.33]); although not very precise, it seems more reliable than the early result: $K_{HTO}^{298} = 6.25$ obtained by *Black* and *Taylor* [3.56]. No experimental investigation of this surprisingly high equilibrium isotope effect seems to have been done so far.

Particularly interesting is a comparison of the equilibrium isotope effects between the bulk of the lattice and the surface, i.e., between dissolved and *chemisorbed hydrogen*. Considering the rather low-lying energy levels of the strong atomic chemisorption at positions 3 in Fig. 3.6b, higher vibrational frequencies are to be expected in these states compared to the interstitial sites. In consequence, the separation factor between gas and low-lying surface states, α_S^G, should be smaller than the one between gas and bulk states, $\alpha_S^G < \alpha_{Pd}^G$.

Since the ratio of dissolved to chemisorbed hydrogen decreases with decreasing pressure, the separation factor should also decrease when the hydrogen pressure changes to low values. The first indication that this actually happens has been obviously obtained by *Botter* [3.48] in a study of H/D isotope effects at room temperature on dispersed Pd and a 10% Pt–Pd alloy precipitated on a porous ceramic carrier. Starting with separation factors of about 2 in the $(\alpha+\beta)$ two-phase region, the values decreased to an average of about 1 when the hydrogen pressure was lowered to the range 5 to 1 Torr. *Botter* herself, however, did not take into account the influence of chemisorption, but considered the small values of the separation factor to be a property of α phase Pd(H, D).

A systematic study of the influence of chemisorbed hydrogen was undertaken by *Gans* [3.55]. He determined H/T separation factors at 80 °C on Pd and Pd black samples of different dispersity and down to hydrogen pressures of about 1 Torr. Since the total amount of hydrogen taken up by the sample was measured, the fraction dissolved in the bulk lattice, φ_b, could be calculated by means of known values of Sieverts' constant. Figure 3.10 demonstrates the separation factors obtained as function of φ_b. In the range $\varphi_b \to 1$ (Pd sheet sample) the curve approaches the value for β phase Pd(H, T): $\alpha''(H, T) = 2.3$ (see Fig. 3.8); in the range $\varphi_b \to 0$ the curve levels out to a limiting value that can be taken, although not very precisely, to be $\alpha_S^G(H, T) \approx 0.1$ at 353 K.

This value has been used to make an estimation of the vibrational frequencies of the H and T atoms in the chemisorption potential wells by means of (3.17). With the assumption $\omega_T = \omega_H / \sqrt{3}$, the estimation yields

$$\hbar\omega_H \approx 125 \text{ meV}; \quad \hbar\omega_T \approx 72 \text{ meV}.$$

This result is in agreement also with the measurements of *Botter* [3.48] since with $\hbar\omega_H \approx 125$ meV (and $\omega_D = \omega_H / \sqrt{2}$), (3.17) yields $\alpha(H, D) \approx 1$ as mentioned already on p. 93.

Recently *Ratajczykowa* [3.57] studied the surface of evaporated Pd films charged with hydrogen by means of *IR reflection-absorption spectroscopy*. No

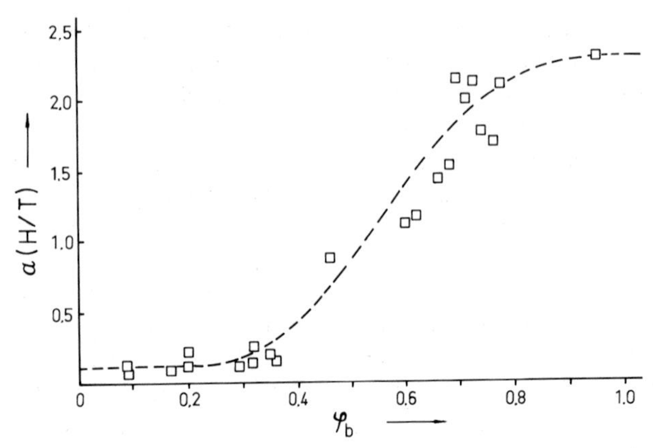

Fig. 3.10. H/T separation factor at 80 °C measured with Pd samples of different surface to bulk ratio. From the total amount of hydrogen taken up by the sample represents: φ_b the fraction dissolved in the bulk lattice, $1-\varphi_b$ the fraction chemisorbed at the surface. Measurements by *Gans* [3.55]

absorption band could be observed from chemisorbed hydrogen. However, after formation of the β phase, two bands occurred at 94 and 109 meV, the intensities of which increased with hydrogen pressure according to the logarithmic course of the $p(n)$ isotherm in the β phase region (see Fig. 3.9). The author ascribes the absorption bands to hydrogen atoms located in surface interstitial sites on the (100) planes (94 meV) and on the (111) planes, similar to Fig. 3.6a (109 meV). The bands are, however, certainly not due to chemisorbed hydrogen atoms in position 3, Fig. 3.6b, because these chemisorption states are saturated at hydrogen pressures far below the formation of the β phase, and do not increase in coverage with hydrogen pressure in the same manner as the hydrogen content n in the bulk β phase. Obviously the absorption observed is connected with an intermediate state—between the molecular chemisorption (position 1, in Fig. 3.6b) and the dissolved state in the bulk—that keeps equilibrium with the hydrogen in the lattice. From the values of the frequencies, a separation factor between this intermediate state and the gas phase is to be expected which corresponds to an enrichment of H on the surface. This agrees with the findings of *Lynch* and *Flanagan* [3.12] whereafter the "weak" chemisorption (see p. 88) shows up separation effects—deuterium favoring the gas phase—similar to the dissolution equilibrium. Hence the IR absorbing states of *Ratajczykowa* are to be correlated likewise to atomic chemisorption at the passages from surface to bulk interstitial sites.

3.2.5 Statistical Thermodynamics of Solute-Solute Interactions

The phase diagrams of many metal-hydrogen systems display a separation into more concentrated and more dilute phases below a critical temperature just as a van der Waals gas. Since the interstitial atoms have a relatively high mobility in the fixed host lattice, the term "lattice gas" was coined to draw attention to the analogy [3.58–61]. The maximum occupancy of one interstitial per site limits the concentration in the lattice gas model in a way similar to the rigid sphere approximation in a van der Waals type gas.

At present, it is still a matter of debate whether the condensation of the lattice gas is brought about by short-range interactions as in a real gas, e.g., by an attraction due to nearest neighbor interactions, or by long-range interactions.

In this presentation, we shall use the *quasichemical model* to obtain a mathematical relationship between the concentration and the chemical potential of the dissolved H, exclusive of the Fermi energy contribution. This model was independently introduced by *Guggenheim* [3.62] and, in a different form, by *Bethe* [3.63]. It belongs to a class of models in which the particles are confined to certain sites on a lattice and in which only particles on adjacent sites interact energetically ("Ising problem") [3.59].

If the number of pairs is assumed to follow the random distribution, a zeroth order solution of this problem is obtained, as first derived by *Gorski* [3.64] and later developed by *Bragg* and *Williams* [3.65].

The quasichemical approach is a first-order solution in which the deviation from randomness of the number of pairs is taken into account in an approximate manner. If an attraction exists between like particles, an excess of clusters over and above the random number is predicted, and the size of these clusters has a tendency to increase at lower temperatures. If like particles repel each other, a tendency to form a superstructure or an ordered solid solution is predicted.

The temperature dependence of the number of pairs is a most welcome feature of the quasichemical approach. In many alloys and mixtures studied in detail, there is experimental evidence of a pronounced temperature dependence of the thermodynamic excess functions, and the quasichemical model is able to describe this temperature dependence in spite of its simplicity and inherently approximate nature.

In a long-range interaction model, the deviations from ideality are formally described in the same way as in the Bragg-Williams approximation, implying a random distribution and an ideal configurational entropy of the interstitial atoms. The observed temperature dependence of the excess potential is either neglected, or accounted for in some way other than the change of the number of energetically favored pairs.

In order to show the relationship of the nearest neighbor interactions to the other contributions to the *chemical potential*, the various parts of (3.1, 1a) are schematically represented in Fig. 3.11,

$$\mu_{\mathrm{H}}=\mu^0_{\mathrm{H_{vib}}}+\mu^0_{\mathrm{H^+}}+\mu^0_{\mathrm{e}}+RT\ln\frac{n_{\mathrm{H}}}{1-n_{\mathrm{H}}}+\Delta\mu_{\mathrm{H^+}}+\Delta\mu_{\mathrm{e}}\,. \tag{3.24}$$

The standard potential has been split up into vibrational, electronic, and protonic contributions $\mu^0_{\mathrm{H_{vib}}}$, μ^0_{e}, and $\mu^0_{\mathrm{H^+}}$, respectively. In Fig. 3.11 the

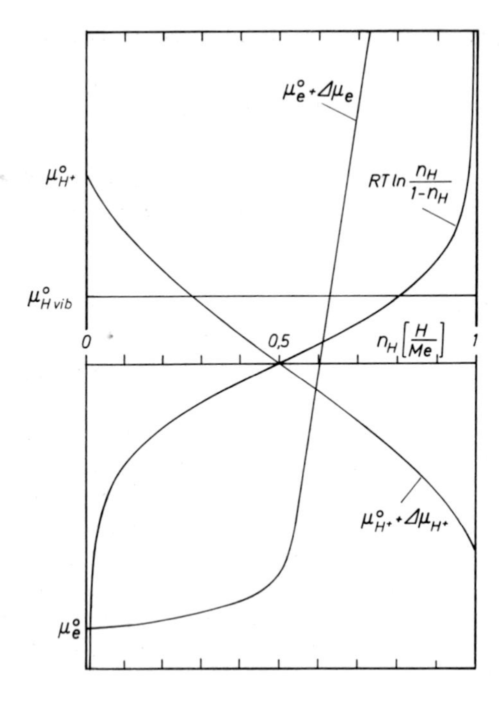

Fig. 3.11. Constituent parts of the chemical potential of H in Pd [3.66]

electronic and protonic parts have been lumped together with their corresponding excess potentials $\Delta\mu_e$ and $\Delta\mu_{H^+}$.

The electronic contribution $\mu_e^0 + \Delta\mu_e$ delineates the gentle rise of the Fermi energy in the 4*d* band regime and the steep rise in the 5*s* region, as already shown in Fig. 3.1. The protonic contribution $\mu_{H^+}^0 + \Delta\mu_{H^+}$ has a slightly curved, inflected shape, calculated according to the quasichemical approximation with a pair interaction energy of $w_{HH} \approx -kT$. It degenerates to a straight line in the Bragg-Williams limit $-w_{HH}/kT = 0$.

On thermodynamic grounds alone, only the total excess potential can be determined experimentally, and not its protonic and electronic parts. However, in Pd alloys, and to some extent in Ni alloys, a very good separation can be brought about due to a fortunate circumstance, the slow rise of the Fermi energy at low H concentrations. That is why Pd and Ni alloys are particularly useful to demonstrate these effects, which are probably important in other alloys, too, but which cannot be separated with equal ease in the other systems. A small error in the initial rise of the *electronic excess potential* will not introduce a great relative uncertainty in the slopes either of $\Delta\mu_{H^+}$ or of the 5*s* band part of $\Delta\mu_e$.

In earlier evaluations [3.9], the initial rise of the electronic excess potential was taken to be zero. The pair interaction energy necessary to calculate the whole curve $\Delta\mu_{H^+}(n_H)$ was obtained from the initial slope of the total excess potential $d(\Delta\mu_H)/dn_H$. In a second step, the electronic excess potential was obtained according to (3.1a) from the experimental values of $\Delta\mu_H$ and the calculated values of $\Delta\mu_{H^+}$.

In later evaluations [3.66], the initial Fermi energy rise was set equal to 1050 cal/mol, obtained from a discussion of the standard potentials $\Delta\mu_H^0$ in a series of Pd/Ag alloys (Sect. 3.4.1).

As an approximation, the Fermi energy rise can be calculated from electronic specific heat data on the isoelectronic Pd/Ag alloys [3.67, 68] or on the Pd/H system proper [3.69], which are of limited use, however, due to the miscibility gap below $n_H = 0.6$. Within the rigid band framework, the slope of $\Delta\mu_e$ is equal to the reciprocal value of the density of states. *Simons* and *Flanagan* [3.70] used an electronic excess potential calculated in this way in conjunction with a Bragg-Williams contribution to interpret the shape of the H absorption isotherms.

As will be discussed in Section 3.4.3, the shortcomings of the rigid band model do not permit a good calculation of the actual Fermi energy rise in these systems, yielding too large values of $\Delta\mu_e$ by a factor of 2 to 3. That is why the empirical way outlined above was adopted. The method is justified by the possibility of using the same electronic excess potential, in good approximation, for all isotherms on pure Pd and on a series of Pd alloys with electron donors as well [3.66, 71–73], see (3.60) p. 124. The electronic excess potential turns out to be rather independent of temperature, except in the transition range between the 4*d* and 5*s* band regimes (see Fig. 3.20 and 3.21).

In a critical comment on this method, *Burch* [3.74] arrived at the opposite conclusion of a prohibitively high variation of $\Delta\mu_e$ with temperature. The method was not properly presented, however. These seems to be a most unfortunate misunderstanding of the electronic excess potential introduced here.

In a discussion of their results on the H absorption capacity of Pd/B and Pd/Ag alloys, in which the H concentration was defined as N_H/N_{Pd} instead of $N_H/(N_{Pd}+N_{Ag})$, *Burch* and *Lewis* [3.75] concluded that B was about seven times as effective as Ag in reducing the H solubility in the 5*s* region instead of three times. The authors suggested, accordingly, that electron band effects were not responsible for limiting the H absorption capacity. (With the concentration variables used in this paper, however, the H absorption capacity and the susceptibility are influenced in the expected way, see Figs. 3.18 and 3.22.)

Consequently, *Burch* [3.76] introduced the idea of a repulsion of H atoms on next-nearest neighbor sites, which would overcompensate the attraction of nearest neighbor H atoms at high concentrations. He obtained a good fit of calculated and experimental curves on the binary Pd/H system; it remains to be seen, however, whether the scheme can be extended to interpret H absorption isotherms in Pd alloys.

All of these models are greatly indebted to the preceding investigation of *Lacher* [3.8], who introduced the Bragg-Williams approximation to account for the two-phase phenomena, but assumed that only part of the octahedral sites were available for H occupation ($n_{max} \approx 0.60$) to account for the asymmetry of the absorption isotherms.

In order to make the underlying assumptions of the quasichemical approach more explicit, it is useful to set up and evaluate the *partition function* of the system. We shall largely follow the presentation given by *Hill* [3.59].

The vibrational partition functions q_{Pd} and q_H are assumed to be independent of concentration. Similarly, the nearest neighbor interaction energies are taken to be independent of concentration and cluster size. There is no interaction between particles which are not nearest neighbors.

Apart from a mathematical approximation described below, these three assumptions involve the most stringent simplifications of the model with respect to real systems. Deviations from the first assumption, however, have often been found to be of minor importance. Comparison with experimental results will have to show how serious the other two assumptions are.

The octahedral sites occur in one of two states: empty, or occupied by H. The occupancy of two adjacent sites may be characterized by the symbols [HH] if both are occupied by H atoms, [H0] if one is occupied by an H and the other is empty, and [00] if both are empty.

The rearrangement of pairs which is formulated in analogy to a chemical reaction

$$2[\mathrm{H0}] \rightleftarrows [\mathrm{HH}] + [00] \tag{3.25}$$

is accompanied by the energy change w_{HH}, the *pair interaction energy*. Usually, this energy is an internal energy independent of temperature, but the problem can be restated to allow for a temperature-dependent energy [3.77].

Let N_H denote the number of occupied sites and $N_0 = N_{Pd} - N_H$ the number of empty sites. Between the pair numbers N_{HH}, N_{00}, and N_{H0} there exist the restrictions

$$2N_{HH} = zN_H - N_{H0}\,, \qquad 2N_{00} = zN_0 - N_{H0} \tag{3.26}$$

such that only one independent pair number, e.g., N_{H0}, remains. z is the coordination number, or maximum number of nearest neighbors; in the Pd/H system $z = 12$.

With the vibrational contributions of the individual Pd and H atoms, q_{Pd} and q_H, on their lattice or interstitial sites, respectively, the partition function is

$$Q(N_{Pd}, N_H, T) = q_{Pd}^{N_{Pd}} \cdot q_H^{N_H} \sum_{N_{H0}} g(N_0, N_H, N_{H0}, T) \exp[-E(N_H, N_{H0})/kT] \tag{3.27}$$

to be summed over all possible pair numbers N_{H0}.

The energy $E(N_H, N_{H0})$ is assumed to be a function of the number of H atoms and of the numbers of pairs only, independent of the particular way of arrangement of pairs,

$$E(N_H, N_{H0}) = -\frac{z}{2} w_{HH} N_H + w_{HH} N_{HH} = -w_{HH} N_{H0}/2\,. \tag{3.28}$$

The energy is defined in such a way, that a "relative" quantity is obtained, vanishing at both $N_H=0$ and $N_H=1$. The inclusion of the term $-(z/2)w_{HH}N_H$, which is sometimes left out, implies an initial energy change, when an H atom enters the lattice far from any neighbor. This energy is successively lowered as more and more pairs are formed (see also Sect. 3.4.1).

In particular, if two isolated interstitials move into adjacent sites, their combined energy decreases from $-zw_{HH}$ to $-(z-1)w_{HH}$, i.e., by the "energy of pair formation" w_{HH}. The ratio of energy per isolated particle to energy per pair is $(-zw_{HH}/2)/w_{HH}=-z/2$.

The weight factor $g(N_0, N_H, N_{H0}, T)$ in (3.27) is the number of arrangements compatible with the particle numbers N_{Pd} and N_H and a given pair number N_{H0}. The task is to find the correct weight factor and then to determine the set of equilibrium pair numbers.

No exact expression has yet been formulated for the weight factor or for the equilibrium pair numbers. If a random arrangement of particles is assumed, the number of pairs is readily stated, and the weight factor is just the combinatorial number, independent of temperature,

$$g_{\text{ideal}}=N_{Pd}!/N_0!N_H!\,, \tag{3.29}$$

leading to an ideal configurational entropy. This is the treatment of the zeroth or Bragg-Williams approximation.

The approach used here, which is more or less explicit in the original papers of *Guggenheim* [3.62] and of *Bethe* [3.63], is to treat the pairs as mutually independent in a first step, as if the relation (3.25) were representing reactions between molecules (quasichemical approach), and to introduce a normalization factor in a second step to remove some effects of the overcounting. The result, not to be derived here, is

$$g(N_0, N_H, N_{H0})=\left[\frac{N_{Pd}!}{N_0!N_H!}\right]^{1-z}\cdot\frac{(zN_{Pd}/2)!}{N_{HH}!N_{00}![(N_{H0}/2)!]^2}\,. \tag{3.30}$$

The equilibrium value of the pair number N^*_{H0} is obtained by the maximal term method, i.e., by differentiating (3.27) with respect to N_{H0} at constant N_H and N_{Pd}.

Substituting $\alpha_{H0}=N^*_{H0}/zN_{Pd}$ and $n_H=N_H/N_{Pd}$, one obtains the equation, reminiscent of a chemical equilibrium,

$$\frac{(n_H-\alpha_{H0})(1-n_H-\alpha_{H0})}{\alpha^2_{H0}}=\exp\left(-\frac{w_{HH}}{kT}\right). \tag{3.31}$$

After solving for α_{H0} and the *equilibrium pair number* N^*_{H0}, respectively,

$$N^*_{H0}=zN_{Pd}\frac{2n_H(1-n_H)}{1+\beta}\,;\quad \beta=\left[1+4n_H(1-n_H)\left(\exp\frac{-w_{HH}}{kT}-1\right)\right]^{1/2}, \tag{3.32}$$

the chemical potential per g-atom of H, exclusive of the Fermi energy part, is obtained from the partion function (3.27) according to

$$\mu_{\mathrm{H}} = -RT(\partial \ln Q/\partial N_{\mathrm{H}})_{N_{\mathrm{Pd}},T}:$$
$$\mu_{\mathrm{H}} = -RT\ln q_{\mathrm{H}} + RT\ln\frac{n_{\mathrm{H}}}{1-n_{\mathrm{H}}} + \frac{z}{2}RT\ln\frac{\beta-1+2n}{\beta+1-2n}\cdot\frac{1-n}{n}, \tag{3.33}$$

which can be rewritten, in view of (3.31) and (3.32), to

$$\mu_{\mathrm{H}} = -RT\ln q_{\mathrm{H}} - \frac{z}{2}N_{\mathrm{L}}w_{\mathrm{HH}} + RT\ln\frac{n_{\mathrm{H}}}{1-n_{\mathrm{H}}} + zRT\ln\frac{2-2n}{\beta+1-2n}. \tag{3.34}$$

In molar quantities, $W_{\mathrm{HH}} = N_{\mathrm{L}} \cdot w_{\mathrm{HH}}$, the last term of (3.34) is the protonic excess potential

$$\Delta\mu_{\mathrm{H}^+} = zRT\ln\frac{2-2n_{\mathrm{H}}}{\beta+1-2n_{\mathrm{H}}} \tag{3.35}$$

while

$$-RT\ln q_{\mathrm{H}} = \mu^0_{\mathrm{H_{vib}}} \tag{3.36}$$

and

$$-\frac{z}{2}W_{\mathrm{HH}} = \mu^0_{\mathrm{H}^+} \tag{3.37}$$

are the vibrational and protonic standard potentials, respectively, shown in Fig. 3.11.

From (3.28) and (3.32) the configurational energy per g-atom Pd is calculated to be

$$E = -\frac{z}{2}W_{\mathrm{HH}}\cdot\frac{2n_{\mathrm{H}}(1-n_{\mathrm{H}})}{1+\beta}. \tag{3.38}$$

3.2.6 Analogous Binary Alloys

The two effects invoked to explain the nonideal behavior of the Pd/H system, the Fermi energy rise, and the clustering tendency, are quite general and should be useful in a number of other systems. In many respects, the applicability of this hypothesis to related alloys can be regarded as a necessary test of the underlying assumptions.

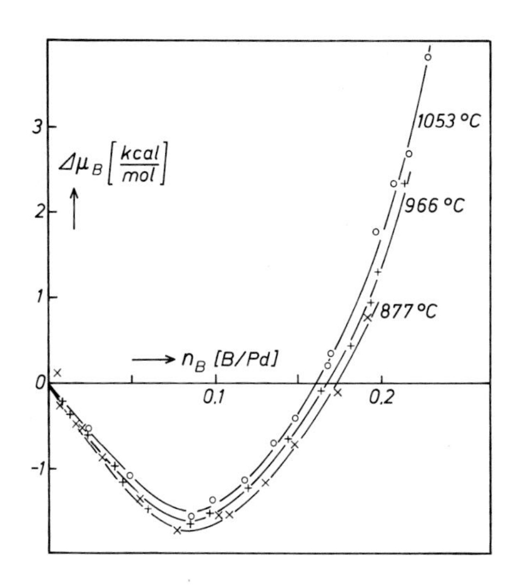

Fig. 3.12. Excess potential of B in Pd [3.78]

Due to the particular band structure, Pd and Ni alloys with electron donators lend themselves remarkably well to such a test. The clustering tendency shows up in the *d* band regime and can often be separated from the Fermi energy effect in this concentration range, while the transition from filling of the *d*-band to the filling of the *s*-band shows up as a landmark in a number of thermodynamic, kinetic, and spectroscopic properties. The systems Pd/B and Pd/Ag may serve as examples for many others to illustrate the point.

At high temperatures, *boron* forms a wide range of homogeneous solid solutions in Pd, up to an atomic ratio $n_B \approx 0.24$ [3.78]. A two-phase region exists just as in the Pd/H system, but with a critical point at 410 °C and at $n_B = 0.0654$ [3.79]. B donates its three valence electrons to the electron bands of the metal as deduced from susceptibility measurements and H absorption isotherms [3.72, 73] on Pd/B alloys. Within the error limits, the susceptibility curves of B and Ag are identical, if the valence electron concentrations are chosen to be $3n_B$ and x_{Ag}, respectively. *Mahnig* and *Toth* [3.80] found a similarly good agreement between the electronic specific heats of the two systems.

The relative lattice expansion $d \ln a/dn_B = 0.173$ is about three times as large as that due to H ($d \ln a/dn_H = 0.056$ from the $\alpha \to \beta$ transition, see p. 77); the activation energy of diffusion is about four times as large (24.1 kcal/mol) [3.78, 81]. For these reasons, it may safely be judged that B occupies the octahedral sites just as H.

Figure 3.12 shows the excess potential $\Delta\mu_B$ of three isotherms. The data were obtained by equilibrating Pd samples in the presence of B_2O_3 in an H_2–H_2O stream, which fixes the B activity, and by subsequently measuring the weight gain [3.78]. A prominent minimum is evident at $n_B \approx 0.1$, with a negative slope in the 4*d* band region and a steep rise in the 5*s* band region. There is a pronounced temperature effect. The initial slope, for example, changes from

-32.0 to -34.3 to -36.0 kcal/mol at the temperatures of 1053, 966, and 877 °C, respectively. This increase of slopes can largely be accounted for by the quasichemical approximation with pair interaction energies W_{BB} between -1.67 and -1.76 kcal/mol.

An evaluation of the two-phase region confirms the temperature effect down to 300–400 °C ($\partial \Delta\mu_B/\partial n_B = -54.5$ kcal/mol). This tremendous temperature dependence is also, in the main part, predicted by the quasichemical approximation, and to a smaller extent by an increase, in absolute terms, of the pair interaction energy ($W_{BB} = -1.83$ kcal/mol).

The ability of the quasichemical theory, despite its approximate nature, to predict the large temperature dependence of the excess potentials both in the system Pd/B [3.79] and in the system Pd/H [3.9, 66] including the specific heat anomaly at 55 K is supporting evidence for the validity of this approach.

In order to compare the excess potential of an interstitial solute like H in Pd with that of a substitutional solute like *silver* in Pd, two aspects have to be considered [3.82].

First, the chemical excess potential of H, $\Delta\mu_H$, is to be compared to the difference of the partial molar excess free energies of Ag and Pd, $\Delta\overline{G^E_{Ag}} - \Delta\overline{G^E_{Pd}}$. By this device, the changes of the free energies due to the introduction of one H or Ag atom, respectively, are in both cases considered at constant number of lattice sites and at constant number of occupiable electronic states.

The concentration variables corresponding to each other are the atom ratio n_H and the mole fraction x_{Ag}, respectively. These concentration variables denote the mole fraction of occupied octahedral sites or of differently occupied lattice sites (influencing the configurational and quasichemical contributions) and they also happen to be equal, in the univalent case, to the valence electron concentration (influencing the Fermi energy contribution, see Sect. 3.4.3).

Second, the excess functions of Pd/Ag alloys are usually presented as "relative" quantities, the pure components serving as states of reference. The activity coefficients are related to Raoult's law. For gaseous solutes such as H_2, however, the usual state of reference is the infinitely dilute solution, and the activity coefficients are related to Henry's law.

The second difference involves an additive constant in the functions $\Delta\mu_H$ or $\Delta\overline{G^E_{Ag}} - \Delta\overline{G^E_{Pd}}$. It can be removed by subtracting $\Delta\overline{G^E_{Ag}}(x_{Ag}=0)$ from $\Delta\overline{G^E_{Ag}} - \Delta\overline{G^E_{Pd}}$, making this function start at the origin like $\Delta\mu_H$. Alternatively, it can be removed by subtracting $\int_0^1 \Delta\mu_H dn_H$ fro n $\Delta\mu_H$, making a relative function out of $\Delta\mu_H$. [The relative character of the difference of the partial molar excess free energies of Ag and Pd shows up in the condition $\int_0^1 (\Delta\overline{G^E_{Ag}} - \Delta\overline{G^E_{Pd}}) dx_{Ag} = 0$.] The last alternative has been adopted to juxtapose the corresponding excess functions of the two systems in Fig. 3.13.

Results on the Pd/Ag alloy were obtained by *Schmahl* and *Schneider* [3.83] and by *Myles* [3.84] with two different methods around 700 and around

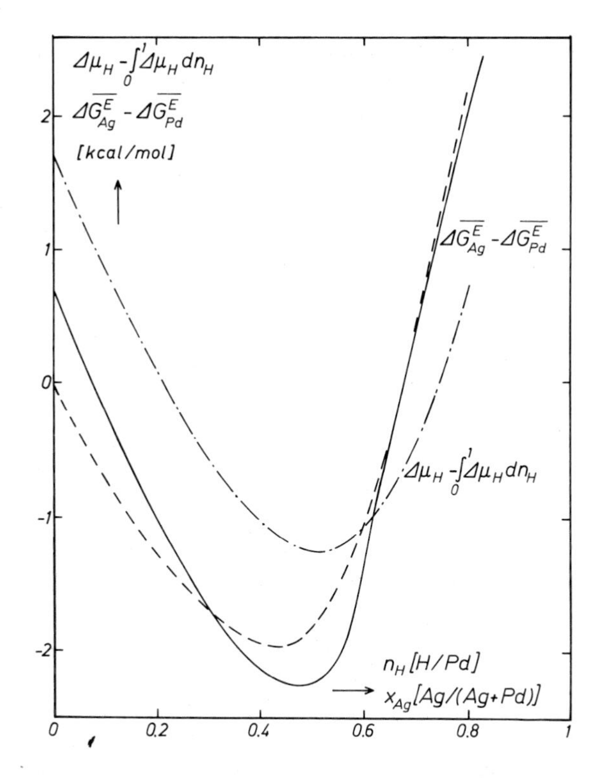

Fig. 3.13. Excess potential of H in Pd and difference of the partial molar excess free energies of Ag and Pd [3.82]

940 °C, respectively. The shift between the curves is fully accounted for by the excess entropy [3.83], which was also measured by *Myles* [3.84].

On the whole, the two sets of curves have the same shape, a negative slope in the 4*d* band region and steep rise in the 5*s* band region. A pair interaction energy of $W_{\mathrm{AgAg}} = -660$ kcal/mol is calculated from the initial slope, if the Fermi energy rise in this region is neglected, or of -721 kcal/mol, if an initial rise of the Fermi energy of 1050 kcal/mol is adopted. The Fermi energy rise in the 5*s* band region is given by 24.5 or 25.5 kcal/mol, respectively.

The extended negative slope of the excess functions $\overline{\Delta G^{\mathrm{E}}_{\mathrm{Ag}}} - \overline{\Delta G^{\mathrm{E}}_{\mathrm{Pd}}}$ in Fig. 3.13 suggests that a *miscibility gap* exists at lower temperatures on the Pd-rich side of the phase diagram. Extrapolations of the high-temperature data by various methods yield two-phase formation below about 300 °C, similar to the one in the Pd/H system but not quite so wide [3.85].

Unfortunately, diffusion rates are too low to bring about such a phase separation in the bulk alloy. However, there are indications in the literature that phase separation probably did occur in thin films. From an x-ray investigation, *Moss* and *Thomas* had an indication of two coexisting phases at 240 °C, which they disbelieved, however, since in a system "showing large negative enthalpies of formation, ... clearly a miscibility gap is not expected" [3.86]. Of course, a positive enthalpy of formation is a sufficient but not a necessary condition for phase separation to occur under equilibrium.

The Pd/Ag system is a rare example, where an indentation exists in the low temperature $\Delta G(x_i)$ curves without a concomitant positive $\Delta H(x_i)$ curve. Meanwhile, the contour lines of the miscibility gap were traced out by x-ray measurements on samples obtained by two different methods: annealing of evaporated films and direct electrolytical codeposition or cementation of the alloy [3.85]. The shape is curiously similar to the one in the Pd/H system.

Other alloys to be mentioned in this context which were found to have the same causes of nonideality are Pd/Cd [3.87], Pd/In, and Pd/Sn [3.88].

3.3 Interpretation of Nonideal Solution Behavior

3.3.1 Lattice Strain Energy

Nearest neighbor interactions were introduced into the Bragg-Williams approximation or the quasichemical model above to account for part of the deviations from ideal solution behavior of H in Pd. The nature of these interactions does not have to be specified, as far as the statistical treatment is concerned, and cannot be elucidated without recourse to other methods. Two hypotheses were advanced to explain the existence of pair interactions in solid alloys: the oscillatory decay of the potential around a solute atom [3.89] and lattice strain.

According to the first hypothesis, an attraction between two solute atoms will occur if the wavelength of the oscillations around each solute atom is such as to produce the opposite potential at a nearest neighbor site. The wavelength is strongly dependent on the valence electron concentration. Therefore, the pair interaction energies should be extremely sensitive to changes of the valence electron concentration, both with respect to magnitude and to sign, which has not been observed. Also, the model is restricted to metals and semiconductors and cannot explain similar behavior in nonmetallic solid solutions.

Therefore, the *strain energy model* seems more likely, which implies a difference of the lattice strain energy about two nearest neighbor solute atoms on the one hand, and about two farther separate solute atoms on the other hand. If the total strain energy is lower in the case of nearest neighbor site occupancy, an attraction between like atoms results.

On the basis of the elastic displacement of individual solvent atoms, *Fisher* [3.90] calculated an energy of attraction for H atoms on certain neighboring sites in bcc lattices. The tendency to form extra pairs is supported by internal friction measurements on dilute interstitial solutions of bcc metals [3.91, 92]. The trapping of H at interstitial impurities is an analogous case [3.92, 93].

Recently, *Zimmermann* determined an energy of attraction between H atoms in Pd, evaluated from the orientation of H–H pairs in internal friction experiments [3.94]. *Zimmermann* was even able to assess the concentration dependence of the pair interaction energy. Using inelastic neutron scattering experiments on α phase Pd/H, *Drexel* and co-workers found strong evidence for the existence of clusters [3.34]. They could best interpret their results in terms of the screened proton model (Sect. 3.4.3).

On quite different grounds, using continuum elastic considerations, *Lawson* calculated the molar *misfit energy* of infinitely dilute solute atoms in a host lattice [3.95]. This misfit energy corresponds to the strain energy part of the partial molar excess free energy of the solute at zero concentration: $\Delta\overline{G_i^E}(x_i=0)$. While *Lawson* originally deduced a free misfit energy, *Heumann* stressed the entropy part [3.96]. By combining the two contributions, the corresponding enthalpy is obtained [3.88, 97–99].

With slight modifications, Lawson's argument can be presented as follows: The work of expanding the volume of a host atom V_1^0/N_L to the volume $\bar{V}_2/N_L$, large enough to accommodate a solute atom on a lattice site, supposing the partial molar volume $\bar{V}_2$ of the solute component is larger than V_1^0, is set equal to the partial molar excess free energy

$$\Delta\overline{G_2^E}(x_2=0)/N_L = -\int_{V_1^0/N_L}^{\bar{V}_2/N_L} P\,dV, \tag{3.39}$$

where the (negative) hydrostatic pressure P is given by

$$P=(V_1^0-V)/\kappa_1 V_1^0 . \tag{3.40}$$

κ_1 is the isothermal compressibility of pure solvent 1. Thus the molar free misfit energy of substitutional solute atoms 2 on the lattice of solvent atoms 1 is

$$\Delta\overline{G_2^E}(x_2=0)=(\bar{V}_2-V_1^0)^2/2\kappa_1 V_1^0 . \tag{3.41}$$

Due to the square in the numerator, the misfit energy may be regarded as the work of either compressing pure solvent from volume $\bar{V}_2$ to V_1^0 or dilating the solvent from volume V_1^0 to $\bar{V}_2$ (supposing $\bar{V}_2>V_1^0$).

Lawson's model implies work to be done on the volume of the inserted particle only and not on the host matrix, which is certainly a rough approximation. A more complete treatment of misfit particles in a coherently distorted lattice based on *continuum elasticity theory* is given by *Wagner* in [Ref. 3.100, Chap. 2].

Wagner showed, in particular, that a *long-range interaction* exists in these alloys. It should be borne in mind, however, that the applicability of the continuum elasticity theory is restricted to regions of small strain outside the sphere of close neighbors. An *atomistic approach* is necessary to describe the behavior of adjacent solute atoms such as the method used by *Fisher* [3.90], which happens to predict pair formation. While the experimental evidence does not exclude the existence of long-range interactions, it seems to indicate a preponderance of *pair interactions* [3.34, 71, 72, 78, 79, 91–94].

If both pure components crystallize in the same lattice, suitable interpolation formulae can be found to describe various excess functions in the

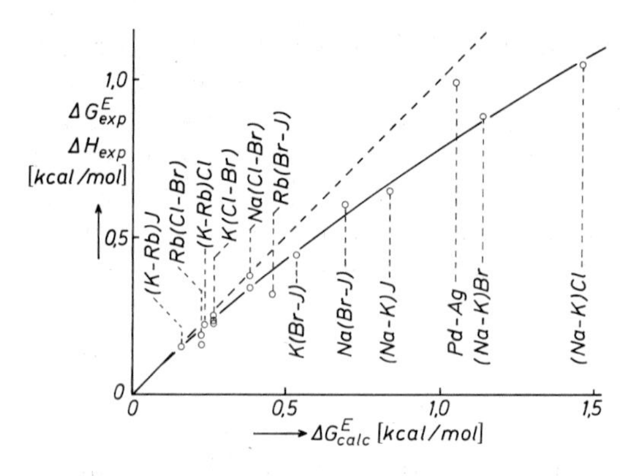

Fig. 3.14. Comparison of experimental and calculated relative molar excess free energies or enthalpies of mixing, respectively, of various equimolar alkali halide mixtures [3.99] and of an equimolar Pd/Ag alloy (strain energy part only): (K/Rb)J [3.103]; Rb(Cl/Br) [3.104]; (K/Rb)Cl [3.105, 106]; K(Cl/Br) [3.99, 107, 108]; Na(Cl/Br) [3.104, 108, 110]; Rb(Br/J) [3.111]; K(Br/J) [3.110]; Na(Br/J) [3.110]; (Na/K)J [3.110]; (Na/K)Br [3.110]; (Na/K)Cl [3.110, 112]; Pd/Ag, this chapter

intermediate concentration range from the misfit energy at vanishing concentrations in (3.39). With the abbreviations $\Delta\overline{G_2^{\mathrm{E}}}(x_2=0)\equiv L$ and $\Delta\overline{G_1^{\mathrm{E}}}(x_2=1)\equiv M$, one obtains [3.99]

$$\Delta\overline{G_2^{\mathrm{E}}}=(2M-L)(1-x_2)^2+2(L-M)(1-x_2)^3 \tag{3.42}$$

$$\Delta\overline{G_1^{\mathrm{E}}}=(2L-M)x_2^2-2(L-M)x_2^3 \tag{3.43}$$

$$\Delta\overline{G_2^{\mathrm{E}}}-\Delta\overline{G_1^{\mathrm{E}}}=L(1-2x_2)-(L-M)(2x_2-3x_2^2) \tag{3.44}$$

$$\Delta\overline{G^{\mathrm{E}}}=Lx_2(1-x_2)-(L-M)x_2^2(1-x_2)\,. \tag{3.45}$$

Experimental data and calculated values of $\Delta\overline{G^{\mathrm{E}}}(0.5)=(L+M)/8$ of a number of alkali halides are compared in Fig. 3.14. Molar volumes were used in (3.41) wherever partial molar volumes are not available to account for the size factor. A remarkable correlation is observed, with calculated values systematically low by 10 to 20%.

While the misfit energy may be the only major cause of nonideality in alkali halide mixtures, Fermi energy effects have certainly to be considered in alloys. For Pd/Ag alloys a separation of the two influences was brought about in a way similar to the Pd/H system [3.82] (see also Sects. 3.2.6 and 3.4.1). The strain energy part of the partial molar excess free energies was evaluated to be 3.96 kcal/mol (4.48 kcal/mol if the Fermi energy rise in the 4*d* band is not neglected) compared to 4.31 kcal/mol (Ag in Pd) and 4.10 kcal/mol (Pd in Ag) as calculated from x-ray data [3.101] according to (3.41).

For interstitial solutions such as Pd/H, the upper limit of the integral in (3.39) is $(V_{\mathrm{Pd}}^0+\bar{V}_{\mathrm{H}})/N_{\mathrm{L}}$, and the misfit energy of H in Pd, in view of (3.37), is

$$\mu_{\mathrm{H}^+}^0=-6W_{\mathrm{HH}}=\bar{V}_{\mathrm{H}}^2/2\kappa_{\mathrm{Pd}}V_{\mathrm{Pd}}^0\,, \tag{3.46}$$

which relates the misfit energy to the pair interaction energy[6].

[6] Within the framework of the quasichemical approximation, the pair interaction energy may equally well be treated as dependent or as independent of temperature [3.62, 77].

From β phase lattice expansion data [3.14], the strain energy parts of the standard potentials of H and D are $\mu_{H^+}^0 = 3.75$ kcal/g-atom H and $\mu_{D^+}^0 = 3.69$ kcal/g-atom D according to (3.46), compared to 4.03 and 3.92 kcal/g-atom, respectively, from the evaluation of H absorption isotherms [3.71] (4.55 and 4.44 kcal/g-atom, respectively, if the initial Fermi energy rise is included [3.66]).

Extending this scheme to the concentration-dependent part of the partial molar free excess energy, one obtains, in the Bragg-Williams approximation

$$\mu_{H^+}^0 + \Delta\mu_{H^+} = -6W_{HH}(1-2n_H) = \frac{\bar{V}_H^2}{2V_{Pd}^0 \kappa_{Pd}}(1-2n_H)\,. \tag{3.47}$$

The slightly curved lines representing $\mu_{H^+}^0 + \Delta\mu_{H^+}$ in Fig. 3.11 were calculated according to the quasichemical approximation.

3.3.2 Excess Functions at Constant Volume and at Constant Pressure

In a notable paper, *Wagner* investigated the significance and the consequences of the assumption of constant pressure, implicit in most evaluations of the excess potential of interstitial solutions [3.102]. He threw light upon the implications involved by studying the influence of the alternative assumption of constant volume.

Briefly, using the present nomenclature, his argument runs as follows:

$$d\Delta\mu_H = \left(\frac{\partial \Delta\mu_H}{\partial n_H}\right)_{T,V} dn_H + \left(\frac{\partial \Delta\mu_H}{\partial V}\right)_{T,n_H} dV \tag{3.48}$$

$$\left(\frac{\partial \Delta\mu_H}{\partial n_H}\right)_{T,P} = \left(\frac{\partial \Delta\mu_H}{\partial n_H}\right)_{T,V} + \left(\frac{\partial \Delta\mu_H}{\partial V}\right)_{T,n_H} \left(\frac{\partial V}{\partial n_H}\right)_{T,P} \tag{3.49}$$

where $(\partial V/\partial n_H)_{T,P} = \bar{V}_H$ and

$$\left(\frac{\partial \Delta\mu_H}{\partial V}\right)_{T,n_H} = \left(\frac{\partial \Delta\mu_H}{\partial P}\right)_{T,n_H} \bigg/ \left(\frac{\partial V}{\partial P}\right)_{T,n_H} = -\frac{\bar{V}_H}{V_{Pd}^0 \kappa_{Pd}}\,.$$

Therefore,

$$\left(\frac{\partial \Delta\mu_H}{\partial n_H}\right)_{T,V} = \left(\frac{\partial \Delta\mu_H}{\partial n_H}\right)_{T,P} + \frac{\bar{V}_H^2}{V_{Pd}^0 \kappa_{Pd}}\,. \tag{3.50}$$

Comparing the two terms on the right-hand side of (3.50), *Wagner* found that they are opposite in sign and practically cancel for a number of widely varying temperatures. Thus, at constant volume, *Wagner* obtained the remarkable result that within the limits of error the chemical potential is that of an *ideal solution:*

$$V = \text{const.}\,; \qquad \mu_H - \mu_H^0 - RT\ln(n_H/(1-n_H)) \approx 0\,. \tag{3.51}$$

He concluded that, since the lattice far away from H atoms at infinite dilution is unaffected at constant volume, there is no pair formation either at constant volume or at constant pressure, the nonvanishing excess potential at constant pressure being entirely due to the lattice expansion.

While the thermodynamic relations of (3.50, 51) are straightforward, different interpretations seem to be possible. The assertion that the distances of Pd atoms remain unaffected by the introduction of H atoms at fixed average lattice constant has certainly to be qualified. Actually, the Pd atoms around an occupied octahedral site will be pushed apart and the ones about empty sites will be pushed together.

This may be clarified more quantitatively in the analogous case of a system forming an uninterrupted series of substitutional solid solutions. If x_2 moles of substance 2 of molar volume $V_2^0 > V_1^0$ and $(1-x_2)$ moles of substance 1 of molar volume V_1^0 are jointly compressed from $V_1^0 + x_2(V_2^0 - V_1^0)$ to the common volume V_1^0 prior to mixing, the distances between like particles in both unmixed samples will certainly decrease and remain so after mixing occurs at constant volume. In this case, the volumes shrink to $V_2(P) = V_2^0 - x_2(V_2^0 - V_1^0)/V_i^0\kappa_i$ and $V_1(P) = V_1^0 - (1-x_2)\cdot(V_2^0 - V_1^0)V_i^0\kappa_i$ and the work necessary to bring this compression about is $(V_2^0 - V_1^0)^2 x_2^2/2V_i^0\kappa_i$. Here $V_i^0\kappa_i$ is the weighted average value $(1-x_2)V_1^0\kappa_1 + x_2V_2^0\kappa_2$.

An interpretation holding for a larger concentration range beyond the limiting case of very dilute solutions is desirable. Experimentally, the slope of $\Delta\mu_H$ at constant pressure is found to be roughly constant right into the region of $n_H \approx 0.3$, where every Pd atom has on the average about 1.8 H atoms next to it. Indeed, the change of slope at higher concentrations $n_H > 0.3$ is most certainly due to Fermi energy effects rather than to a breakdown of the nearest neighbor model. The applicability of the pair interaction energy w_{HH} obtained from α phase measurements to describe the β phase heat peak at 55 K suggests that the model may approximately hold to the highest H concentrations (see Sect. 3.3.4). Thus any more general interpretation should not necessarily be confined to the infinitely dilute case, in which most Pd atoms are far away from dissolved H atoms.

One more point deserves special attention, namely the meaning of the *standard potential* in the two cases. Formally, the standard potentials at $P = \text{const.}$ and $V = \text{const.}$ are entirely equivalent and they are indeed identical in the limiting case of $n_H = 0$.

Figure 3.15 illustrates the change of the function

$$\mu_H^0 + \Delta\mu_H = \mu_H - RT\ln n_H/(1-n_H)$$

upon charging Pd with H along various pathways.

Path $\overline{AB}$ denotes the change of $\mu_H^0 + \Delta\mu_H$ under ambient pressure $P \approx 0$ as obtained experimentally upon addition of n_H^* g-atoms H to 1 g-atom Pd. If $\overline{BC} = \bar{V}_H^2\, n_H^*/V_{Pd}^0\kappa_{Pd}$ is added to $\overline{AB}$, the path $\overline{AC}$ is obtained, indicating a change

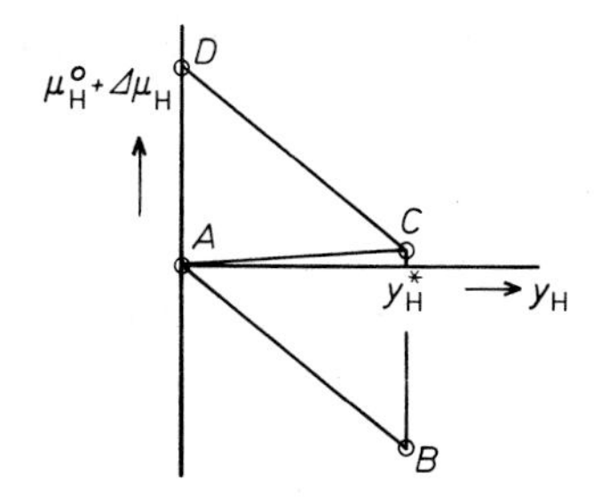

Fig. 3.15. Change of the function of $\mu_H^0 + \Delta\mu_H$ upon H absorption at constant pressure and at constant volume

of the function $\mu_H^0 + \Delta\mu_H$ of about zero upon addition of n_H^* g-atoms H to 1 g-atom Pd at constant volume V_{Pd}^0.

Significantly, point C may be reached from point A on an alternative pathway via point D. Path $\overline{DC}$ involves charging of H at constant pressure $P \gg 0$ such that at the atomic ratio n_H^* (point C) the volume of the system is V_{Pd}^0. The pressure is calculated according to (3.40) to be $P = \bar{V}_H n_H^* / V_{Pd}^0 \kappa_{Pd}$. The volume of the pure Pd at pressure P (point D) is $V_{Pd}^0 - \bar{V}_H n_H^*$. Therefore, the work of compression of one mole of Pd by $\bar{V}_H n_H^*$ is $\bar{V}_H^2 n_H^{*2} / 2V_{Pd}^0 \kappa_{Pd}$. The path $\overline{AD}$ is equal to the corresponding partial molar quantity $\bar{V}_H^2 n_H^* / V_{Pd}^0 \kappa_{Pd}$.

The equivalence of pathways $\overline{AC}$ and $\overline{AD} + \overline{DC}$ suggests that a more complicated situation may quite possibly be involved in the seemingly ideal behavior of inserting H atoms into a Pd lattice at constant volume. In particular, it should be borne in mind that the standard potential at constant volume implies insertion of H atoms at increasingly higher pressure. This effect is separately shown in the alternative path via point D, where the initial increase of the standard potential and the subsequent decrease of the excess potential more or less cancel. It might be argued that the negative slope of path $\overline{DC}$ leaves room for the nearest neighbor hypothesis even in the constant volume case.

Actually, similar opposing influences on the standard potential and on the excess potential are built-in features of various types of thermodynamic models. It is also exhibited, e.g., in the strain energy part of the partial molar excess free energies $\Delta\overline{G_2^E} - \Delta\overline{G_1^E} = -6W_{ii}(1-2x_2)$ of the analogous substitutional mixtures described in (3.44), supposing $L = M = -6W_{ii}$, or in the corresponding expression (3.47) for interstitial alloys.

3.3.3 The Specific Heat Anomaly at 55 K

For a system with *cluster formation*, the most natural way to reduce its configurational entropy at low temperatures is to segregate into two phases. In the Pd/H system, this happens only in or near the 4*d* band region. The energy requirement for complete phase separation above $n_H \approx 0.6$ is too high due to the steep Fermi energy rise in this concentration range. The formation of large clusters might be an alternative way to get rid of the configurational entropy.

The rise of the Fermi energy is also the cause for the incomplete phase separation observed in He^3–He^4 *mixtures* below 1 K [3.113], which are in many ways analogous to the system Pd/H [3.114]. Even at the lowest temperatures, there remains a residue of 6.4% He^3 dissolved in He^4 in equilibrium with pure He^3. Some mechanism other than phase separation must operate in this system, too, to bring its entropy to zero at vanishing temperatures.

If clustering does indeed exist in the β phase of Pd/H, a specific heat anomaly can be postulated: At very high temperatures, the number of pairs is that of the random configuration, and remains so upon a change of $\Delta T = 1$ K; similarly, at very low temperatures, the number of pairs is maximum and remains so upon a change of 1 K. At intermediate temperatures, however, there occurs a finite change of the number of pairs with a specific heat peak at the temperature of maximum change [3.66, 115].

The reasons for postulating a specific heat peak to exist in a system of clustering solute atoms are not unlike those that bring about a Schottky anomaly in a system with two energy levels. The lower level is represented by [HH] pairs, the upper level is represented by [OH] pairs. Of the total number of z bonds around the H atoms, only the part $zN_H(N_{Pd} - N_H)$ enters into the formation of extra clusters, which is not involved in clusters in the random arrangement at high temperatures.

A requirement for the anomaly to be actually observed in a system with cluster formation is the mobility of at least one component, sufficient to bring about equilibrium in the time available. Metal hydrides seem to be the only systems, besides liquid mixtures of the He isotopes 3 and 4, where this condition is fulfilled at the low temperatures of the expected anomalies [3.114].

Nace and *Aston* found anomalies of the specific heat in β phase and various two-phase samples of Pd/H and Pd/D [3.116]. The extra specific heat per g-atom H of the systems with respect to pure Pd, Δc_p, is plotted versus temperature in Fig. 3.16.

Einstein model vibrational specific heats c_{vib} can be fitted into the curves, such that for characteristic temperatures $\theta_H/\theta_D = \sqrt{2}$, there remains a residue $c_p^E = \Delta c_p - c_{vib}$ identical for the two isotopes. The anomaly is also plotted in Fig. 3.16.

The frequencies $\hbar\omega_H = 60.3$ meV and $\hbar\omega_D = 42.6$ meV corresponding to the characteristic temperatures of 700 and 495 K are in agreement with recent results of neutron inelastic scattering on β-PdH_n and β-PdD_n (57 ± 2 and 40 ± 2 meV, respectively) quoted on p. 94, although these values are slightly lower.

Equilibrium was only partially achieved in the specific heat experiments, as evidenced by two observations of *Nace* and *Aston* [3.116] and *Mitacek* and *Aston* [3.117]. They obtained a zero point entropy of 0.59 ± 0.18 cal/g-atom H and a continuing heat evolution at temperatures below the anomaly, also investigated by *Mackliet* and *Schindler* [3.118]. *Manchester* and co-workers noticed an ageing of rapidly cooled samples of Pd/D and Pd/H by means of

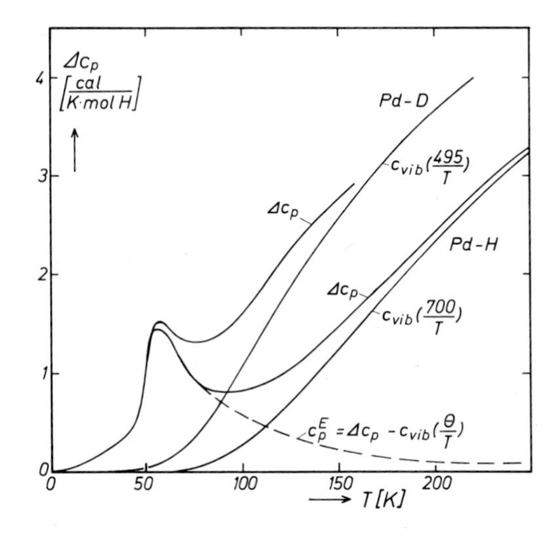

Fig. 3.16. Specific heat anomaly of Pd/H and Pd/D [3.66, 116]

electrical resistance measurements [3.119], as well as by internal friction measurements [3.120, 121].

In addition to the spontaneous heat emission at liquid helium temperatures, *Aston* and co-workers discovered similar diffusion controlled processes in the 150–200 and 200–250 K temperature ranges on several Pd/H samples [3.116, 117]. Both are evidence for the existence of sites of lower and higher energy, a strong indication of nearest neighbor interactions. The formation of an ordered structure as in the analogous system Pd/B [3.79] could also be responsible for these effects, but indications of superstructure lines or of additional miscibility gaps have yet to turn up, in spite of extensive research.

An expression for the *excess specific heat* c_p^{E}, can be derived from the quasichemical approximation. Differentiation of the configurational energy (3.38) of n_{H} g-atom H in 1 mole Pd with respect to T and division by n_{H} leads to the specific heat per mole H

$$\frac{1}{n_{\mathrm{H}}}\frac{\partial E}{\partial T}=c_p^{\mathrm{E}}=24R\left(-\frac{w_{\mathrm{HH}}}{kT}\right)^2\frac{n_{\mathrm{H}}(1-n_{\mathrm{H}})^2}{\beta(1+\beta)^2}\exp\left(-\frac{w_{\mathrm{HH}}}{kT}\right). \tag{3.52}$$

At half occupancy, $n_{\mathrm{H}}=0.5$, this reduce to

$$c_p^{\mathrm{E}}(n_{\mathrm{H}}=0.5)=\frac{3}{4}\left(\frac{w_{\mathrm{HH}}}{kT}\right)^2\frac{1}{\cosh^2\left(\dfrac{w_{\mathrm{HH}}}{4kT}\right)}. \tag{3.53}$$

Pair energy and temperature occur in the combination $kT/-w_{\mathrm{HH}}$ only, i.e., the specific heat can be plotted vs the reduced temperature $kT/-w_{\mathrm{HH}}$ for all values of w_{HH}. The magnitude of the effect is independent of the pair energy. Table 3.2 shows the peak temperatures T_{m} and the corresponding heat peak $c_p^{\mathrm{E}}(T_{\mathrm{m}})$ calculated for a number of H concentrations.

Table 3.2. Dependence of reduced peak temperature and peak height on H concentration

$n_H \frac{H}{Pd}$	$\frac{T_m}{-w_{HH}/k}$	$\frac{c_p^E(T_m)}{R}$
0.5	0.209	5.27
0.6	0.208	4.29
0.7	0.206	3.39
0.8	0.201	2.52
0.9	0.190	1.57

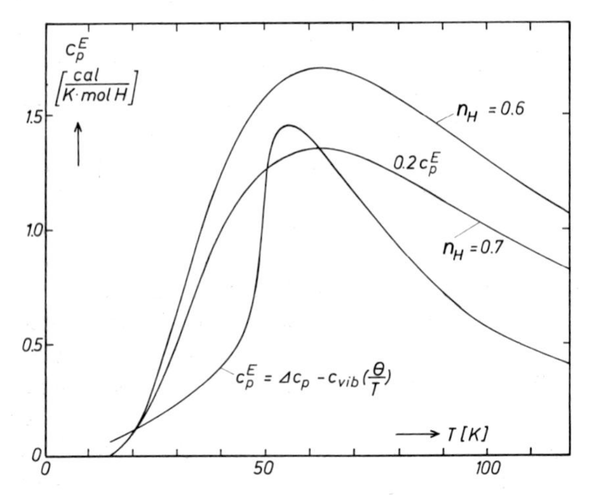

Fig. 3.17. Calculated and experimental specific heat anomaly [3.66]

The pair formation part of the specific heat per mole H, c_p^E, is included in Fig. 3.17 for a pair energy of $W_{HH} = -600$ cal/mol pairs and for two atomic ratios, $n_H = 0.6$ and 0.7, close to the ones that occurred in the investigation of *Mitacek* and *Aston* [3.117] (0.75 in the homogeneous sample and about 0.62 in the β phase of the two-phase samples).

Unfortunately, most of the samples where in the two-phase region, with part of the material in the α-phase almost devoid of H. The proportionality between peak height and H concentration in two-phase samples commented upon by some authors is quite trivial. Measurements of the true concentration dependence of the anomaly on homogeneous β phase samples with as small a zero point entropy as possible would be particularly valuable.

In view of the approximations involved in the quasichemical approach, which is more suitable to describe small deviations from randomness at temperatures $T \gg -w_{HH}/k$, the comparison with experimental data will necessarily be only qualitative in the temperature region of the anomaly $T < -w_{HH}/k$.

The peak temperature of 62.1 K calculated with $W_{HH} = -600$ cal/mol pairs is surprisingly close to the one obtained experimentally (55 K) (molar quantities are used to state the experimental values). The agreement is not bad even with a pair energy of $W_{HH} = -752$ cal/mol pairs extrapolated to 55 K from α phase activity data at higher temperatures [3.9] ($W_{HH} = -764 + 0.259\, T/\mathrm{K}$ [cal/mol pairs]).

The magnitude of the anomaly is much greater in the calculated curves than in the experimental curves c_p^E. (Note the scaling factor of 0.2 in Fig. 3.17). The discrepancy is not fully removed by making allowance for the observed zero point entropy of 0.59 ± 0.18 cal/g-atom H [3.116, 117].

In spite of the poor prediction of the magnitude of the anomaly by the quasichemical model, clustering seems to be a more likely explanation than alternative explanations by *Nace* and *Aston* [3.116] and by *Ferguson* et al. [3.122]. *Nace* and *Aston* suggested an awakening of rotational modes of moleculelike PdH_4 groups in the metal [3.116, 117]. *Ferguson* et al. connected the anomaly with a partial transition of H atoms from octahedral to tetrahedral sites at low temperature, according to their neutron diffraction results [3.122]. Aside from the difficulties involved in the energy and entropy changes implied in such a transition, the occupation of tetrahedral sites was not confirmed by later low-temperature neutron diffraction work by *Somenkov* [3.123] and by *Mueller* et al. [3.124].

Wallace and co-workers discovered several specific heat peaks in the system Ta/H [3.125]. While the double peak at 333 K is associated with the first-order phase transition between bcc and the orthorhombic structures, the peak at 306 K might be accounted for by the kind of clustering discussed here. According to this interpretation, a pair interaction energy of $W_{HH} \approx -3000$ cal/mol pairs is estimated. This is in rough agreement with values derived from H absorption isotherms in Ta [3.126] and of the same order of magnitude as values obtained from internal friction data on the related system Nb/H (-1490 cal/mol) [3.91]. *Wallace* and co-workers found that the configurational entropy of their $TaH_{0.5}$ sample is practically zero below the anomaly at 306 K [3.125]. This is in accordance with a clustering model. The same would be expected for the anomaly in Pd/H if the H mobility in Pd at 55 K were as high as in Ta at 306 K. With the limited mobility, equilibrium is only partially achieved in the time available.

Two recent theoretical efforts to understand the causes of specific heat anomalies in metal hydrides deserve attention. *Jacobi* and *Vaughan* discussed the 55 K heat peak of Pd/H in terms of a *Schottky* anomaly [3.127]. By adjusting two parameters, the energy difference and the ratio of degeneracy of two levels, they were able to arrive either at the correct peak temperature and peak height or at the correct temperature and entropy change. The parameters are not related, however, to any other data on the Pd/H system.

Stafford and *McLellan* obtained specific heat anomalies as a corollary of their statistical treatment of interstitial solid solutions [3.128]. A repulsive pair interaction also leads to an anomaly, which they found to be applicable to the

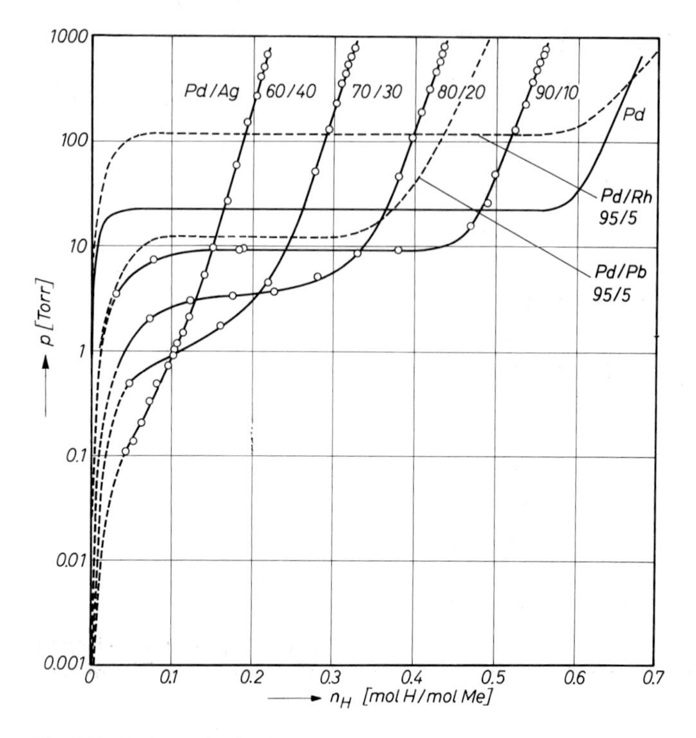

Fig. 3.18. H absorption isotherms at 50 °C in Pd [3.11], and in Pd alloys with Ag [3.71], Pb [3.129], and Rh [3.129]

Ta/H system. For the Pd/H system they adopted the clustering model. A pair interaction energy of −600 cal/mol pairs leads to a heat peak temperature of about 90 K.

3.4 Palladium Alloys

3.4.1 Fermi Energy Rise and Nearest Neighbor Effects in Ternary Systems

Numerous examples have been studied of how the addition of alloy components to Pd influences the H solubility [3.2]. Figure 18 shows absorption isotherms of Pd [3.11] and of a series of Pd/Ag alloys [3.71] as well as of a Pd/Pb and of a Pd/Rh alloy at 50 °C [3.129]. It is noteworthy that at pressures above 100 Torr, the solubility decreases with addition of Ag or Pb, while at pressures below 1 Torr, the solubility increases in the same alloys compared to pure Pd. Additions of Rh have the opposite effect of Ag or Pb.

The particular pressure, which separates the areas of greater and smaller solubility with respect to Pd, is a function of temperature. At elevated

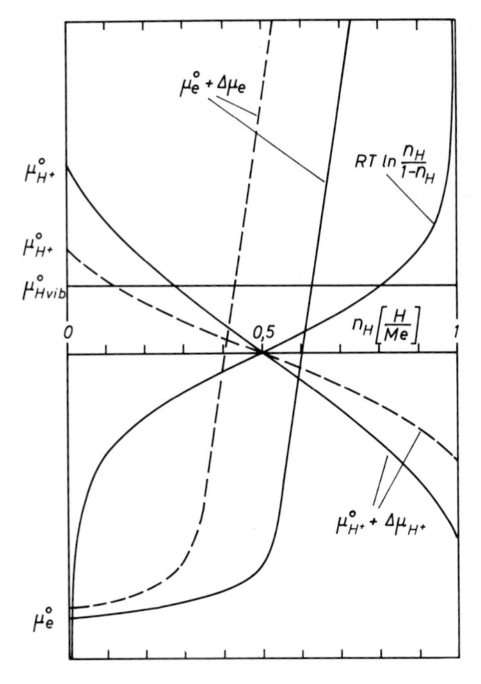

Fig. 3.19. Constituent parts of the chemical potential of H in Pd and in a Pd/20% Ag alloy

temperatures, very high pressures are needed to demonstrate a reduction of solubility in a Pd-Ag alloy.

This conflicting behavior puzzled early investigators and led to some controversy about the influence of alloy components on H absorption. Actually, it is readily understood in terms of band filling and lattice strain effects.

The various constituent parts of the *chemical potential* μ_H of H in a Pd alloy containing about 20 at.% Ag are shown in Fig. 3.19 to illustrate the point, together with the corresponding curves already shown in Fig. 3.11. The vibrational part of the standard potential, $\mu^0_{H_{vib}}$, is regarded to be equal in pure Pd and in the alloy and so is, of course, the configurational term of an ideal solution, $RT \ln n_H/(1-n_H)$. Actually, the small frequency change of H in α- and β-Pd/H as well as in several alloys has been measured by neutron scattering, and the assumption of constant $\mu^0_{H_{vib}}$ is no longer necessary.

The dashed line marked $\mu^0_e + \Delta\mu_e$ indicates the *Fermi energy rise* in the alloy. The transition from the gentle rise in the 4*d* band region to the steep rise in the 5*s* band region is shifted from about $n_H = 0.5$ in Pd and to about 0.3 in the 20 at.% Ag alloy, since part of the 4*d* band gap has already been filled by the valence electrons of Ag.

Other electron donors such as Pb [3.129], Sn [3.72, 129] or B [3.72, 73] behave likewise, while an addition of 5% Rh increases the number of unfilled 4*d* states available for H electrons [3.129]. Magnetic susceptibility measurements

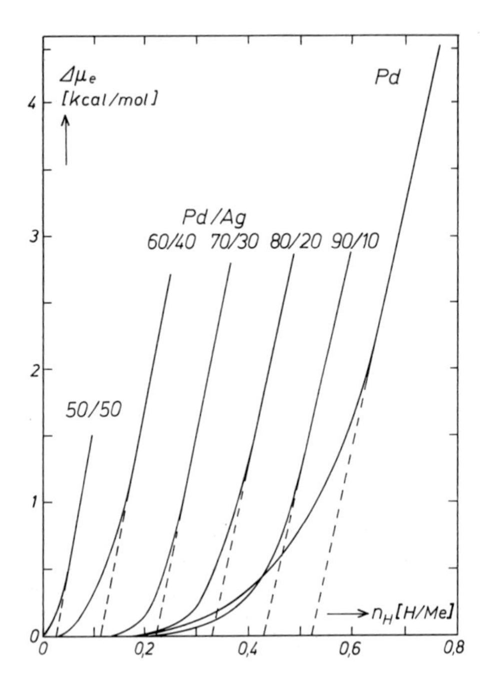

Fig. 3.20. Fermi energy part $\Delta\mu_e$ of the excess potential of H in Pd and in Pd/Ag alloys [3.66, 71]

on a number of ternary Pd/Ag/H [3.72, 132] and Pd/Sn/H [3.72] alloys confirm the notion of a strictly additive effect of two electron donors in filling up the empty 4*d* states. It is this filling of part of the empty 4*d* states by Ag or Pb valence electrons which reduces the H solubility in the 5*s* band region, or high pressure region, mentioned above.

The solid and dashed lines marked $\mu^0_{H^+} + \Delta\mu_{H^+}$ indicate the nearest neighbor interaction part of the chemical potential μ_H, calculated according to the quasi chemical approach [3.9]. As stated above, the pair interaction energy is related to the corresponding part of the standard potential by $\mu^0_{H^+} = -6w_{HH}$ (3.37).

In other alloys with lattice expansion, e.g., Pd/Sn [3.72, 129], Pd/Au [3.130], Pd/Pb or Pd–B [3.72], a similar effect of *solubility enhancement* is usually observed. In alloys with lattice contraction, on the other hand, e.g., Pd/Rh [3.129] or Pd/Ni [3.129, 131] the opposite effect of an increase of μ^0_H is observed. The reduction of $\mu^0_{H^+}$, corresponding to a solubility increase at low H concentrations, is attributed to the lowering of the *strain energy* due to the larger octahedral sites of the Pd/Ag alloy.

The filling of the 4*d* band and the rise of the Fermi energy in the 5*s* band is demonstrated by the *electronic excess potentials* $\Delta\mu_e$ of H in Pd/Ag alloys (Fig. 3.20). The pair interaction energies W_{HH} of the corresponding elastic contributions to the excess potential $\Delta\mu_{H^+}$ were evaluated to be -672, -465, -360, -264, and -208 cal/mol pairs for pure Pd and alloys containing 10 through 40 at.% Ag, respectively [3.66, 71].

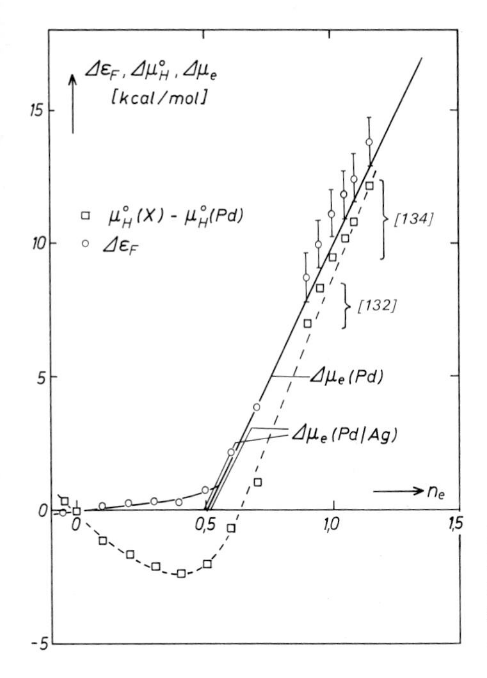

Fig. 3.21. Electronic excess potential $\Delta\mu_e$, Fermi energy shift $\Delta\varepsilon_F$, and change of the standard potential of H, $\Delta\mu_H^0$, in various alloys relative to pure Pd [3.66, 71, 129, 133–135]. Details in the text

The separation of the experimental excess potential $\Delta\mu_H$ into its constituent parts was effected by the approximation $(\partial/\partial n_H)\Delta\mu_e = 0$ at $n_H < 0.1$, i.e., by neglecting the Fermi energy rise in the $4d$ band. In the 50% Ag alloy, the Fermi energy rise could no longer be neglected, and an extrapolated value of the pair interaction energy was used to calculate $\Delta\mu_{H^+}$ ($W_{HH} = -200 \pm 50$ cal/mol pairs).

The similar slopes of about 20 ± 2 kcal/g-atom H of the steep rise of the excess potentials indicate a similar effective density of the $5s$ states in these alloys, and the regular shifts of 0.1 of the intercepts by an addition of 10% Ag indicate a competitive $5d$ band by H and Ag.

This is shown more directly in Fig. 3.21, where the Fermi energy parts of the excess potentials in the $4d$ band regime are plotted vs the *valence electron concentration*

$$n_e = z_H n_H + z_{Ag} x_{Ag}, \tag{3.54}$$

where x_{Ag} is the atom fraction of Ag in the absence of H, $N_{Ag}/(N_{Ag} + N_{Pd})$, while $z_H = 1$ and $z_{Ag} = 1$ are the numbers of electrons donated to the conduction bands by H and Ag, respectively. [A more general definition of the valence electron concentration is introduced by (3.63), p. 127.]

The short lines redrawn from Fig. 3.20 crowd around the long one marked $\Delta\mu_e$(Pd), which has a slope of 20 kcal/g-atom H and an intercept at $n_e = 0.5$. Only the two lines $\Delta\mu_e$(Pd/Ag) with the largest deviation from the one for pure Pd are shown in the drawing.

In order to provide an additional test of the consistency of this approach, the Fermi energy rise $\Delta\varepsilon_F$ in these alloys with respect to the Fermi level in pure Pd is independently traced by the change of the electronic parts of the standard potentials

$$\Delta\varepsilon_F = \mu_e^0(\text{alloy}) - \mu_e^0(\text{Pd}) \tag{3.55}$$

obtained by (3.24). The elastic part of the standard potential is taken from (3.37), while the vibrational part is assumed to be independent of alloy composition and cancels.

Figure 3.21 shows the Fermi energy rise vs valence electron concentration n_e in the range between -0.05 (Pd+5% Rh) [3.129], 1.0 (Ag) and 1.15 (Ag+15% Cd) [3.133–135]. The points show the general features of 4*d* and 5*s* band filling depicted in Fig. 3.19, in particular a gentle slope of about 1.05 kcal/g-atom H in the 4*d* band regime and the steep rise expected in the 5*s* band regime.

The difference of the standard potentials

$$\Delta\mu_H^0 = \mu_H^0(\text{alloy}) - \mu_H^0(\text{Pd}) \tag{3.56}$$

also included in Fig. 21 does show the same steep rise at $n_H > 0.5$. In the 4*d* band range, however, $\Delta\mu_H^0$ has a negative slope, brought about by the increasing lattice constant of the alloys and the concomitant reduction of the strain energy.

The possibility of turning the pronounced negative initial slope of the $\Delta\mu_H^0$ vs n_e curve (-8.2 kcal/mol) around into an initial slope of the $\Delta\mu_e^0$ vs n_e curve close to zero (actually $+1.05$ kcal/mol), as expected for the Fermi energy rise in the *d* band, by calculating the protonic standard potentials from the pair interaction energies

$$\mu_{H^+}^0 = -6W_{HH} \tag{3.57}$$

may be regarded as experimental proof of (3.37).

The initial slope of $\Delta\varepsilon_F$ of 1.05 kcal/g-atom H is a better approximation of the effective reciprocal density of states in the 4*d* band than the zero slope approximation. As a matter of fact, the short stubs of $\Delta\mu_e$ of the series of alloys included in Fig. 3.21 as well as the line for pure Pd were calculated using 1.05 kcal/g-atom H as the initial slope of the electronic potential.

The two series of points $\Delta\mu_H^0(n_e)$ and $\Delta\varepsilon_F(n_e)$ calculated for the Ag-rich Pd and Cd alloys both lie close to the line of the electronic excess potential of pure Pd, $\Delta\mu_e = 20 \cdot (n_e - 0.50)$ kcal/g-atom H. The values of $\Delta\mu_H^0$ were calculated for a common temperature of 350 K from the original data, and an estimated value of $W_{HH} = -250 \pm 150$ cal/mol pairs was used to arrive at the shifts of the Fermi energy $\Delta\varepsilon_F$. The error bars introduced by the uncertainty of this estimate do not seriously detract from the good fit of values derived from excess potentials and standard potentials of H in the series of alloys studied. The slope of close to 20 kcal/g-atom H, in particular, is obtained by both sets of data [3.66, 135].

The sign of this slope, i.e., the increase of H solubility in Ag-rich Pd alloys with respect to the solubility in pure Ag, was pointed out in two early papers by *Himmler* [3.133] and by *Wagner* [3.136], who already related this effect to the shift of the Fermi energy.

3.4.2 Statistical Thermodynamics of Ternary Systems

The solubility enhancement, brought about by the larger size metal atoms, may be interpreted in two different ways. In a 1971 paper, *Wagner* considered the limiting case of vanishing Ag concentration, which will produce an increase of the average lattice constant everywhere, enlarging the octahedral sites far away from dissolved Ag atoms [3.102]. According to this *long-range interaction model*, Ag atoms will alleviate the lattice strain around H atoms on sites far away from them.

Of course, at moderate or high Ag concentrations, there are practically no octahedral sites without one or more Ag atoms on nearest neighbor positions on the host sublattice. That is why an alternative model has to be seriously considered, in which a nearest neighbor interaction energy between solute atoms on different sublattices will be negative between interstitial atoms and substitutional components which expand the host lattice, like Ag, Sn, or Pb, and positive if the host lattice is contracted, as by Rh and Ni. The situation in real systems may perhaps be regarded as an intermediate case between the two extreme models of exclusively long-range and exclusively short-range interactions.

In a 1928 paper, taken up in his book *Thermodynamics of Alloys*, *Wagner* himself considered *nearest neighbor interactions* of a dilute solution of a substance $3 (n_3 \approx 0)$ in a binary alloy of components 1 and 2 in a random mixture [3.137, 138]. If a solute atom 3 has z nearest neighbors, r of component 1 and $(z-r)$ of component 2, there are $(z+1)$ kinds of sites for this atom, each with a relative abundance of $[z!/r!(z-r)!](1-x_2)^r x_2^{(z-r)}$. The probability of finding atom 3 on this site is proportional to the Boltzmann factor $\exp[-rE_{13}-(z-r)E_{23}/zRT]$ where E_{13}/N_{L} and E_{23}/N_{L} are the interaction energies of an atom 3 with z nearest neighbors of 1 and 2, respectively. The amount of substance 3 dissolved in the alloy at a certain activity is proportional to the sum of these $z+1$ products, which happens to reduce to the binomial expression

$$C \cdot [(1-x_2)\exp(-E_{13}/zRT) + x_2 \exp(-E_{23}/zRT)]^z ,$$

where C is a proportionality constant. At equal activity of component 3, the amount dissolved in pure substance 1 is

$$C \cdot \exp(-E_{13}/\mathrm{RT}) .$$

The ratio of concentrations at equal activity in the alloy and in the pure substance 1 is

$$[(1-x_2)+x_2\eta]^z; \quad \eta=\exp[(-W_{23}/RT)]; \quad W_{23}=[(E_{23}-E_{13})/z].$$

This may be rephrased, in the ideally dilute case, for the ratio of activities or for the difference of chemical potentials at constant concentration n_3

$$\mu_3(x_2)=\mu_3^0+RT\ln\frac{n_3}{1-n_3}-zRT\ln[1+x_2(\eta-1)]+E_{13} \tag{3.58}$$

$$\mu_3(x_2=0)=\mu_3^0+RT\ln\frac{n_3}{1-n_3}+E_{13}. \tag{3.58a}$$

Applying these expressions to the case of H (at vanishing concentration) in a Pd alloy, it is to be remembered that the number of metal atoms surrounding an octahedral site is $z=6$ and that only vibrational effects ($\mu^0_{\mathrm{Hvib}}=\mu_3^0$) and nearest neighbor effects ($\mu^0_{\mathrm{H}^+}=E_{13}$ and $\mu^0_{\mathrm{H}^+}(x_2)=-zRT\ln[1+x_2(\eta-1)]+E_{13}$ for pure Pd and for the alloy) have been included so far. Thus, the change of the *strain energy* part of the standard potential between Pd and Pd alloys is

$$\mu^0_{\mathrm{H}^+}(x_2)-\mu^0_{\mathrm{H}^+}(\mathrm{Pd})=-zRT\ln[1+x_2(\eta-1)]. \tag{3.59}$$

The Fermi energy shift in the standard potential $\mu_e^0(x_2)-\mu_e^0(\mathrm{Pd})$ can be obtained from the *electronic excess potential* calculated from the curve for pure Pd shown in Fig. 3.20 (plus an initial slope of 1050 cal/g-atom H): $n_e<0.656$:

$$\Delta\mu_e(n_e)=1402\,n_e-7093\,n_e^2+42924\cdot n_e^3-83065\,n_e^4+70419\,n_e^5(\mathrm{cal/g\text{-}atom\ H}) \tag{3.60}$$

$$n_e>0.656: \Delta\mu_e(n_e)=19250\,(n_e-0.492). \tag{3.60a}$$

The change of the standard potential is

$$\mu_{\mathrm{H}}^0(x_2)-\mu_{\mathrm{H}}^0(\mathrm{Pd})=-zRT\ln[(1+x_2(\eta-1)]+\Delta\mu_e(n_e). \tag{3.61}$$

Correspondingly, the changes of the partial molar enthalpies and standard entropies of dissolved H, equal but of opposite sign to the difference of enthalpies and entropies of desorption between alloy and pure Pd, are

$$\bar{H}_{\mathrm{H}}(x_2)-\bar{H}_{\mathrm{H}}(\mathrm{Pd})=-zW_{2\mathrm{H}}x_2/[1+x_2(\eta-1)]+\Delta\mu_e(n_e) \tag{3.62}$$

$$\bar{S}^0_{\mathrm{H}}(x_2)-\bar{S}^0_{\mathrm{H}}(\mathrm{Pd})=zR\ln[1+x_2(\eta-1)]-zW_{2\mathrm{H}}x_2\eta/[1+x_2(\eta-1)]\cdot T. \tag{3.62a}$$

Experimental and calculated values of these changes for a number of Pd/Ag and Pd/Sn alloys are shown in Table 3.3.

Table 3.3. Experimental and calculated thermodynamic values of H in Pd and Pd alloys at 350 K

	x_2	ΔH^0 $\left[\frac{\text{kcal}}{\text{g-atom H}}\right]$	ΔS^0 $\left[\frac{\text{cal}}{\text{K}\cdot\text{g-atom H}}\right]$	$\Delta\mu_e(n_e)$ $\left[\frac{\text{kcal}}{\text{g-atom H}}\right]$	W_{HH} $\left[\frac{\text{cal}}{\text{mol pairs}}\right]$	$\mu_H^0(x_2)-\mu_H^0(\text{Pd})$ $\left[\frac{\text{kcal}}{\text{g-atom H}}\right]$		$\bar{H}_H(x_2)-\bar{H}_H(\text{Pd})$ $\left[\frac{\text{kcal}}{\text{g-atom H}}\right]$		$\bar{S}_H^0(x_2)-\bar{S}_H^0(\text{Pd})$ $\left[\frac{\text{cal}}{\text{K}\cdot\text{g-atom H}}\right]$		Ref.
						exp	calc	exp	calc	exp	calc	
Pd	0	2.25 ± 0.1	12.55 ± 0.3	—	−672	—	—	—	—	—	—	[3.9]
Pd/Ag	0.2	4.67	14.8	0.23	−360	−1.63	−1.63	−2.42	−2.49	−2.25	−2.46	[3.71]
	0.3	5.50	11.8	0.44	−264	−2.11	−2.11	−3.25	−3.02	−3.25	−2.60	[3.71]
	0.4	5.08	15.7	0.77	−208	−2.48	−2.38	−3.58	−3.24	−3.15	−2.46	[3.71]
	0.5	5.07	14.75	1.30	—	−2.05	−2.36	−2.82	−3.12	−2.20	−2.17	[3.66]
	0.6	4.43	16.8	2.27	—	−0.69	−1.86	−2.18	−2.49	−2.45	−1.79	[3.66]
	0.7	2.82	17.2	4.12	—	−1.06	−0.43	−0.82	−0.90	−4.65	−1.37	[3.66]
Pd/Sn	0.025	2.70	13.1	0.09	−5.88	−0.26	−0.26	−0.45	−0.56	−0.55	−0.85	[3.72]
	0.0478	3.23	14.0	0.17	−555	−0.48	−0.47	−0.98	−0.98	−1.45	−1.46	[3.129]
	0.05	3.13	13.7	0.18	−525	−0.48	−0.49	−0.88	−1.02	−1.15	−1.51	[3.72]
	0.075	3.37	13.8	0.33	−478	−0.68	−0.642	−1.12	−1.34	−1.25	−2.01	[3.72]
	0.10	3.85	14.7	0.55	−365	−0.85	−0.70	−1.60	−1.54	−2.15	−2.40	[3.72]
	0.125	4.15	15.5	0.87	−281	−0.87	−0.64	−1.90	−1.58	−2.95	−2.96	[3.72]
	0.15	4.16	16.3	1.37	−240	−0.60	−0.39	−1.91	−1.41	−3.75	−2.91	[3.72]

The number of electrons donated to the electron bands by Sn has been set equal to $z_{Sn} = 3.4$, after an evaluation of the $\Delta\mu_e$ vs n_H curves [3.72, 129] and also of the susceptibility curves of these alloys [3.72, 139]. In these measurements, the effective valence turns out to be slightly lower than 4.

Remarkably good agreement between calculated and experimental values of these alloys is obtained, except at the large concentrations, by adjusting one parameter for each alloy:

$$W_{HAg} = \tfrac{1}{6}(E_{HAg} - E_{HPd}) = -926(\text{cal/mol pairs})$$
$$W_{HSn} = \tfrac{1}{6}(E_{HSn} - E_{HPd}) = -1043(\text{cal/mol pairs}).$$

A more general scheme has been developed for fcc lattices, in which the restrictions of random distribution of the host metal atoms and of vanishing interstitial concentration were dropped [3.140]. The method is equally applicable to reciprocal salt mixtures of the NaCl type, where mixtures occur in both intersecting fcc sublattices.

The starting point for writing the partition function for these alloys is the recognition that for every nearest neighbor pair on one sublattice there is one and only one "conjugated" pair on the other sublattice. The conjugated pairs form diagonals on squares, four of which comprise a face of the unit cell.

There are nine ways to occupy the corners of the square, with corresponding energies obtained as the sum of the interaction energies of every particle with every other particle on the square. The total energy of the system is a function of the numbers of squares only. Three square numbers are eliminated by conditions analogous to (3.26). The equilibrium values of the remaining six square numbers are obtained by differentiating the partition function with respect to these square numbers (maximum-term method), very much like the procedure of finding the equilibrium number of pairs in the binary system.

The method was applied to the Pd/Ag/H system. Besides the known pair interaction energies $W_{HH} = -705$ and $W_{AgAg} = -660$ cal/mol pairs [3.82] of the two binary subsystems, one additional pair energy W_{HAg} has to be determined for the ternary system. With $W_{HAg} = -900$ cal/mol pairs, a good fit was obtained between all the experimental and calculated isotherms of four Pd–Ag alloys [3.40].

Not surprisingly, this value is close to, but a little less exothermic than, the one obtained from the standard potentials of assumed random Pd/Ag alloys. In the real alloy, the number of Ag clusters will be in excess of that in a random configuration, and the H atoms will seek these clusters preferentially. That is why a less exothermic value of W_{HAg} will do in the clustering model to account for the observed increase of solubility and decrease of entropy of the H atoms.

The partition function has also been set up for an entirely different ternary system such as Pd/B/H [3.73, 141]. In this case there is only one kind of particle (Pd) on the host sublattice, while the lattice points of the other sublattice assume one of three states: empty, occupied by H, or occupied by B. Compared to the Pd/Ag/H case, matters are not as complicated. Only pairs have to be considered. Of the six kinds of pairs, three pair numbers can be eliminated due to restrictions analogous to (3.26), but three independent pair numbers remain, e.g., N_{H0}, N_{B0}, and N_{HB}.

Application of the maximum-term principle to the partition function yields three simultaneous quadratic "quasichemical" equations in the equilibrium pair numbers. These degenerate to the three equations of the subsystems, if N_{H}, N_{B}, or N_0 are zero. The pair interactions of two of the subsystems $W_{\mathrm{HH}} = -705$ and $W_{\mathrm{BB}} = -1830$ cal/mol pairs are known [3.9, 78, 79] (based on an initial Fermi energy rise of 1050 cal/g-atom H), while the third value W_{HB}, pertaining to the subsystem without empty sites, is adjusted from H absorption isotherms on Pd/B alloys [3.73].

Using the electronic excess potential (3.60) with $n_{\mathrm{e}} = n_{\mathrm{H}} + 3n_{\mathrm{B}}$ and the pair interaction energy $W_{\mathrm{HB}} = -700$ cal/mol pairs, good fit could be obtained for a number of isotherms on alloys with $n_{\mathrm{B}} = 0.062$ [3.73], 0.033, 0.067, and 0.10 [3.142].

3.4.3 The Rigid Band Model and Its Limitations

The rigid band model, introduced by *Mott* and *Jones* [3.143], is based on the concept that alloy components or "impurities" contribute electrons to the energy bands of the host metal—or withdraw electrons from there—but do not change the band structure. The effect of alloying, according to this model, will be merely a shift of the Fermi level to higher or lower energies, and to higher or lower densities of electronic states. Electronic properties that depend on the Fermi density of states (like the magnetic susceptibility, or the electronic specific heat) or change with the energetic position of the Fermi level (like the chemical potential of dissolved hydrogen as discussed in Sect. 3.4.1, or the isomer shift of Mössbauer resonances) can therefore serve as indications of the validity of the rigid band model.

Gerstenberg [3.144], in a fundamental investigation of the magnetic behavior of substitutional alloys of Pd with other transition metals, demonstrated that the molar susceptibilities of numerous alloys of this type can be represented in the low concentration range as a uniform function of the number of d and s electrons from incomplete shells brought into the lattice by the alloy components (for higher concentrated alloys see [3.145, 146]). In order to account for interstitial solutes like H and B, too, the definition of the valence electron concentration (VEC) was extended by *Brodowsky* [3.82, 135] to the two-termed expression

$$n_{\mathrm{e}} = \sum_{i,j} (z_i x_i + z_j n_j), \tag{3.63}$$

where the $x_i = N_i / (\sum N_i + N_{\mathrm{Pd}})$ represent the atomic fractions of the substitutional components, the $n_j = N_j / (\sum N_i + N_{\mathrm{Pd}})$ the atomic ratios of the

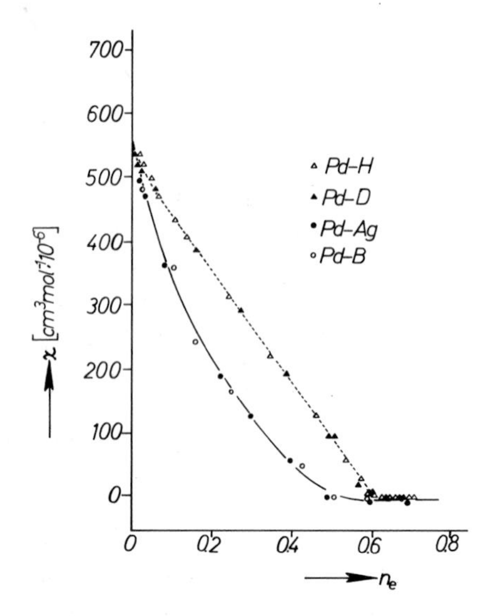

Fig. 3.22. Susceptibility of Pd/Ag and Pd/B alloys as well as of Pd/H and Pd/D at 20 °C as a function of the VEC (*Husemann* et al. [3.72, 73])

interstitial solutes; the z_i and z_j mean the valencies of these components, respectively. This definition relates the VEC to the number of lattice points $\sum N_i + N_{Pd}$ and thereby implies that the substitutional components provide not only electrons, but also electronic states to the energy bands of the host metal, while the interstitial solutes contribute only electrons.

Figure 3.22 demonstrates the conception of the rigid band model, illustrated by the *magnetic susceptibilities* of Pd/Ag and Pd/B alloys, after measurements of *Husemann* [3.72, 73]. Although Ag is a substitutional, and B an interstitial solute, the curves of $\chi(n_e)$ coincide to a uniform course if Ag is taken as univalent, and B as trivalent in (3.63). The values of Pd/H and Pd/D coincide with the bent curve in the α and β phase regions only; the linear decrease in between represents the straight line connecting the coexistent states of the α and β phase across the two-phase region, as in the measurements of *Frieske* (see Fig. 3.2). The obvious lack of differences between the Pd/H and Pd/D values confirms that electronic isotope effects can be neglected as mentioned already p. 90. If Pd/Ag or Pd/B alloys are charged by hydrogen, their susceptibilities decrease along the bent curve, and the dissolved hydrogen turns out to be equivalent to Ag with respect to susceptibility [3.72].

Frieske and *Mahnig* [3.147] made use of this equivalence when they studied the magnetic behavior of the ternary system Pd/Fe/H. They represented the measured susceptibilities as composed of the contribution χ_{Fe} of the iron *local moments*, and of an electron *band contribution* χ_B,

$$\chi = \chi_{Fe} + \chi_B(n_e, T) \tag{3.64}$$

with $n_e = 3x_{Fe} + n_H$, assuming trivalent iron. Numerical data of χ_B were obtained from experimental results on Pd/Ag alloys, used as empirical standard of the band contribution

$$\chi_B(n_e, T) = \chi_{Pd\text{-}Ag}(x_{Ag}, T) . \tag{3.64a}$$

By this method *Frieske* and *Mahnig* separated $\chi_{Fe}(T)$ from values of the binary system Pd/Fe measured by *Gerstenberg* [3.144] and obtained a Curie-Weiss relationship with a local moment of $m_{Fe} = 5.9\,\mu_B$, confirming the valency 3 of iron in the Pd host lattice. When the measurements on the ternary system Pd/Fe/H were evaluated accordingly, the separated iron contribution χ_{Fe} turned out to decrease with increasing hydrogen content, thereby showing up the reduction of the magnetic coupling ability of the Pd host lattice by the dissolved hydrogen. At high hydrogen contents, χ_{Fe} levels out to constant values, the temperature dependence of which yields again $m_{Fe} = 5.9\,\mu_B$, i.e., trivalent iron, as *Burger* et al. [3.148] have shown.

The $\chi(n_e)$ curve of Pd/Ag and Pd/B in Fig. 3.22 reveals the completion of the 4*d* band by its bending off to the limiting value in the range $0.5 \leqq n_e \leqq 0.6$. The kink in the χ curve of Pd/H (and Pd/D), on the other hand, does not occur before $n_e \gtrsim 0.6$ at 20 °C. With increasing temperatures the kink slides along the bent curve to lower n_e values, thereby tracing out the β_{min} limit of the two-phase region as represented in Fig. 3.3. Only a loose connection exists, therefore, between the position of this kink and the completion of the 4*d* band: the steep rise of the electronic excess potential $\Delta\mu_e$, after the 4*d* band has been filled up, sets an upper bound at $n_e \approx 0.6$ that can be overstepped only slightly by the coexistence limit β_{min}, even at low temperatures (see Fig. 3.19); there is no analogous restriction, however, at lower n_e values and higher temperatures.

The relation between the chemical potential of the dissolved hydrogen and the rigid band model showed up already in the Figs. 3.18 and 3.21. Particularly the well-proportioned shifts of the logarithmic isotherm branches in Fig. 3.18 demonstrate the exchangeability of Ag and H in filling up the electron bands of the host lattice. Similar results have been obtained with systems containing trivalent components like Pd/B/H [3.72], Pd/Fe/H [3.149], and Pd/Al/H [3.150].

The *electronic specific heats* of Pd/Ag alloys [3.67, 68, 151] and of Pd charged with hydrogen [3.69, 152] also indicate that the 4*d* band is complete at $n_e \approx 0.6$, irrespective of the origin of the electrons (although the γ coefficients of β-Pd/H are higher than those of the corresponding Pd/Ag alloys, due to the electron-phonon interaction in β-Pd/H that promotes the superconductivity, see *Wolf* et al. [3.152]). According to measurements of *Mahnig* [3.153] with Pd/V alloys, V at low concentrations up to 5 at.% can replace Ag in Pd with respect to the γ value of the electronic heat capacity; the plot of γ against n_e reveals V to behave as a pentavalent "impurity". Even with such high valent impurity atoms the rigid band concept remains applicable up to $n_e \approx 0.25$; at higher V concentrations, however, deviations occur that may be attributed to localized 3*d* states at the V atoms [3.153]. The same feature of V

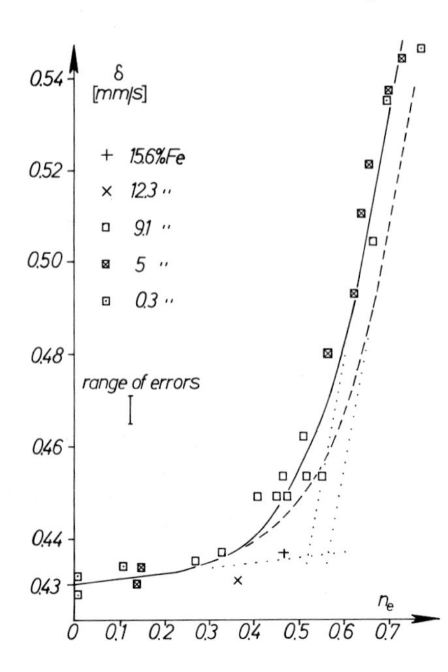

Fig. 3.23. Isomer shift of ^{57}Fe Mössbauer resonances at 35 °C in Pd/Fe/B alloys (*Wanzl* [3.154]). Dashed curve: measurements on the system Pd/Fe/H (*Mahnig* [3.155]). Dotted lines: linear extrapolations

in Pd/V alloys shows up in measurements of magnetic susceptibility [3.144], hydrogen solubility [3.131], and Mössbauer ^{57}Fe spectroscopy [3.153, 154] on these alloys.

The *isomer shifts* δ of ^{57}Fe Mössbauer resonances in Pd alloys were investigated by *Mahnig* [3.155] and by *Wanzl* [3.154]. Figure 3.23 represents δ values obtained with ^{57}Fe in Pd/Fe/H and Pd/Fe/B alloys. Several runs were made, each starting with a Pd/Fe alloy of fixed Fe content between 0.3 and 15.6 at-%, to which then different amounts of H or B were added. For intensity reasons the 0.3% Fe samples were prepared with an enriched ^{57}Fe probe (91 at-%); the other samples contained natural Fe (2.2 at-% ^{57}Fe). The curves in Fig. 3.23, plotted against the VEC ($n_e = 3x_{Fe} + 3n_B$, or $n_e = 3x_{Fe} + n_H$, respectively), show surprising similarity to the course of the electronic excess potential $\Delta\mu_e$ (Figs. 3.19–21). Extrapolations of the steeply rising branches intersect the extrapolation of the gradual initial increase at n_e values between 0.5 and 0.6 as indicated in the figure. These results can be understood by applying the concept of virtually bound states of *Friedel* [3.89] to the broadened localized 3*d* states of the $^{57}Fe^{3+}$ ions in the lattice. The lower half of the 3*d* states is occupied almost completely by localized electrons—responsible for the local moment of 5.9 μ_B—and the small deficiency of *d* electrons will be filled up according to the rise of the Fermi level [3.155]. The isomer shift δ, as a measure of the *d* electron concentration in the shell of the ^{57}Fe nuclei, traces, therefore, the increase of the Fermi energy with increasing VEC, similar to the excess potential $\Delta\mu_e(n_e)$. Analogous results have been obtained by *Wanzl* [3.154] with Pd/Ag/H alloys

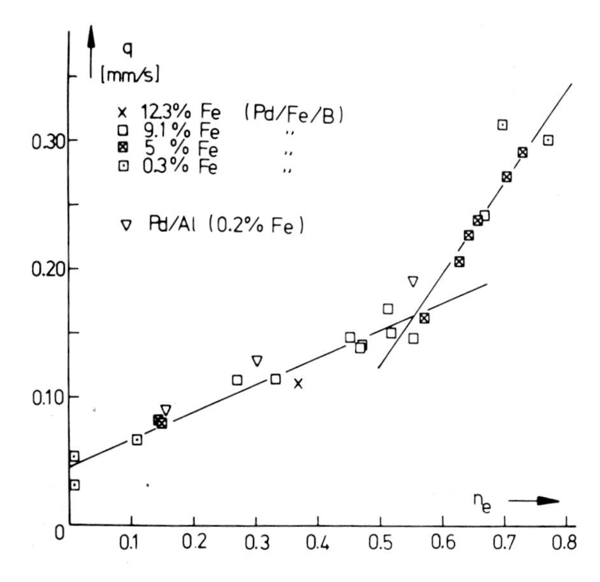

Fig. 3.24. Quadrupole splitting of ^{57}Fe Mössbauer resonances in Pd/Fe/B and Pd/Fe/Al at 35 °C dependent on the VEC (*Wanzl* [3.153, 154])

that contained 0.3% ^{57}Fe as indicator, and by *Mahnig* [3.155] with Pd/Sn/H alloys, studying the isomer shift of ^{119}Sn.

Mössbauer spectroscopy, besides the isomer shift, provides as a second information the *quadrupole splitting q*. q values observed on Pd/Fe/B alloys by *Wanzl* [3.153, 154] are represented in Fig. 3.24 as function of $n_e = 3x_{Fe} + 3n_B$. At $n_e \approx 0.55$ a rather sharp change of slope shows up. Since the quadrupole splitting results from field gradients at the position of the Mössbauer nucleus, brought about by deviations from cubic symmetry in the neighborhood, this kink indicates a change of the screening ability of the band electrons (see also *Longworth* [3.157]). In the scope of *Friedel*'s model [3.89, 156] for screening of extra charges in alloys, the simplest approximation leads to an exponential decay of the screening charge density with increasing distance, characterized by a decay length, or *screening radius* of

$$\varrho = 1/\sqrt{4\pi e_0^2 (dN/d\varepsilon)_F}\,, \tag{3.65}$$

where $(dN/d\varepsilon)_F$ is the density of states (number per eV and volume of unit cell) at the Fermi limit. Thus the screening radius is small when the density of states is high, as in pure Pd, where even nearest neighbors of an impurity atom do not feel that their neighbor is different from a matrix atom. The change of slope in Fig. 3.24, therefore, indicates the decrease in the density of states when the Fermi level passes the upper border of the 4d band.

The experimental evidence listed so far seems to provide an imposing confirmation of the applicability of the rigid band model to Pd alloys, and of the existence of between 0.5 and 0.6 holes in the 4d band of Pd. However, as

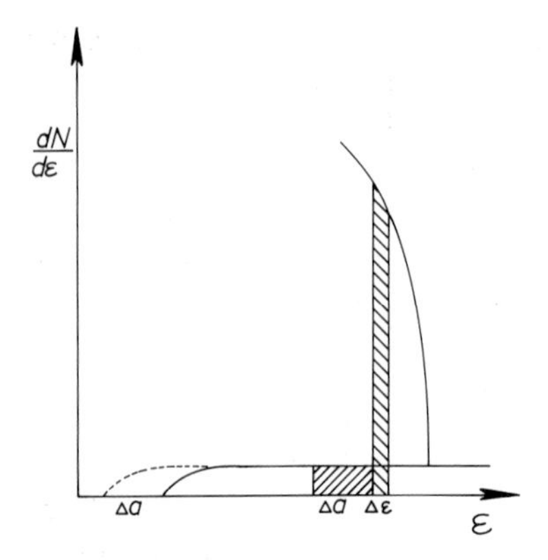

Fig. 3.25. Shift of the s band beneath the d band by screening effects of solute ions (*Brodowsky* [3.135])

Vuillemin and *Priestley* [3.158] demonstrated by measurements of the de Haas-van Alphen effect on pure Pd, there are only 0.36 ± 0.01 electron holes in the $4d$ band[7].

In order to reconcile this contradiction, it seems necessary in a certain sense to modify the meaning of the expression "rigid band model". The statement that quite a number of electronic properties of Pd alloys behave as uniform functions of the VEC after (3.63) cannot be invalidated. But this does not mean that the structure of the electron bands remains unchanged; instead, it may be that the structure changes also as a uniform function of n_e. As a matter of fact, just this seems to be the case.

When an "impurity" ion is screened by band electrons, the screening charge is proportional to the nuclear charge number or valency z of the impurity ion. This is true as long as the density of states at the Fermi level is high enough and, accordingly, the screening radius after (3.65) small enough that no overlapping of the screening clouds occurs. The solute ions by their coulomb fields pull down occupied electron states from the conduction bands of the host lattice to low energy levels in order to provide for their screening charges. These screening states are broadened appreciably by strong exchange interactions with the band states. From the viewpoint of band structure, these low-lying states for nonlocalized electrons, induced by the solute ions, show up like a *shift of the* broad $5s$ *band* of the Pd host lattice towards lower energies beneath the $4d$ band, as sketched in Fig. 3.25. When, by alloying, the Fermi level is raised by the amount $\Delta\varepsilon$, the s band shifts simultaneously by Δa; therefore, not only the $4d$ holes have to be filled up by electrons of the solute, but also the $5s$ states that shifted beneath the Fermi level. As about 0.55 electrons are needed to raise the Fermi level to the top of the $4d$ band—i.e., to fill the 0.36 d holes—the s band

[7] From isochromat-spectroscopic investigations of binary alloys of the series Rh–Pd–Ag, *Eggs* and *Ulmer* [3.159] concluded, in apparent agreement with this result, that the $4d$ band of Pd is filled up if more than 38 ± 5 at-% of Ag are alloyd to Pd. This statement, however, is based upon a misinterpretation of the measurements; in fact, the low energy peak in the course of the isochromates is representative of the d-band holes, and this vanishs between 50 and 75 at. % Ag in Pd.

shift, obviously, holds responsible for the additional 0.2 electrons. For the same reason the electronic excess potential $\Delta\mu_e$ turns out to be smaller, i.e., the increase of $\Delta\mu_H(n_e)$ at $n_e \gtrsim 0.6$ is less steep, as could be expected from the density of states in the 5*s* band (see [3.66–69]). Due to the narrow connection of the shift with the screening of the solute ions, i.e., with their valencies and concentrations, this change of electron band structure, as represented in Fig. 3.25, can be expected in fact as a uniform function of the VEC[8]. This statement implies, particularly with regard to the uniform course of $\chi(n_e)$ and $\gamma(n_e)$, that the enhancement effects of electron-phonon interactions and of exchange interactions are also uniform functions of n_e (exception: γ of β-PdCH, see p. 129).

The concept of band shifting in the Pd host lattice was forwarded first by *Dugdale* and *Guénault* [3.160] for explaining the concentration dependence of the resistivity and the low-temperature thermopower of Pd/Ag alloys, and by *Montgomery* et al. [3.68] with regard to the electronic heat capacity of these alloys. *Brill* and *Voitländer* [3.161] discussed the idea in connection with susceptibility measurements on Pd/H samples (from Pd black).

The concept gained evidence and new aspects by the results of extended APW *band calculations* performed for Pd hydride [3.40, 162] as well as for Ni hydride [3.40]. *Switendick* [3.40], in order to demonstrate the effect of increasing hydrogen content, made calculations for a number of ordered hydride phases with fixed compositions: Pd_4H, Pd_4H_3, PdH, and the hypothetical PdH_2. The density of states distributions of PdH, compared with Pd metal, are only little affected for states with predominant *d* character. At the low-energy side of the 4*d* band, however, new *s*-like states occur, pulled down below the top of the *d* band by the dissolved hydrogen. The number of these states can only be roughly estimated to be between 0.5 and 0.7 per hydrogen atom [3.40, 41]. The experimental values, 0.55 electrons for 0.36 *d* holes, yield for comparison: $(0.55-0.36)/0.55=0.35$ states pulled down per hydrogen atom. These will be, however, no Pd–H bonding states as suggested previously [3.41], similarly as the screening states of any other solute X cannot be expected to represent Pd–X bonding states.

Zbasnik and *Mahnig* [3.162] performed APW calculations on nonstoichiometric Pd hydride phases PdH_n, $0.6 \leqq n \leqq 1$, with the hydrogen randomly distributed among the octahedral sites. The result of *Switendick* concerning the low-lying *s* states induced by the hydrogen was confirmed, and it could be ascertained that their positions shift downward from the Fermi level approximately linearly with hydrogen concentration.

Hence the model of "screening induced band shifts" (instead of "rigid bands") appears to be well founded also by theory. It implies that electronic properties of Pd alloys can be represented as functions of the VEC only, irrespective of the individual nature of the solutes. The model is applicable up to such VEC values, where the screening spheres of the solute ions begin to

[8] Deficiencies in the VEC occur in case of high valent solutes, such as Sn or Pb for which an effective valency of about 3.5 instead of 4 was observed (see p. 126 and Fig. 3.18).

overlap. Apart from very high-valent solutes, such as V, this limit is reached at $n_e \gtrsim 0.6$, where the sharp decrease of the density of states—when the Fermi level passes the top of the 4d band—let the screening radius increase strongly, according to (3.65). The steep rises of the electronic excess potential $\Delta\mu_e$, as well as of the ^{57}Fe isomer shift, are thus already outside the range of the model. As a matter of fact, individual influences of the solutes show up in the steepness of these branches. The same is true for the increase of the activation energy of hydrogen diffusion in Pd alloys at $n_e \gtrsim 0.6$ that originates from the repulsion between the screening spheres of the protons, when passing over the energy barrier, and the screening spheres of the neighboring solute ions [3.163].

3.5 Hydrogen Mobility in Palladium and Palladium Alloys

3.5.1 Surface Kinetics and Stationary Permeation

In 1932, *Wagner* [3.164] stated that the transfer of hydrogen from the gas phase into the Pd lattice is composed of at least two steps: the *dissociative chemisorption* of hydrogen molecules at the Pd surface (I), and the *passage* of hydrogen atoms from surface sites into the lattice (II):

$$\underset{\text{(I)}}{H_2^g \rightleftharpoons 2H_{ad}}, \qquad \underset{\text{(II)}}{H_{ad} \rightleftharpoons H_{Me}}.$$

When hydrogen is removed from the lattice it has to pass the same steps in the reverse direction. If the first step is rate determining, the rate of dissolution of hydrogen from the gas phase is expected to be proportional to hydrogen pressure p_{H_2}, the rate of hydrogen removal from the lattice back to the gas phase proportional to the square of hydrogen content in the lattice n^2 (second-order process). In cases where the second step is rate determining, the rate of hydrogen uptake is expected to be proportional to $\sqrt{p_{H_2}}$, and the rate of removal proportional to n (first-order process). *Wagner* tried to verify these two limiting rate laws by measuring the change of hydrogen content of Pd wires with time at constant p_{H_2}. He used the electrical resistance of the Pd wires as the measure of their hydrogen content, and found that the particulars of the rate law depend largely on the prehistory and the accidental state of the surface. A few years later *Wagner* [3.165] inserted an intermediate step between (I) and (II) mentioned above: a *migration* of the chemisorbed hydrogen atoms along the surface until they find a passage suitable for entering the lattice. By this idea *Wagner* anticipated an important point of the later developed concepts of hydrogen transfer catalysis [3.16, 17], hydrogen spillover [3.166], and of more detailed mechanisms of surface kinetics [3.167].

The problem was attacked again recently by *Auer* and *Grabke* [3.167]. In order to be sure of measuring the *surface processes* only, they used very thin Pd

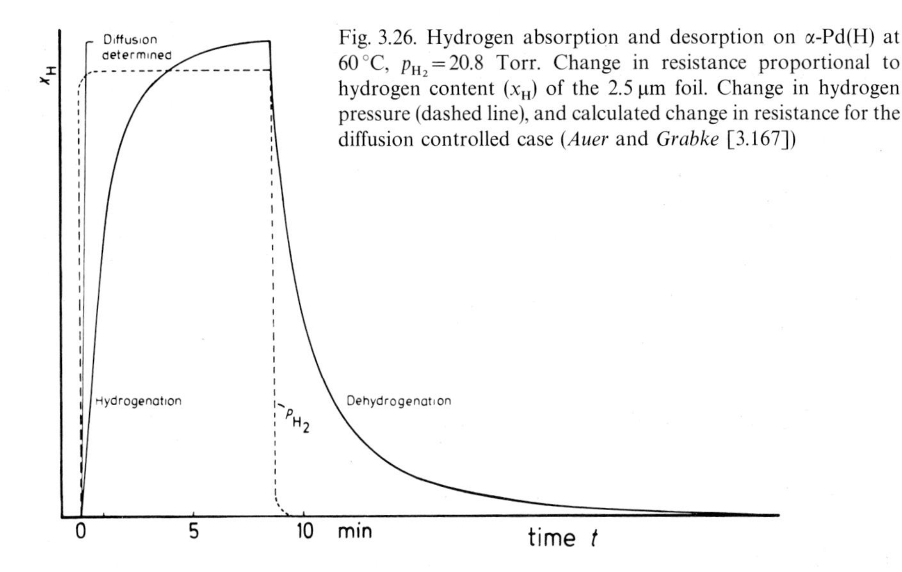

Fig. 3.26. Hydrogen absorption and desorption on α-Pd(H) at 60 °C, $p_{H_2}=20.8$ Torr. Change in resistance proportional to hydrogen content (x_H) of the 2.5 μm foil. Change in hydrogen pressure (dashed line), and calculated change in resistance for the diffusion controlled case (*Auer* and *Grabke* [3.167])

foils (2.5 and 5 μm). These were pretreated by annealing for 2 hrs at 500 °C in an highly pure helium flow. Helium served also as a carrier gas for providing the hydrogen to be taken up by the foils or to be removed. The hydrogen content of the foils was followed by resistance measurements. Figure 3.26 represents examples of a hydrogenation and a dehydrogenation run, initiated by a step up and a step down of hydrogen pressure, respectively. If diffusion controlled, the changes would be much faster, as also shown in Fig. 3.26. The measurements, carried out between 20 and 150 °C, yielded at low hydrogen concentrations the rate law

$$\frac{dn}{dt}=kp_{H_2}\cdot\frac{1}{1+Kn}-k'\cdot\frac{Kn^2}{1+Kn}, \tag{3.66}$$

valid in the range of Sieverts' law, i.e., of ideal solution. The two terms represent the rates of hydrogenation and of dehydrogenation, the proportionality with p_{H_2} and with n^2 indicates step (I) as rate determining, i.e., the dissociation of hydrogen molecules and, in the reverse direction, the recombination of hydrogen atoms in the chemisorbed state. The denominator $1+Kn$ points to a Langmuir-type chemisorption of H atoms, the surface sites of which are moderately occupied under the conditions of measurement. If H atoms from these sites, however, would recombine to H_2 molecules, the denominator would be expected to be $(1+Kn)^2$. The time law (3.66) actually observed needs for explanation a second type of H atom chemisorption, rather weak, i.e., in the linear range of adsorption isotherm. *Auer* and *Grabke* themselves denoted the two types of chemisorption by β and α; in order to avoid confusion with other β and α states we shall use instead the notation $H_{ad}(B)$ and $H_{ad}(A)$, respectively (going back to *Wagner* [3.165]). With the quantities θ_B and θ_A as the fractions of surface sites occupied, respectively, the rate law (3.66) reads:

$$\frac{dn}{dt} = kp_{H_2} \cdot (1 - \theta_B) - k'' \cdot \theta_A \theta_B \,. \tag{3.66a}$$

An activation energy of 6.8 kcal/mol of H_2 was measured for the rate constant of hydrogenation k in (3.66, 66a). From the temperature dependence of the coefficient K in (3.66), the enthalpy of the *segregation equilibrium* $H_{ad}(B) \rightleftharpoons H_{Me}$ was determined as 1.9 kcal/mol of H_2, with $H_{ad}(B)$ on the lower energy level. Considering the enthalpy of dehydrogenation of α-Pd(H), 4.6 kcal/mol of H_2 (p. 79) the enthalpy of adsorption $H_2^g \rightarrow 2H_{ad}(B)$ is obtained as -6.5 kcal/mol of H_2. According to *Grabke*, the type B adsorption sites are located in such a way that hydrogen atoms at these sites block up (reversibly) the passages between the surface and the interior of the lattice. Therefrom result the retardation terms Kn in the denominators of (3.66), diminishing equally the rates of hydrogenation and of dehydrogenation.

Regarding the results of chemisorption research, dealt with in Sections 3.2.3 and 3.2.4, it is obvious to connect this mechanism with the model of chemisorption represented in Fig. 3.6a, b [considering again the (111) lattice plane as the most frequent one in a polycrystalline sample]. The empty tetrahedral sites in Fig. 3.6a, correlated to the "weak" chemisorption investigated by *Lynch* and *Flanagan* [3.12] (p. 88) and to the infrared absorption observed by *Ratajczykowa* [3.57] (p. 97f.), suggest themselves once more as the mentioned passages between surface and interior interstitial sites, i.e., as the loci of the $H_{ad}(B)$ atoms after *Grabke*. The enthalpy of adsorption, -6.5 kcal/mol of H_2 as derived above, can still be taken as compatible with the values 11 to 8.5 kcal obtained by *Lynch* and *Flanagan* [3.12], considering the different type of samples used (annealed Pd foils compared to highly dispersed Pd black, respectively) and the different methods of investigation applied [3.168]. As a matter of fact, hydrogenation and dehydrogenation proceed via rather flat chemisorption wells on a Pd surface, where the deepest potential holes have been filled up by strong preadsorption. The *rate determining step* of hydrogenation is the dissociation of H_2 molecules, chemisorbed in position 1 of Fig. 3.6b, according to

$$H_{2_{ad}} \rightarrow H_{ad}(A) + H_{ad}(B) \,, \tag{III}$$

with the $H_{ad}(A)$ atoms remaining in position 1 or migrating along the surface. The true activation energy of this step is expected to be about 10 to 15 kcal/mol of H_2, if one considers the activation energy of 6.8 kcal, measured by *Auer* and *Grabke* [3.167], and the enthalpy of molecular adsorption in position 1.

The dissociation of chemisorbed molecules into atoms at different sites and energy levels according to (III), supposed originally by *Wagner* [3.165] and verified by *Grabke*, seems to be of more general importance. In recent investigations of the nitrogenation and denitrogenation of iron, *Grabke* [3.169] obtained a rate law formally identical with (3.66); the step (III) mechanism of N_2 dissociation could be supported by LEED studies.

The dissociation of H_2 has been confirmed as the rate determining step in hydrogenation of widely different types of Pd samples. *Bucur* et al. [3.170] verified this statement recently in a study with thin Pd layers, electrodeposited on a piezoelectric quartz single crystal, used as an oscillator microbalance; similar results were obtained earlier by *Suhrmann* et al. [3.171] with evaporated Pd films.

The *permeation* of hydrogen through a Pd foil is composed of the surface processes at the entrance and the exit faces of the foil, and the diffusion through the lattice. The surface kinetics at the gas/metal interfaces can be studied, therefore, also by permeation measurements. The method of stationary permeation, applied by *Meyer* [3.172] and by *Fehmer* [3.173], is based upon the relation

$$j = \frac{D}{V_{Me}} \cdot \frac{n_0 - n_s}{s} \tag{3.67}$$

for the diffusion flux density, with D = Fick's diffusion coefficient; n_0, n_s = H/Pd atomic ratios at the entrance and the exit faces of the metal foil, respectively; s = foil thickness; and V_{Me} = volume of the foil per g-atom of metal. The diffusion flux j is suitably chosen rather small, and the hydrogen pressure p_0 at the entrance side rather high. Under such conditions the solution equilibrium is always established at this side, and

$$n_0 = \sqrt{p_0^*}/K_s\,, \tag{3.68}$$

where K_s is Sieverts' constant and p_0^* differs from p_0 by the corrections accounting for the deviations from Sieverts' law (for details see [3.172]). At the exit face of the foil the diffusion flux, coming out from the lattice, enters the regime of surface kinetics, according to (3.66),

$$j = k' \cdot \frac{K n_s^2}{1 + K n_s} - k p_s \cdot \frac{1}{1 + K n_s}\,. \tag{3.69}$$

If the hydrogen pressure p_s is chosen very small, the second term on the right-hand side may be neglected, likewise the retardation term $K n_s$. The equilibrium condition, applied to (3.66), yields $k'K = k \cdot K_s^2$; hence (3.69) reduces to

$$j = k K_s^2 n_s^2\,, \quad \text{or} \quad n_s = \frac{1}{K_s} \cdot \sqrt{\frac{j}{k}}\,. \tag{3.70}$$

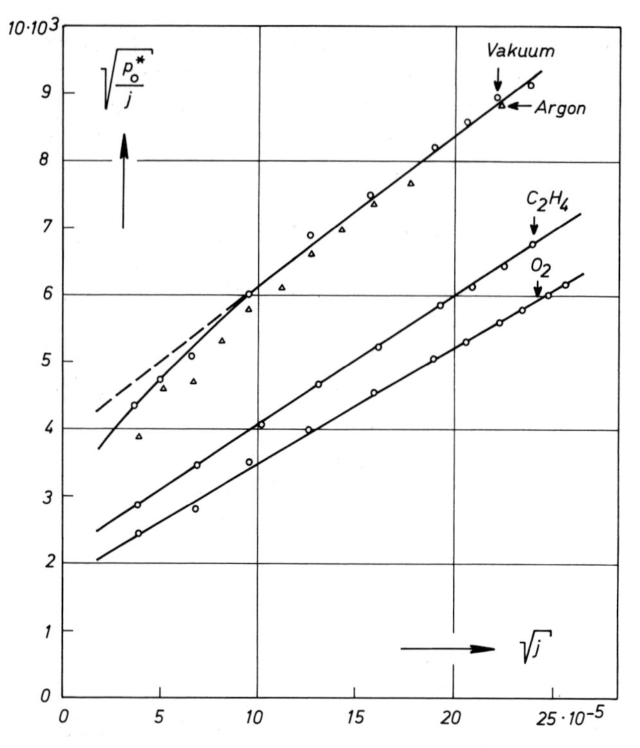

Fig. 3.27. Permeation data measured on 0.05 mm Pd foil (tube wall) at 98 °C under different conditions (see text), plotted according to (3.71). p_0^*: Torr; j: g-atom of H per cm² and s (*Meyer* [3.172])

Introducing (3.68) and (3.70) into (3.67) results in

$$\sqrt{\frac{p_0^*}{j}} = K_s \cdot \frac{s \cdot V_{\mathrm{Me}}}{D} \cdot \sqrt{j} + \frac{1}{\sqrt{k}}. \tag{3.71}$$

Figure 3.27 presents experimental data plotted according to (3.71). The values where measured by *Meyer* [3.172], who used 3 mm Pd tubes, wall thickness $s = 0.05$ mm, activated on both sides by Pd black coating. The inside hydrogen pressures varied within $0.01 < p_0 < 10$ Torr; the external ones were kept very low ($p_s \lesssim 10^{-5}$ Torr) by steady evacuation or by removing the hydrogen with a bypassing gas flow which acted as an inert carrier (argon) or as a reacting agent (ethylene, oxygen). The curves in Fig. 3.27 show up the linear shape predicted by (3.71) (at higher p_0^* values) and the coincidence of the vacuum and the argon curve that could be expected. In measurements between 30 and 100 °C the slopes, proportional to $1/D$, yielded reliable values of the diffusion coefficients, and the intercepts at the ordinate, equal to $1/\sqrt{k}$ after (3.71), resulted in an activation energy of 7.6 (± 0.6) kcal/mol of H_2, to be compared with the value of 6.8 kcal by *Auer* and *Grabke* [3.167]. The same value, 6.8 kcal, was obtained by *Fehmer* [3.173], who applied the method of *Meyer* to Pd tubes with polished bright surfaces, i.e., without Pd black coating, in the range 200 to 350 °C. Hence the "activation" by Pd black is not an effect of lowering the activation energy, but an effect of enlarging the surface area for the H_2 dissociation.

Ethylene and oxygen at the exit side react immediately with the hydrogen atoms as soon as they pass over from the lattice to the surface. Therefore, no congestion at all of adsorbed H atoms—as usually occurs before recombination to H_2 molecules—is necessary in these cases. For that reason, the diffusion coefficients determined by the slopes are a bit higher than in the case of applying vacuum or argon flow. The ordinate intercepts, $1/\sqrt{k}$, however, are appreciably smaller, i.e., the rate constants k, connected by $k'K = k \cdot K_s^2$ with the removal of the H atoms from the surface, are appreciably larger.

The process of *removal* of H atoms from the Pd surface *by chemical reaction* (instead of recombination) can be described by the mechanism

$$H_{Me} + M_{ad} \rightleftharpoons HM_{ad}$$

$$\swarrow \qquad \text{or} \qquad \searrow$$

$$2HM_{ad} \rightarrow M_{ad} + H_2M^g, \quad HM_{ad} + H_{Me} \rightarrow H_2M^g.$$

In the cases discussed here, M_{ad} means adsorbed O atoms or C_2H_4 molecules, HM_{ad} are OH or C_2H_5 groups in chemisorbed state, and H_2M^g the desorption products H_2O or C_2H_6. The characteristic features of the action of C_2H_4 and O_2 as presented in Fig. 3.27 can be evaluated for obtaining indications on the nature of the chemisorbed states M_{ad} (*Meyer* [3.172]). The reverse phenomenon—retardation of hydrogen removal by chemisorbed species—is also possible, as *Huber* [3.174] found in a study with carbon monoxide that blocks up the hydrogen passage through the gas/metal interface by strong chemisorption.

The results of the investigations on surface kinetics lead to the general conclusion that the influence of the transfer resistance at the interfaces decreases not only with decreasing diffusion flux—this is trivial—but also with increasing hydrogen pressure in the gas phase at these interfaces. As a matter of fact, when measuring the permeation through Pd foils—surface activated by Pd black—at elevated pressures in the β phase region, these resistances are of minor importance even at normal temperatures, i.e., "permeation" can be taken equal to "diffusion". This was the basis of the early diffusion measurements by *Bohmholdt* [3.175] applying stationary permeation in the range 20 to 80 °C at hydrogen pressures between 1 and 20 atm. He observed for the first time the *inverse* H/D *isotope effect* as represented in Fig. 3.28, simultaneously with *Holleck* [3.176] who worked with electrochemical methods. The results of *Bohmholdt* and of *Holleck* in the β phase region of PdH_n were confirmed quantitatively in recent NMR studies—T_1 measurements and pulsed-field-gradient method—by *Davis* et al. [3.177].

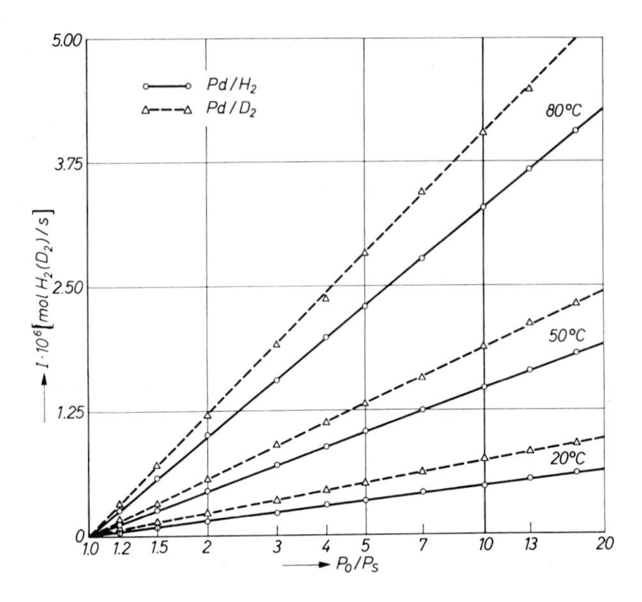

Fig. 3.28. Diffusion fluxes of H and D through the 0.22 mm wall of Pd tube, 3 mm ∅, in the β phase region of Pd(H) and of Pd(D), respectively (*Bohmholdt* [3.175]). Pressure within tube: $p_0 = 1$ to 20 atm; external pressure: $p_s = 1$ atm

3.5.2 Time-Lag Methods

In time-lag techniques for studying hydrogen diffusion or permeation in metals, the central part of the experimental setup is the diffusion foil that separates two vessels representing the "entrance" and the "detection" sides from one another. A change in hydrogen concentration or in hydrogen influx rate at the entrance side—starting from a uniform hydrogen distribution within the foil—appears with a certain time lag in the measuring signal at the detection side. The time lag depends on the mobility of the diffusing species in the metal as well as, in general, on the rate of transfer through phase boundaries. Fortunately, Pd and most of its alloys possess high catalytic activity for equilibration with hydrogen molecules. Thus small impediments at the phase boundaries, which could retard the hydrogen passage especially at normal and lower temperatures, can be removed almost completely[9] by "activation" of the surface with a coating of Pd black [3.16, 17].

Although quite a number of time-lag techniques have been developed, differing by the boundary conditions at both the entrance and the detection sides (for a survey, see [3.178]), the *methods for evaluating* the measured time-lag curves are rather uniform. The measuring signal may indicate the change with time of hydrogen concentration c or hydrogen flux j at the detection side

[9] Small remaining effects of surface impediments will be discussed later on.

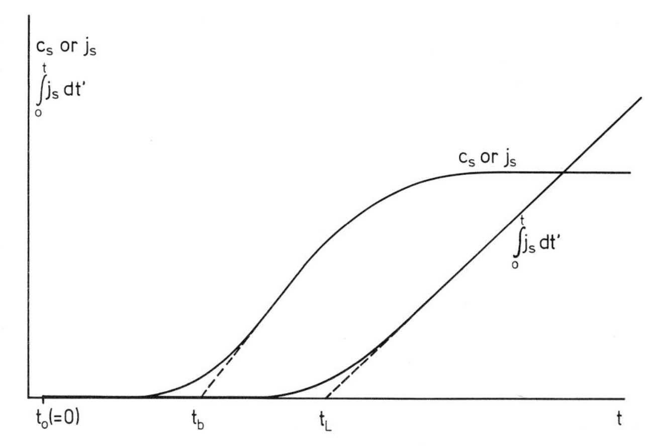

Fig. 3.29. Typical time-lag curves

$x=s$. Typical time-lag curves of the concentration $c_s(t)$ or the flux $j_s(t)$ are shown in Fig. 3.29, as well as the time integral of the flux $\int_0^t j_s(t')dt'$.

In Fig. 3.29, t_0 indicates the time of change of the hydrogen concentration at the entrance side $x=0$, which is taken as the zero point of the time scale. The intersection of the tangent at the inflection point of the S-shaped time-lag curve with the initial concentration level yields the "breakthrough" time t_b. The intercept of the extrapolated straight line branch of the integral curve determines the so-called "time lag" t_L. Both characteristic time intervals t_i are proportional to s^2/D,

$$t_i = \varkappa \cdot s^2/D\,, \tag{3.72}$$

where the coefficient $\varkappa$ depends on the boundary conditions to be considered for the measurement in question. This relationship and the proper values of the constant $\varkappa$ can be derived from the solution $c(x, t)$ of the diffusion equation

$$\frac{\partial c}{\partial t} = D\frac{\partial^2 c}{\partial x^2} \tag{3.73}$$

with the appropriate boundary conditions. In this equation the diffusion coefficient represents a mean value, averaged over the range of concentration Δc, covered by the actual measurement. It is a particular advantage of the time-lag technique to permit the choice of such small values of Δc that the concentration dependence of D can be neglected within this range.

The flux of hydrogen passing momentarily through the surface of the metal foil at the detection side $x=s$ can be obtained from the solution $c(x, t)$ by differentiating

$$j_s(t) = -D \cdot q \cdot [\partial c(x, t)/\partial x]_s = -D \cdot q \cdot c'_s(t)\,. \tag{3.74}$$

The total amount of hydrogen that emerged from the foil is, accordingly,

$$Q(t) = -Dq \int_0^t c_s'(t')dt' . \tag{3.75}$$

Time-lag methods were developed and first used by means of gas-volumetric techniques as early as 1920 (*Daynes* [3.179]) and 1940 (*Barrer* [3.180]) and were later transferred to the field of electrochemistry. For hydrogen in palladium the *electrochemical techniques* [3.178, 181–186] were preferred, in general, because of their more simple procedure and their flexibility in varying experimental conditions. Moreover, these techniques allow measurements at very low equilibrium pressures of hydrogen which are inaccessible to gas-volumetric and most other methods. The basis of electrochemical time-lag measurements are the following relationships.

At the surface of a Pd electrode, activated by a coating of Pd black, the equilibrium

$$\tfrac{1}{2}\mathrm{H_2} + n\mathrm{H_2O} \rightleftarrows \mathrm{H^+_{aq}} + \mathrm{e^-_{Me}}$$

is established reproducibly. The electrode potential follows Nernst's equation

$$\eta = \frac{RT}{F} \ln \frac{a_+}{\sqrt{p}} \tag{3.76}$$

(a_+ = hydrogen ion activity, p = hydrogen equilibrium pressure, F = Faraday charge). If the metal dissolves hydrogen, and the equilibrium

$$\mathrm{H_{Me}} + n\mathrm{H_2O} \rightleftarrows \mathrm{H^+_{aq}} + \mathrm{e^-_{Me}}$$

can be established at the metal/electrolyte interface, the hydrogen concentration in the metal near the surface determines the electrochemical potential at the interface. Concentration changes in the metal near the surface can thus be followed by monitoring the electrochemical potential at the interface, measured against a calibrated reference electrode. When small steps of concentration $\Delta c \ll c$ are applied in an experimental run, the proportionality

$$\Delta\eta_s(t) \sim c_s(t) \tag{3.77}$$

can be used in good approximation.

Certain time-lag techniques are based upon the measurement of the unsteady permeation rate, i.e., of the hydrogen flux that leaves the foil at the detection side during the non-steady-state process. The anodic current, measured by means of a potentiostat that maintains a fixed hydrogen concentration at the foil's exit surface, is by Faraday's law proportional to the momentary rate

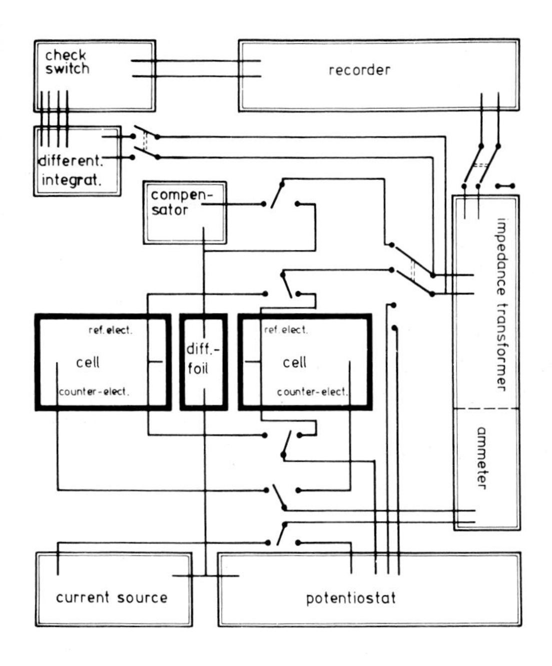

Fig. 3.30. Electrical circuit [3.178]

of hydrogen permeation

$$j_s(t) = i_s(t)/F\,. \tag{3.78}$$

The total amount of hydrogen that left the foil during a certain time t is accordingly given by

$$Q = \frac{1}{F}\int_0^t i_s(t')\,dt'\,. \tag{3.79}$$

There are no remarkable differences of equipment and type of measuring cells used for electrochemical time-lag measurements by different authors. A schematic survey of a complete experimental setup and a suitable electrochemical cell are shown in Figs. 3.30 and 3.31, respectively. Detailed descriptions are given elsewhere [3.178].

Devanathan and *Stachurski* [3.181] were the first to apply electrochemical time-lag methods to hydrogen in palladium. They used the non-steady-state technique that consisted of establishing at time $t=0$ a fixed and constant hydrogen concentration at the entrance side of the membrane by means of a potentiostat, while the detection side of the membrane was kept at hydrogen concentration $c_s = 0$ by anodic polarization. The anodic current $i_s(t)$ and/or its integral after (3.79) are recorded, and the diffusion coefficient is determined, corresponding to Fig. 3.29, by

$$t_b = 0.5\cdot s^2/(\pi^2\cdot D)\,, \tag{3.80}$$

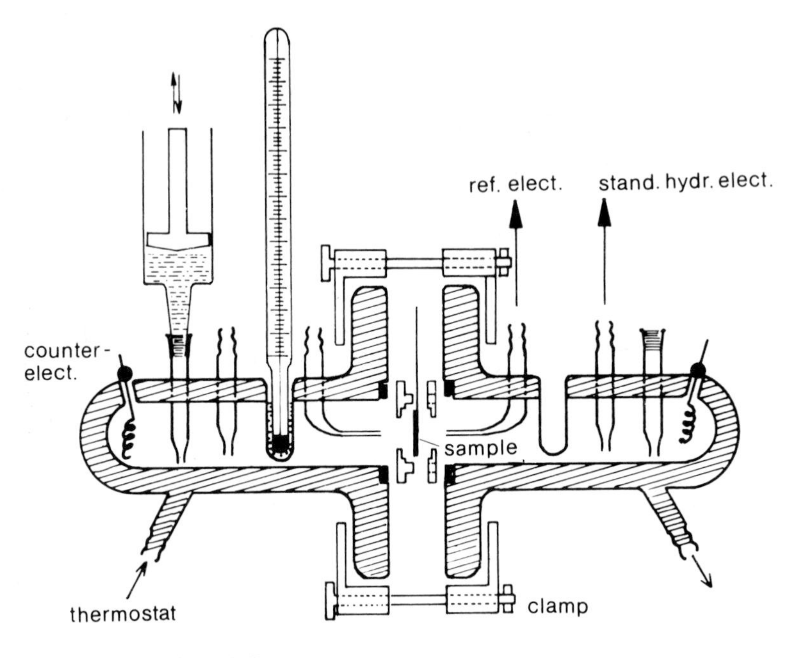

Fig. 3.31. Measuring cell [3.178]

or

$$t_{\mathrm{L}} = \frac{1}{6} \cdot s^2/D\,. \tag{3.81}$$

The measurements of *Devanathan* and *Stachurski* [3.181] reveal good agreement between both methods, when evaluated correctly[10] (see Table 3.4). The absolute value, $D = 1.3 \cdot 10^{-7}\ \mathrm{cm^2\ s^{-1}}$ at room temperature, however, is certainly too small. The reason is that the samples used in this investigation were neither annealed nor activated. As *Boes* and *Züchner* [3.178] showed, the method of anodic current, applied here, is particularly sensitive even to small impediments to the hydrogen passing through the surface of the metal.

For similar reasons the data of *von Stackelberg* and *Ludwig* [3.182], obtained by the same technique (and evaluated with another incorrect coefficient in the formula $t_{\mathrm{b}} = \varkappa s^2/D$) can indicate the order of magnitude only. Besides this, the D values, determined in this investigation at higher hydrogen concentrations, are influenced by the $\alpha \rightarrow \beta$ phase transition of Pd(H).

Küssner [3.183] already stated that the measurements of potentials at $j_s = 0$ instead of permeation rates yield more reliable diffusion data, because the potential values are less affected by surface kinetics. He used, therefore, the so-

[10] The authors used in [3.181] the incorrect relationship $t_{\mathrm{b}} = (1/15.3)s^2/D$. The D values in Table 3.4 were calculated with the correct formula (3.80).

Table 3.4. Diffusion data of H, D, and T in Pd and Pd alloys at $n \to 0$, obtained by electrochemical and radiochemical time-lag methods

System	X_i [at %]	$D \cdot 10^7$ [cm^2 s^{-1}] (25 °C)			$D_0 \cdot 10^3$ [cm^2 s^{-1}]			A_E [kcal g-atom^{-1}]			Remarks	Ref.
		H	D	T	H	D	T	H	D	T		
		1.3										[3.181]
Pd	0	1.2										[3.182]
		5.4	6.0	4.8	5.3	2.7	7.2	5.45	4.9	5.7		[3.189, 195]
Pd/Ag	10	5.2	3.5*	4.2	4.3	2.3*	6.3	5.35	5.2*	5.7		[3.185, 189, 195]
	20	4.1	5.3	3.7	3.4	1.6	5.6	5.35	4.75	5.7*		
	30	1.8	1.7*	1.9	2.3	1.1*	6.5	5.6	5.2*	6.2		
	40	0.23*	0.51*	0.28	1.9*	2.5*	2.7	6.7*	6.4*	6.8		[3.184, 185, 189]
	50	0.020*	0.041*	0.033	1.1*	1.3*	4.7	7.8*	7.5*	8.4		
	60	0.0035*	0.0046*	0.0012	1.0*	3.0*	7.0	8.8*	9.3*	10.6		
Ag	100	(7±3) 10^{-5}									Evaporated thin Ag films	[3.192]
Pd/Au	18.8	3.5									D values are valid for 37 °C; prehistory of the foils is unknown	[3.186]
	26.5	1.7										
	35.1	0.27										
	44.7	0.037			0.6			7.3				
	55.7	0.0074										
Pd/Cu	5			4.3			4.6			5.5		[3.190]
	10			3.1			2.8			5.4		
	15			3.0			3.3			5.5		
	20			2.3			2.8			5.6		
	25			2.3			3.0			5.6		
Pd/Ni	3			4.9			6.1			5.6		[3.190]
	7			3.8			4.8			5.6		
	10			3.4			4.3			5.6		
	15			2.6			3.9			5.7		
	20			2.3			3.4			5.7		
	26			1.6			3.9			6.0		
Pd/V	2			3.1			4.7			5.7		
	4			1.4			2.5			5.8		[3.190]
	6			1.3			2.8			5.9		
	10			0.5			3.4			6.6		
Pd/Rh	10	1.1			10.4			6.8			Prehistory of the foils is unknown	[3.191]

* Foils have not been annealed

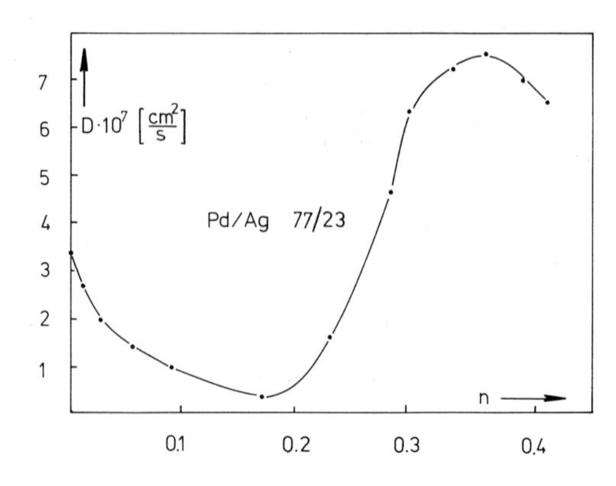

Fig. 3.32. Concentration dependence of the diffusion coefficient of H in Pd/Ag (77/23) at 30 °C (*Küssner* [3.183])

called "step method" for investigating hydrogen diffusion in Pd/Ag (77/23) foils, but, nevertheless, activated their surfaces. This method differs from that of *Devanathan* and *Stachurski* by the boundary conditions at the detection side. Not the anodic current $i_s(t)$ at $c_s=0$ is measured, but instead the course of hydrogen potential with time $\eta_s(t)$ [i.e., of the concentration $c_s(t)$] at $i_s=0$. In this case, the breakthrough time t_b is related to the diffusion coefficient by

$$t_b = 0.755\, s^2/(\pi^2 D)\,. \tag{3.82}$$

The method is suited excellently for studying the *concentration dependence* of the diffusion coefficient by a series of measurements with stepwise increasing concentrations. The concentration level, uniformly distributed throughout the foil after one measurement, represents the starting level for the next measuring run. Results obtained by *Küssner* [3.183] with this procedure for hydrogen concentrations $0 \lesssim n \leqq 0.42$ in Pd/Ag (77/23) alloy at 30 °C are shown in Fig. 3.32; for discussion see below.

The same method was applied by *Züchner* [3.184, 185] for investigating a series of Pd/Ag alloys with different compositions. His results are presented in Fig. 3.33 as diffusion coefficients at 30 °C versus hydrogen potential, $D(\eta)$. By means of Nernst's equation (3.76), $D(\eta) \sim D(-\ln p_{H_2})$, and of the well-known $p-n$ isotherms of Pd/Ag alloys (*Brodowsky* and *Poeschel* [3.71]) it is easy to reduce $D(\eta)$ to $D(n)$; as an example, the results for Pd/Ag (60/40) are presented in Fig. 3.34. The concentration dependence shows up quite similar to the observations of *Küssner* (Fig. 3.32).

The strong dependence $D(n)$ in the region covered by Figs. 3.32 and 3.33 is due to the deviations from ideal solution behavior. Application of Fick's law (3.73) yields Fick's diffusion coefficients $D(n)$, which are connected with the mobility, or Einstein's diffusion coefficients D^*, by the *thermodynamic factor* $f(n)$,

$$D(n) = D^* \cdot f(n) = D^* \frac{d \ln a}{d \ln n}\,. \tag{3.83}$$

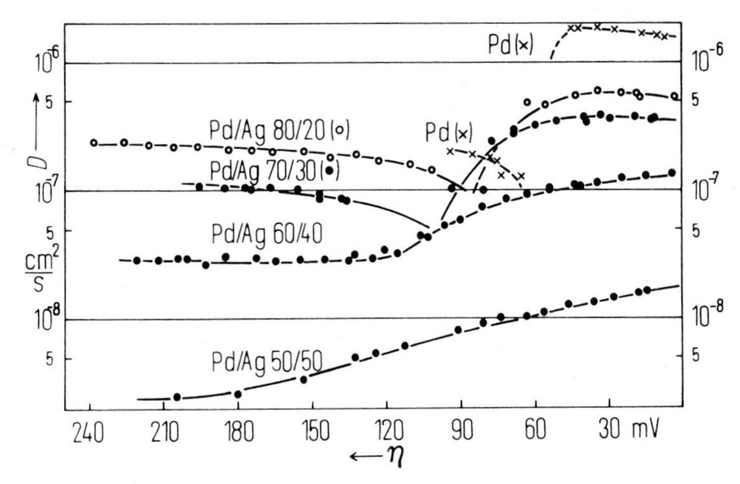

Fig. 3.33. Plots of diffusion coefficients of H in Pd/Ag alloys vs. hydrogen potential η (proportional to the logarithm of H_2 pressure) at 30 °C [3.185]

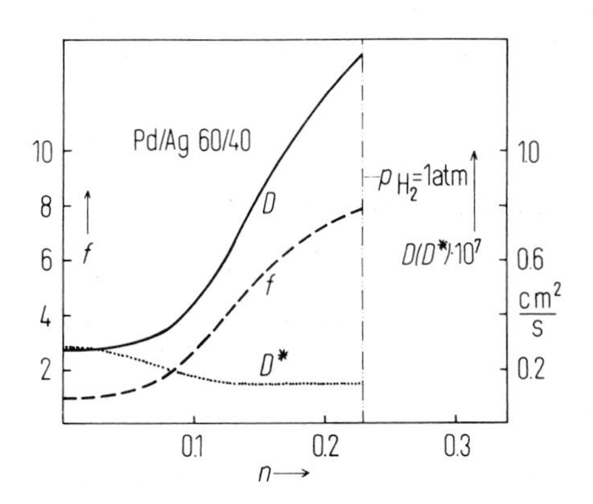

Fig. 3.34. Dependence of Fick's (D) and Einstein's (D^*) diffusion coefficients and of the thermodynamic factor f on atomic ratio H/Me ($=n$) [3.185]

The quantity a represents the activity of the dissolved hydrogen atoms that is related to the chemical potential by

$$\mu_H = \mu_H^0 + R \cdot T \cdot \ln a\,. \tag{3.84}$$

Combination with the equilibrium condition (3.7) yields the relationship

$$\frac{d \ln a}{dn} = \frac{1}{2} \frac{d \ln p_{H_2}}{dn}\,. \tag{3.85}$$

Accordingly, the course of the thermodynamic factor $f(n)$ can be determined from the slope of the $p-n$ isotherm; an example is represented in Fig. 3.34. There results, however, as Fig. 3.34 demonstrates, a concentration dependence of D^*, too. The decrease of D^* with increasing n can be attributed partly to the vacancy factor $(1-n)$ (see *Bohmholdt* [3.175] and *Holleck* [3.176]); the residual influence of concentration on D^* has not been explained so far.

Maestas and *Flanagan* [3.186] applied the step method for investigating the diffusion of hydrogen in a series of Pd/Au alloys at small hydrogen contents, where $D = D^*$ holds. Their values are listed in Table 3.4 and will be discussed together with the Pd/Ag results.

Küssner [3.183] already pointed out that the charging of the double layer capacity of the activated surface at the polarization (entrance) side of the membrane causes a certain relaxation time, which has to be considered as a correction especially at small breakthrough times. The effect was confirmed by *Maestas* and *Flanagan* [3.186]. It can be avoided by using the so-called "pulse method" developed by *Züchner* [3.184, 185] in 1969. This method differs from the step method in the way of charging the foil with hydrogen on the entrance side. Starting again with a uniform hydrogen distribution throughout the foil, a concentration peak is forced into the entrance side of the foil by a short cathodic current pulse. The initially narrow peak smoothes out by diffusion through the sample, and the change of concentration with time at the detection side is followed by recording the hydrogen potential ($i_s = 0$). The diffusion coefficient is obtained from the breakthrough time t_b by applying (3.80) because the derivation of the quantity $\varkappa$ in (3.72) under the boundary conditions to be considered in this case yields the same value $\varkappa = 0.5/\pi^2$ as was given in (3.80).

A special device of the pulse method is the *radiochemical technique* developed by *Sicking* and *Buchold* [3.187] for measuring tritium diffusion. The diffusion foil in this case consitutes the bottom of the electrolyte chamber containing dilute sulfuric acid with small amounts of HTO, and at the same time serves as the window of a proportional counter. The hydrogen/tritium peak at the entrance side is created by a short time cathodic polarization; the β radiation from the tritium decay in the metal near the surface at the detection side is taken as measuring signal. The pulse-time curve looks like the electrochemical time-lag curves (Fig. 3.29) and can be evaluated analogously.

Both the electrochemical and the radiochemical pulse methods are well adapted for measuring diffusion coefficients of H, D, and T *directly* at very small hydrogen-isotope contents, where nonideality can be neglected. An extrapolation to infinite dilution, as is often necessary in *Gorsky* experiments [3.188], is not needed here. Besides this, the radiochemical method is a specific one for measuring *tritium diffusion*. Most other methods must fail in this case, because it is rather difficult to attain sufficiently high concentrations of pure tritium as are required for those experiments.

The electrochemical and the radiochemical methods have been used to investigate a number of palladium alloys (and other metals) with regard to the diffusion of the three hydrogen isotopes. The results obtained—diffusion constants at room temperature, activation energies, and frequency factors—are compiled in Table 3.4. All data refer to infinite hydrogen dilution. Some values

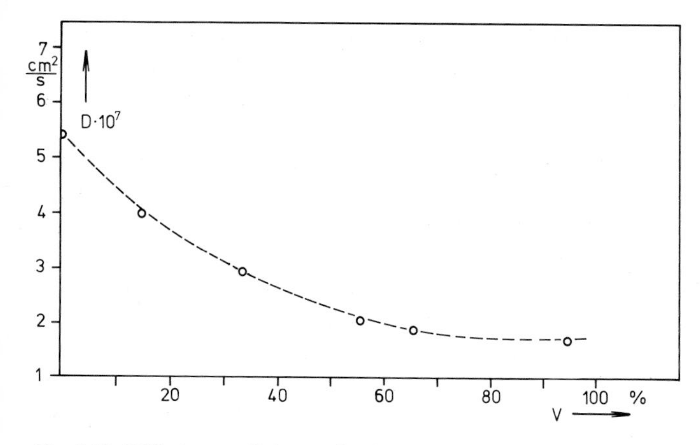

Fig. 3.35. Diffusion coefficients of T in Pd as a function of degree of deformation V at 30 °C (*Buchold* [3.189])

Table 3.5. Diffusion data of H and D in Pd at $n \to 0$

	Method	Ref.	A_E [kcal g-atom^{-1}]	$D_0 \cdot 10^3$ [cm^2 s^{-1}]	$D \cdot 10^7$ [cm^2 s^{-1}]		Temp. range [°C]
					25 °C	400 °C	
H	Gorsky effect	[3.196]	5.21 ± 0.15	2.5 ± 0.5	3.8	510	−40 − +200
	Electrochem. time lag	[3.195]	5.45 ± 0.16	5.3 ± 1.0	5.4	900	0 − +100
	Neutron scattering	[3.197]				960	400
D	Gorsky effect	[3.196]	4.75 ± 0.17	1.7 ± 0.6	5.6		−60 − + 70
	Electrochem. time lag	[3.195]	4.9 ± 0.17	2.7 ± 1.0	6.9		0 − +100

of diffusion constants in the systems Pd/Ag (H, D, T) were reported in earlier publications [3.184, 185, 187] to be somewhat smaller than the corresponding results listed in Table 3.4. The difference is because the diffusion coefficients are rather sensitive to *deformations* of the metal lattice [3.189, 190]. In the earlier investigations the foils were not annealed, but, on the contrary, somewhat deformed. Annealed samples always yield higher values of the diffusion coefficient (see Fig. 3.35) and smaller activation energies.

The diffusion data of H and D in Pd are well suited to compare the results of the time-lag method [3.195] with the results of those methods, where the interphase transfer is excluded, like *Gorsky effect* measurements [3.196]. As Table 3.5 demonstrates, the activation energies obtained with these entirely different methods agree within the limits of error. Only the D_0 values differ by about a factor of 2, the time-lag method yielding the higher D_0.

In this connection it is interesting to consider for comparison also the results of *neutron scattering* studies. Earlier investigations by this method yielded no satisfactory diffusion data as consequence of too low a resolution. Also, the more recent results of *Carlile* and *Ross* [3.198] are characterized by rather high uncertainty. Only the application of high neutron fluxes (HFR-Grenoble) quite recently yielded hydrogen mobility values of high accuracy. The diffusion coefficient determined by *Kley* [3.197] at 400 °C and low hydrogen content $n \to 0$, listed in Table 3.5, agrees excellently with the values extrapolated from time-lag results that have been obtained in the temperature range 0 to 100 °C. These comparisons demonstrate the reliability of time-lag diffusion data; besides this, special attention is called to the general convergence of the results obtained by the different methods of diffusion research.[11]

The influence of *alloy components* on the mobility of hydrogen in Pd is remarkably small at low additions. Ag, Au, Ni, and Pt up to about 25 at.% leave the diffusion coefficient and the activation energy nearly unchanged. At higher additions the activation energies increase and the diffusion constants decline in a nearly logarithmic fashion to quite low values for alloys with 50 or 60 at.% of solute.

The influence of the added components on the hydrogen diffusion and the activation energies in the alloys was discussed by *Buchold* et al. [3.163, 189]. The authors split up the activation energy into a basic contribution, which can be approximately identified with the energy of elastic strain of the host lattice caused by the diffusion jump, and an electronic term. This term is closely connected with the VEC (see p. 131) and can be estimated by means of the repulsion between the screening spheres of the solute ion and the proton.

The experimental techniques employed so far could not be used for diffusion measurements in Pd alloys with more than 60 at.% of solute, such as Ag, Au, or Pt. This is due to the small solubility and mobility of hydrogen in these alloys. Recently, however, *Diederichs* [3.192] succeeded in extending the time-lag method to hydrogen diffusion in pure silver, applying ultrahigh vacuum evaporation techniques. He used thin silver films of some 1000 Å thickness, evaporated onto palladium foils which acted only as a support to get mechanically stable diffusion foils necessary for time-lag experiments. The preliminary result obtained so far is

$$D_{\mathrm{H}}(\mathrm{Ag}, 30\,^{\circ}\mathrm{C}) = (7 \pm 3) \cdot 10^{-13}\,\mathrm{cm}^2\mathrm{s}^{-1}\,.$$

Palladium, owing to its high permeability for hydrogen, is suitable not only as a supporting material for thin diffusion membranes, but also as a permeable *protecting layer* at the surface of less noble metals. By this means it could be made possible to apply time-lag methods also to metals of the Vb group. Normally these metals are not permeable for hydrogen, because oxide layers at their surfaces prevent the hydrogen from passing through. *Boes* and *Züchner*

[11] For a review of these methods see, for example, *Völkl* and *Alefeld* [3.199].

[3.193, 194] succeeded in removing these layers by annealing the metal foils in ultrahigh vacuum, and by protecting the foils against reoxidation in air by a thin palladium film evaporated onto both sides of the cleaned samples. The thinness of the protecting palladium films and their high permeability for hydrogen reduce the three-layer diffusion problem—when applying the electrochemical pulse method—to a one-layer problem easy to solve. The diffusion data of H and D in Vb metals obtained with this technique confirm the results received from Gorsky effect measurements [3.188], and earlier from resistance [3.200] and lattice expansion studies [3.201, 202].

The possibility of building up "sandwich structures" from different metals by means of UHV procedures described above opens new fields of research—referring to the interesting problem of hydrogen passage from an fcc metal lattice into a bcc lattice, for instance—and new fields of application of the palladium/hydrogen system.

References

3.1 T.Graham: Phil. Trans. Roy. Soc. (London) **156**, 415 (1866)
3.2 F.A.Lewis: *The Palladium Hydrogen System* (Academic Press, London, New York 1967)
3.3 T.B.Flanagan: Engelhard Ind. Techn. Bull. **7**, 9 (1966)
3.4 J.E.Worsham,Jr., M.K.Wilkinson, C.G.Shull: J. Phys. Chem. Sol. **3**, 303 (1957)
3.5 G.Löbl, H.-D.Carstanjen, R.Sizmann: DPG-Frühjahrstagung „Festkörperphysik" Münster 1977, Verhandl. DPG(VI) **12**, 315 (1977)
3.6 T.Skośkiewicz; Ber. Bunsenges. Physik. Chem. **76**, 847 (1972); Phys. Status Solidi (a) K **123**, 6 (1972)
3.7 C.Wagner: Z. Physik. Chem. A **193**, 386, 407 (1944)
3.8 J.R.Lacher: Proc. Roy. Soc. (London), A **161**, 525 (1937)
3.9 H.Brodowsky: Z. Physik. Chem. NF **44**, 129 (1965)
3.10 M. von Stackelberg, P.Ludwig: Z. Naturforsch. **19**a, 93 (1964)
3.11 E.Wicke, G.H.Nernst: Ber. Bunsenges. Phys. Chem. **68**, 224 (1964)
G.H.Nernst: Dissertation, Münster (1963)
3.12 J.F.Lynch, T.B.Flanagan: J. Phys. Chem. **77**, 2628 (1973)
3.13 H.O. von Samson-Himmelstjerna: Z. Anorg. Allg. Chem. **186**, 337 (1930)
3.14 J.E.Schirber, B.Morosin: Phys. Rev. B **12**, 117 (1975)
3.15 B.Baranowski, S.Majchrzak, T.B.Flanagan: J. Phys. F.: Metal Phys. **1**, 258 (1971)
3.16 A.Küssner, E.Wicke: Z. Physik. Chem. NF **24**, 152 (1960)
A.Küssner: Dissertation, Hamburg (1959)
3.17 A.Küssner: Ber. Bunsenges. Physik. Chem. **66**, 675 (1962)
3.18 T.Graham: Proc. Roy. Soc. (London) **16**, 422 (1868); Phil. Mag. **36**, 63 (1868)
3.19 N.Boes, H.Züchner: Ber. Bunsenges. Physik. Chem. **80**, 22 (1976)
3.20 J.D.Clewley, T.Curran, T.B.Flanagan, W.A.Oates: J. Chem. Soc. Faraday Trans. I **69**, 449 (1973)
3.21 J.W.Simons, T.B.Flanagan: J. Phys. Chem. **69**, 3773 (1965)
3.22 H.Frieske, E.Wicke: Ber. Bunsenges. Physik. Chem. **77**, 50 (1973)
H.Frieske: Dissertation, Münster (1972)
3.23 W.M.Mueller, J.P.Blackledge, G.G.Libowitz: *Metal Hydrides* (Academic Press, London, New York 1968)
3.24 D.M.Nace, J.G.Aston: J. Am. Chem. Soc. **79**, 3619, 3623, 3627 (1957)
3.25 J.G.Aston: Engelhard Ind. Techn. Bull. **7**, 14 (1966)

3.26 T.B.Flanagan, J.F.Lynch: J. Phys. Chem. **79**, 444 (1975); J. Chem. Soc. Faraday Trans. I **70**, 814 (1974)
3.27 H.Kubicka: J. Catalysis **5**, 39 (1966)
3.28 P.C.Aben: J. Catalysis **10**, 224 (1968)
3.29 P.W.Selwood: *Chemisorption and Magnetization* (Academic Press, New York, San Francisco, London 1975)
3.30 A.W.Aldag, L.D.Schmidt: J. Catalysis **22**, 260 (1971)
3.31 H.Conrad, G.Ertl, E.E.Latta: Surface Sci. **41**, 435 (1974)
3.32 W.H.Weinberg, R.P.Merrill: Surface Sci. **33**, 493 (1972)
3.33 G.Sicking: Ber. Bunsenges. Physik. Chem. **76**, 790 (1972)
3.34 W.Drexel, A.Murani, D.Tocchetti, W.Kley, I.Sosnowska, D.K.Ross: J. Phys. Chem. Sol. **37**, 1135 (1976); The Motion of Hydrogen Isotopes in Metals and Intermetallic Compounds, Part I/II, Commission of the European Communities Rep. EUR 5465c and e (Luxembourg 1976)
3.35 M.Sakamoto: J. Phys. Soc. Japan **19**, 1862 (1964)
3.36 J.A.Pryde, I.S.T.Tsong: J. Chem. Soc. Faraday Trans. I **67**, 297 (1971)
3.37 R.H.Wiswall, J.J.Reilly: Inorg. Chem. **11**, 1691 (1972)
3.38 H.Brodowsky, E.Wicke: Engelhard Ind. Techn. Bull. **7**, 41 (1966)
3.39 L.J.Gillespie, W.R.Downs: J. Am. Chem. Soc. **61**, 2496 (1939)
3.40 A.C.Switendick: Ber. Bunsenges. Physik. Chem. **76**, 535 (1972)
3.41 D.E.Eastman, J.K.Cashion, A.C.Switendick: Phys. Rev. Lett. **27**, 35 (1971)
3.42 B.Stritzker, W.Buckel: Z. Physik **257**, 1 (1972)
3.43 T.Skośkiewicz, A.W.Szafranski, W.Bujnowski, B.Baranowski: J. Phys. C.: Solid State Phys. **7**, 2670 (1974)
3.44 R.J.Miller, C.B.Satterthwaite: Phys. Rev. Lett. **34**, 144 (1975)
3.45 A.Sieverts, W.Danz: Z. Physik. Chem. B **38**, 61 (1937)
3.46 M.R.Chowdhury: J. Phys. F.: Metal Phys. **4**, 1657 (1974)
3.47 H.C.Urey, D.Rittenberg: J. Chem. Phys. **1**, 137 (1933)
3.48 F.Botter: J. Less-Common Metals **49**, 111 (1976); Intern. Meeting on Isotope Effects, Cluj, Rumania 1973
3.49 G.Sicking: Z. Physik. Chem. NF **93**, 53 (1974)
3.50 J.Bergsma, J.A.Goedkoop: Physica **26**, 744 (1960); *Inelastic Scattering on Neutrons in Solids and Liquids* (I.A.E.A. Vienna 1961) p. 501
3.51 D.G.Hunt, D.K.Ross: J. Less-Common Metals **49**, 169 (1976)
3.52 J.M.Rowe, J.J.Rush, H.G.Smith, M.Mostoller, H.E.Flotow: Phys. Rev. Lett. **33**, 1297 (1974)
3.53 W.M.Jones: J. Chem. Phys. **16**, 1077 (1948)
3.54 H.Brodowsky, H.Gibmeyer, E.Wicke: Z. Physik. Chem. NF **49**, 222 (1966)
H.Gibmeyer: Dissertation, Münster (1965)
3.55 E.Gans: Unpublished experimental work
3.56 J.F.Black, H.S.Taylor: J. Chem. Phys. **11**, 395 (1943)
3.57 I.Ratajczykowa: Surface Sci. **48**, 549 (1975)
3.58 T.D.Lee, C.N.Yang: Phys. Rev. **87**, 410 (1952)
3.59 T.L.Hill: *Statistical Thermodynamics* (Addison-Wesley, Reading, London 1960) Chap. 14, pp. 235–255
3.60 G.Alefeld: Ber. Bunsenges. Physik. Chem. **76**, 746 (1972)
3.61 F.D.Manchester: J. Less-Common Metals **49**, 1 (1976)
3.62 E.A.Guggenheim: Proc. Roy. Soc. (London) A **148**, 304 (1935)
3.63 E.Bethe: Proc. Roy. Soc. (London) A **150**, 552 (1935)
3.64 W.Gorski: Z. Physik **50**, 64 (1928)
3.65 W.L.Bragg, E.J.Williams: Proc. Roy. Soc. (London) A **145**, 699 (1934); **151**, 540 (1935)
3.66 H.Brodowsky: Ber. Bunsenges. Physik. Chem. **76**, 740 (1972)
3.67 F.E.Hoare, B.Yates: Proc. Roy. Soc. (London) A **240**, 42 (1957)
3.68 H.Montgomery, G.P.Pells, E.M.Wray: Proc. Roy. Soc. (London) A **301**, 261 (1976)
3.69 C.A.Mackliet, A.J.Schindler: Phys. Rev. **146**, 468 (1966)

3.70 J.W.Simons, T.B.Flanagan: Can. J. Chem. **43**, 1665 (1965)
3.71 H.Brodowsky, E.Poeschel: Z. Physik. Chem. NF **144**, 143 (1965)
E.Poeschel: Dissertation, Münster (1964)
3.72 H.Husemann, H.Brodowsky: Z. Naturforsch. **23**a, 1693 (1968)
H.Husemann: Dissertation, Münster (1968)
3.73 H.Brodowsky, H.Husemann, R.Mehlmann: Ber. Bunsenges. Physik. Chem. **77**, 36 (1973)
3.74 R.Burch: Trans. Faraday Soc. **66**, 736 (1970)
3.75 R.Burch, F.A.Lewis: Trans. Faraday Soc. **66**, 727 (1970)
3.76 R.Burch: Trans. Faraday Soc. **66**, 749 (1970)
3.77 E.A.Guggenheim: *Mixtures* (Clarendon Press, Oxford 1952) pp. 78–81
3.78 H.Brodowsky, H.-J.Schaller: Trans. Met. Soc. AIME **245**, 1015 (1969)
3.79 H.Brodowsky, H.-J.Schaller: Ber. Bunsenges. Physik. Chem. **80**, 656 (1976)
3.80 M.Mahnig, L.E.Toth: Phys. Lett. **32**a, 319 (1970)
3.81 H.-J.Wernicke, Diplomarbeit, Kiel (1974)
H.Brodowsky, H.-J.Schaller, H.-J.Wernicke: To be published
3.82 H.Brodowsky: Z. Naturforsch. **22**a, 130 (1967)
3.83 N.G.Schmahl: Z. Anorg. Chem. **266**, 1 (1951)
N.G.Schmahl, W.Schneider: Z. Physik. Chem. NF **57**, 218 (1968)
3.84 K.M.Myles: Acta Met. **13**, 109 (1965)
3.85 H.Brodowsky, H.Frieske: In preparation
3.86 R.L.Moss, D.H.Thomas: J. Catalysis **8**, 151 (1967)
3.87 H.Brodowsky, Y.Oei, H.-J.Schaller: J. Metals **26**, 38 (1974)
3.88 H.-J.Schaller, H.Brodowsky: Ber. Bunsenges. Physik. Chem. **73**, 915 (1969); **82**, 773 (1978); Z. Metallk. **69**, 87 (1978)
3.89 J.Friedel: Ber. Bunsenges. Physik. Chem. **76**, 828 (1972)
3.90 J.C.Fisher: Acta Met. **6**, 13 (1958)
3.91 H.K.Birnbaum, C.Baker: Ber. Bunsenges. Physik. Chem. **76**, 827 (1972)
3.92 H.K.Birnbaum, C.A.Wert: Ber. Bunsenges. Physik. Chem. **76**, 813 (1972)
3.93 W.Dresler, M.Frohberg, H.G.Faller: Z. Metallk. **63**, 94 (1972)
3.94 G.J.Zimmermann: J. Less-Common Metals **49**, 49 (1976)
3.95 A.W.Lawson: J. Chem. Phys. **15**, 851 (1947)
3.96 T.Heumann: Ber. Bunsenges. Physik. Chem. **57**, 724 (1953)
3.97 L.L.Seigle, M.Cohen, B.L.Averbach: J. Metals **4**, 1320 (1952)
3.98 H.-J.Schaller: Dissertation, Münster (1972)
3.99 H.Brodowsky, W.Kock: Ber. Bunsenges. Physik. Chem. **79**, 985 (1975)
3.100 G. Alefeld, J.Völkl (eds.): *Hydrogen in Metals I. Basic Properties*, Topics in Applied Physics, Vol. 28 (Springer, Berlin, Heidelberg, New York 1978)
3.101 B.R.Coles: J. Inst. Metals **84**, 346 (1955)
3.102 C.Wagner: Acta Met. **19**, 843 (1971)
3.103 A.Makarov, D.Y.Stupin: Zhur. Fiz. Khim. **35**, 743 (1961)
3.104 V.Hovi: Ann. Acad. Sci. Fennicae AI **190** (1955)
3.105 N.Fontell: Soc. Sci. Fennicae AI **55** (1948)
3.106 N.Fontell, V.Hovi, L.Hyvönen: Ann. Acad. Sci. Fennicae AI **65** (1949)
3.107 W.H.McCoy, W.E.Wallace: J. Am. Chem. Soc. **78**, 5995 (1956)
3.108 V.Hovi: Ann. Acad. Sci. Fennicae AI **55** (1948)
3.109 M.A.Finemann, W.E.Wallace: J. Am. Chem. Soc. **76**, 4165 (1948)
3.110 M.W.Lister, N.F.Meyers: J. Phys. Chem. **62**, 145 (1958)
3.111 H.Koski: Thermochim. Acta **5**, 360 (1973)
3.112 W.T.Barrett, W.E.Wallace: J. Am. Chem. Soc. **76**, 370 (1954)
3.113 J.Wilks: *Liquid and Solid Helium* (Clarendon Press, Oxford 1967) p. 242
3.114 H.Brodowsky: Chemie-Ing.-Techn. **44**, 1089 (1972)
3.115 H.Brodowsky: Ber. Bunsenges. Physik. Chem. **72**, 1055 (1968)
3.116 D.M.Nace, J.G.Aston: J. Am. Soc. **79**, 3623, 3627 (1957)
3.117 P.Mitacek, J.G.Aston: J. Am. Soc. **85**, 137 (1963)
3.118 C.A.Mackliet, A.J.Schindler: J. Chem. Phys. **45**, 1363 (1966)

3.119 N.S.Ho, F.D.Manchester: J. Chem. Phys. **51**, 5437 (1969)
3.120 J.K.Jacobs, C.R.Brown, F.D.Manchester: Ber. Bunsenges. Physik. Chem. **76**, 827 (1972)
3.121 J.K.Jacobs, F.D.Manchester: J. Less-Common Metals **49**, 67 (1976)
3.122 G.A.Ferguson,Jr., A.I.Schindler, T.Tanaka, T.Morita: Phys. Rev. **137**, A 483 (1965)
3.123 V.Somenkov: Ber. Bunsenges. Physik. Chem. **76**, 733 (1972)
3.124 M.H.Mueller, J.Faber, H.E.Flotow, D.G.Westlake: Bull. Am. Phys. Soc. **20**, 421 (1975)
3.125 W.G.Saba, W.E.Wallace, H.Sandmo, R.S.Craig: J. Chem. Phys. **35**, 2148 (1961)
3.126 H.Brodowsky, J.Zierath: Ber. Bunsenges. Physik. Chem. **74**, 938 (1970)
J.Zierath: Dissertation, Münster (1969)
3.127 N.Jacobi, R.W.Vaughan: Scripta Met. **10**, 437 (1976)
3.128 S.W.Stafford, R.B.McLellan: Acta Met. **22**, 1391 (1974)
3.129 H.Brodowsky, H.Husemann: Ber. Bunsenges. Physik. Chem. **70**, 626 (1966)
3.130 K.Allard, A.Maeland, J.Simons, T.B.Flanagan: J. Phys. Chem. **72**, 126 (1968)
3.131 Th.Singe: Dissertation, Münster (1976)
3.132 T.Tsuchida: J. Phys. Soc. Japan **18**, 1016 (1963)
3.133 W.Himmler: Z. Physik. Chem. A **195**, 253 (1950)
3.134 W.Siegelin, K.H.Lieser, H.Witte: Ber. Bunsenges. Physik. Chem. **61**, 359 (1957)
3.135 H.Brodowsky: Habilitationsschrift, Münster (1968)
3.136 C.Wagner: J. Chem. Phys. **19**, 626 (1951)
3.137 C.Wagner: Z. Physik. Chem. **132**, 273 (1928)
3.138 C.Wagner: *Thermodynamics of Alloys* (Addison-Wesley, Reading, London 1952) Chap. 2, pp. 40–41
3.139 I.R.Harris, M.Cordey-Hayes: J. Less-Common Metals **16**, 223 (1968)
3.140 H.Brodowsky, R.Mehlmann: Ber. Bunsenges. Physik. Chem. **78**, 1259 (1974)
3.141 K.Hagemark: J. Phys. Chem. **72**, 2316 (1968)
3.142 G.v.Maltzahn: Diplomarbeit, Kiel (1977)
3.143 N.F.Mott, H.Jones: *The Theory of the Properties of Metals and Alloys* (Oxford University Press, London 1936)
3.144 D.Gerstenberg: Ann. Phys. Leipzig (7) **2**, 236 (1958)
3.145 E.Kudielka-Artner, B.B.Argent: Proc. Phys. Soc. **80**, 1143 (1962)
3.146 D.J.Lam, K.M.Myles: J. Phys. Soc. Japan **21**, 1503 (1966)
3.147 H.Frieske, M.Mahnig: Z. Naturforsch. **24**a, 1801 (1969)
3.148 J.P.Burger, E.Vogt, J.Wucher: C.R. Acad. Sci. Paris **249**, 1480 (1959)
3.149 R.Mehlmann: Diplomarbeit, Münster (1970)
3.150 Ch.Lambrou: Diplomarbeit, Münster (1972)
3.151 F.E.Hoare: In *Electronic Band Structure and Alloy Chemistry of Transition Elements*, ed. by P.A.Beck (Interscience/Wiley, New York 1963)
3.152 G.Wolf, M.Zimmermann, K.Bohmhammel: Phys. Status Solidi (a) **37**, 179 (1976)
3.153 A.Obermann, W.Wanzl, M.Mahnig, E.Wicke: J. Less-Common Metals **49**, 75 (1976)
3.154 W.Wanzl: Dissertation, Münster (1973)
3.155 M.Mahnig, E.Wicke: Z. Naturforsch. **24**a, 1258 (1969)
M.Mahnig: Dissertation, Münster (1968)
3.156 J.Friedel: Nuovo Cimento **7**, Suppl. **1**, 287 (1958)
3.157 G.Longworth: J. Phys. C: Metal Phys., Suppl. **1**, S 81 (1970)
3.158 J.J.Vuillemin, M.G.Priestley: Phys. Rev. Lett. **14**, 307 (1965)
J.J.Vuillemin: Phys. Rev. **144**, 396 (1966)
3.159 J.Eggs, K.Ulmer: Z. Physik **213**, 293 (1968)
3.160 J.S.Dugdale, A.M.Guénault: Phil. Mag. **13**, 503 (1966)
3.161 P.Brill, J.Voitländer: Z. Naturforsch. **24**a, 1 (1969)
3.162 J.Zbasnik, M.Mahnig: Z. Physik B **23**, 15 (1976)
3.163 H.Buchold, G.Sicking, E.Wicke: J. Less-Common Metals **49**, 85 (1976); Ber. Bunsenges. Physik. Chem. **80**, 446 (1976)
3.164 C.Wagner: Z. Physik. Chem. A **159**, 459 (1932)
3.165 C.Wagner: Z. Elektrochem. Angew. Physik. Chem. **44**, 507 (1938)
3.166 M.Boudart, M.A.Vannice, J.E.Benson: Z. Physik. Chem. NF **64**, 171 (1969)

3.167 W. Auer, H. J. Grabke: Ber. Bunsenges. Physik. Chem. **78**, 58 (1974)
3.168 H. J. Grabke: Private communication
3.169 H. J. Grabke: Z. Physik. Chem. NF **100**, 185 (1976)
3.170 R. V. Bucur, V. Mecea, T. B. Flanagan, E. Indrea: Surface Sci. **54**, 477 (1976); J. Less-Common Metals **49**, 147 (1976)
3.171 R. Suhrmann, G. Schumicki, G. Wedler: Z. Naturforsch. **19**a, 1208 (1964)
3.172 E. Wicke, K. Meyer: Z. Physik. Chem. NF **64**, 225 (1969)
K. Meyer: Dissertation, Münster (1968)
3.173 D. Fehmer: Dissertation, Münster (1971)
3.174 B. Huber: Diplomarbeit, Münster (1974)
3.175 G. Bohmholdt, E. Wicke: Z. Physik. Chem. NF **56**, 133 (1967)
G. Bohmholdt: Dissertation, Münster (1966)
3.176 G. Holleck, E. Wicke: Z. Physik. Chem. NF **56**, 155 (1967)
G. Holleck: Dissertation, Münster (1966)
3.177 P. P. Davis, E. F. W. Seymour, D. Zamir, W. David Williams, R. M. Cotts: J. Less-Common Metals **49**, 159 (1976)
3.178 N. Boes, H. Züchner: J. Less-Common Metals **49**, 223 (1976)
3.179 H. Daynes: Proc. Roy. Soc. (London) A **97**, 286 (1920)
3.180 R. M. Barrer: Trans. Faraday Soc. **35**, 628 (1939); **36**, 1235 (1940)
3.181 M. A. V. Devanathan, Z. Stachurski: Proc. Roy. Soc. (London) A **270**, 90 (1962)
3.182 M. v. Stackelberg, P. Ludwig: Z. Naturforsch. **19**a, 93 (1964)
3.183 A. Küssner: Z. Naturforsch. **21**a, 515 (1966)
3.184 H. Züchner: Z. Naturforsch. **25**a, 1490 (1970); Dissertation, Münster (1969)
3.185 H. Züchner, N. Boes: Ber. Bunsenges. Physik. Chem. **76**, 783 (1972)
N. Boes: Diplomarbeit, Münster (1971)
3.186 S. Maestas, T. B. Flanagan: J. Phys. Chem. **77**, 850 (1973)
3.187 G. Sicking, H. Buchold: Z. Naturforsch. **26**a, 1973 (1971)
G. Sicking: Dissertation, Münster (1970)
H. Buchold: Diplomarbeit, Münster (1971)
3.188 G. Schaumann, J. Völkl, G. Alefeld: Phys. Status Solidi **42**, 401 (1970)
3.189 H. Buchold: Dissertation, Münster (1974)
3.190 B. Huber: Dissertation, Münster (1977)
3.191 D. Artmann, T. B. Flanagan: J. Phys. Chem. **77**, 2804 (1973)
3.192 H. Diederichs: Dissertation, Münster (1976)
3.193 N. Boes, H. Züchner: Phys. Status Solidi A **17**, K 111 (1973)
3.194 N. Boes, H. Züchner: Z. Naturforsch. **31**a, 754, 760 (1976)
N. Boes: Dissertation, Münster (1974)
3.195 N. Boes, H. Züchner: To be published
3.196 J. Völkl, G. Wollenweber, K.-H. Klatt, G. Alefeld: Z. Naturforsch. **26**a, 922 (1971)
3.197 W. Kley, W. Drexel: Commission of the European Communities, Physical Sciences, EUR 5466e (1976)
3.198 C. J. Carlile, D. K. Ross: Solid State Commun. **15**, 1923 (1974)
3.199 J. Völkl, G. Alefeld: In *Diffusion in Solids: Recent Developments*, ed. by A. S. Nowick, J. J. Burton (Academic Press, New York 1975)
3.200 B. A. Merisov, V. I. Khotkevich, A. I. Karnus: Phys. Metals Metall. **22**, 163 (1966)
3.201 H. Züchner, E. Wicke: Z. Physik. Chem. NF **67**, 154 (1969)
H. Züchner: Z. Physik. Chem. NF **82**, 240 (1972)
3.202 E. Wicke, A. Obermann: Z. Physik. Chem. NF **77**, 163 (1972)

4. Metal-Hydrogen Systems at High Hydrogen Pressures

B. Baranowski

With 26 Figures

In recent decades increasing attention has been devoted to high pressure research in physics and chemistry. Various factors are responsible for this development. High pressure range means, first of all, the region of behavior where strong intermolecular interaction is present, due to the increasing numerical density of the particles. This causes several specific consequences.

In gaseous systems, large deviations from ideality and a continuous approach to liquids may be observed. As a convincing example of this tendency, the appearance of two-phase equilibria in high pressure gaseous systems may be mentioned. The increase of density in solids is very often followed by a rearrangement of the lattice, leading to new space configurations, usually of a more closely packed character when increasing the pressure. In fact the variety of phases observed due to the changes of pressure is richer than that caused by the variation of temperature.

Which contributions can be expected from high pressure research in metal-hydrogen systems? The increase of pressure in solid hydrides with a constant hydrogen concentration will surely lead to the appearance of new crystallographic structures. This is an uncommon and poorly investigated phenomenon. For example, the prediction of a pressure-induced transformation in lithium hydride [4.1] has not been confirmed in static compressibility measurements up to 40 kbar [4.2]. Recently a systematic investigation of the electrical resistivity in the rare earth, alkaline earth, and borohydrides was reported [4.3]. In several cases the changes observed can be explained by probably phase transitions. Even at normal conditions the interatomic distances between the hydrogen particles are rather small in solid hydrides. The application of high pressure will improve the interaction between these particles, because of the decrease of distances. It seems quite probably that a metallic behavior of the hydrogen involved can be achieved, if the required pressure and temperature conditions are realized in suitable solid hydrides. Very little has been done so far in this respect. In dynamic conditions an experiment was performed [4.4] which was ultimately determined to be misleading, because it was not confirmed at hydrostatic pressures. The results mentioned are surely the very beginning of extended investigations which will show the variety of phase transformations in solid hydrides caused by the increase of the hydrostatic pressure.

A different class of changes may occur if the metallic phase is brought into contact with high pressures of gaseous hydrogen[1]. Only this aspect will be considered in this chapter. Here the pure hydrostatic influence, as discussed above, may be negligible as compared with the effects caused by the high thermodynamic activity of gaseous hydrogen. The condition of thermodynamic equilibrium requires the equality of the chemical potential of hydrogen in the gaseous and the solid phases. The common effect to be expected, when increasing the pressure of gaseous hydrogen, will be the growth of hydrogen concentration in the solid phase. This can lead to new hydride phases which require a high pressure of gaseous hydrogen under equilibrium conditions. Historically the discovery of nickel hydride in our laboratory by electrochemical methods in 1959 [4.5] was the very beginning for an extended and systematic study of this metal in contact with high pressure gaseous hydrogen. After special devices were developed, other metallic systems were treated by gaseous hydrogen under high pressure conditions in other laboratories. A review of all results achieved so far is presented below. Some of these were previously summarized in three review papers [4.6–8].

Investigations carried out with active hydrogen but without the high pressure technique will be mentioned in this chapter only briefly. Preference will be given to measurements carried out in situ in high pressure gaseous hydrogen; for example, the ab- and desorption isotherms, the electrical resistance, thermopower, magnetic moment, lattice constant, thermal conductivity, diffusion, and the Hall effect can be treated in this way.

Let us finally remark that only gaseous hydrogen as the reference phase guarantees reproducible equilibrium conditions in the systems studied. The supplies of active hydrogen particles by electrochemical processes, gas discharges, or implantation are in most cases far from those of any comparable equilibrium state. They may serve as simple and effective methods for preparative purposes but are usually of limited importance for basic research. These aspects were discussed in more detail previously [4.7].

4.1 High Pressure Devices

4.1.1 First Experiments with High Pressures of Gaseous Hydrogen

In the early history of high pressure research [4.9] the only experiments with high pressure gaseous hydrogen were carried out by *Amagat* [4.10]. These were limited to the compressibility of this gas, measured up to 3000 atm and in a temperature range of 200 °C. The same property of gaseous hydrogen was later investigated by *Bridgman* [4.11], who extended the pressure range to

[1] High pressure of gaseous hydrogen means, in this chapter, pressures of the order of magnitude 10^3 bar. Results achieved at lower pressures will be discussed mainly only if related to this pressure range.

13,000 atm. The starting pressure was about 3000 atm, and the previous data of *Amagat* were taken over as correct in Bridgman's calculations. Manganin resistance served as a pressure gauge and the gas investigated filled an internal steel vessel, surrounded by kerosene as the transmitting medium.

A considerable improvement of the PVT data for gaseous hydrogen was achieved in the van der Waals Laboratory in Amsterdam in 1959 [4.12]. Measurements were carried out at temperatures ranging from -175 to 150 °C and at densities up to 900 Amagat, corresponding to pressures below 3 kbar. All three variables were determined to an accuracy of one part in 30,000. This allowed presentation of the experimental values of PV in the form of a power series up to the 5th power of the densities (in Amagat units). All values of the hydrogen fugacity used below are either taken over from this paper [4.12] or are results of numerical extrapolations based on the constants tabulated there.

Recently the equation of state of gaseous hydrogen was investigated in a slightly extended pressure range (0.5–5 kbar and sometimes to 6.5 kbar) in the temperature range 25–150 °C [4.13]. As can be verified, these results coincide with the extrapolated data mentioned above [4.12].

First experiments with high pressure gaseous hydrogen acting on metals were described in 1952 by *Frumkin* and co-workers [4.14]. The solubility of hydrogen in palladium hydride, that is, in samples with initial concentrations higher than 0.6 in atomic ratios n_H/n_{Pd}, was investigated in the temperature range from -78 to 100 °C and at pressures from 10^{-2} to 1.7×10^3 atm. A linear relationship was found between the concentration of hydrogen in the metal and the logarithm of hydrogen pressure,

$$\frac{n_H}{n_{Pd}} = a + b \ln p_{H_2} \,. \tag{4.1}$$

n_H/n_{Pd} denotes the atomic ratio of hydrogen to palladium, a, b are constants, and p_{H_2} is the pressure of gaseous hydrogen. The maximal hydrogen content reported in this paper was 0.92, in atomic ratio, and it was achieved at -78 °C and a gaseous hydrogen pressure of 1000 atm. This result clearly contradicted the earlier assumptions about the limiting concentration of hydrogen in this metal around 0.6 in atomic ratio. Relation (4.1) was established earlier [4.15] and electrochemical measurements in particular led many times to a linear dependence between the concentration of hydrogen and the electrode potential of the hydride phase, determined relative to the normal hydrogen electrode [4.16]. Because this potential is proportional to the logarithm of pressure (correctly to the logarithm of fugacity), the correspondence with (4.1) is obvious. In some cases the constants a and b of (4.1), as determined by absorption in gaseous hydrogen and in electrochemical measurements, were compared and satisfactory agreement was found [4.14]. As we shall see later, the validity of (4.1) is limited and its interpretation in simple terms is not possible.

In supercritical temperatures (from 326–477 °C) the isotherms of the Pd/H_2 system were determined up to 990 atm in 1959 by *Levine* and *Weale* [4.17]. A continuous increase of sorption with hydrogen pressure was observed with no indication of a limiting value. The highest atomic ratio n_H/n_{Pd}, equal to 0.69, was achieved at 990 atm and 326 °C. Isosteric heats of sorption were calculated for the concentration range from $n_H/n_{Pd}=0.225$ to 0.65, showing a strong increase with higher hydrogen contents from -11.0 kcal/mole H_2 to -2.8 kcal/mole H_2 in the concentration range indicated. An inverse course seemed to be indicative of temperatures higher than 437 °C and atomic ratios n_H/n_{Pd} above 0.5. Simultaneously the relative electrical resistance was measured at three temperatures (366°, 396°, and 456 °C) and at hydrogen pressures up to 1000 atm. A maximum was observed at about $n_H/n_{Pd}\approx 0.4$ for all three temperatures. The interpretation that this behavior is caused by co-conduction of the absorbed hydrogen, in contrast to the so-far observed continuous increase of the electrical resistance at room temperature, will be shown later to be incorrect.

As already mentioned, the origin of the high pressure experiments with gaseous hydrogen in our laboratory goes back to the discovery of nickel hydride in 1959 [4.5]. This new phase was observed by suitable cathodic saturation of metallic nickel; the details of this procedure as well as some kinetic characteristics may be found in earlier papers [4.18–22]. The necessity of high pressure work appeared as soon as we asked for the free energy of formation of nickel hydride from the elements. The only method which seemed to be applicable for this purpose was the determination of the equilibrium conditions between gaseous hydrogen and the solid hydride phase. Any electrochemical method could not be taken into consideration, because the formation of nickel hydride takes place here under typical nonequilibrium conditions and any thermodynamic information extracted from it can be very doubtful [4.7]. Preliminary experiments indicated that even 2000 atm of gaseous hydrogen at room temperature is not sufficient for the stability of nickel hydride and first one-dimensional hydrostatic pressures of about 10 kbar led to a clear inhibition of the decomposition kinetics of this phase [4.23]. This raised the serious problem of construction of a high pressure apparatus whose main purpose was to maintain stationary pressures of gaseous hydrogen of the order of several kbar for even several weeks. The decomposition pressure of nickel hydride was finally determined by putting compact tablets of this phase into a copper container, supported from outside by steel elements and closed by a sealed piston including a manganin gauge. At 25 °C, this pressure equals 3400 ± 70 atm of gaseous hydrogen. The details of the high pressure device and the results are published elsewhere [4.24, 25]. These successful experiments have shown that the sealing of high pressure gaseous hydrogen in the dry way—that is without any liquid phase—is possible in long time experiments, and this result encouraged us to further development of the high pressure technique. The next important step was the construction of a device enabling the continuous change of the hydrogen pressure to 18 kbar and the direct synthesis of nickel hydride from the elements [4.26].

4.1.2 Devices Applying Liquid Transmitting Media

Experiments with gaseous hydrogen in the high pressure region involve specific problems. As one deals here with the lightest gas, special attention has to be paid to all leakage. The destructive action of gaseous hydrogen on all steel elements represents a completely different problem. This danger was recognized very early, and an extensive description of the effects observed can be found in *Bridgman*'s paper [4.11]. The common result of an action of high pressure gaseous hydrogen on steel is formation of rifts, explosive ruptures, and related destruction, leading to an unforeseen if not dangerous end to a high pressure experiment. This also explains the fact that high pressure work with gaseous hydrogen did not belong to conventional and often practiced procedures. The only solution to the problem mentioned is to avoid in all steps of an experiment any contact between steel and the high pressure gaseous hydrogen. Separation of the materials by a thick layer of an organic liquid does not give the required guarantee. Therefore, since 1970 the apparatus which will be described in some detail [4.6, 30] has been used in our laboratory.

The pressure vessel is the internal conical steel cylinder (1), mechanically supported by two steel rings (2). The high pressure volume consists of a cylinder of about 3.2 cm^2 basis. Hydrogen (3) fills up a special container (4) which is supplied with an initial pressure of 300–600 atm of gaseous hydrogen from a separate multiplier outside the high pressure vessel. The details of the filling procedure are described elsewhere [4.6, 30]. The samples to be investigated are placed on the internal surface of the mobile piston (5) which seals the initial hydrogen pressure. The hydrogen container (4) is connected with the steel stopper (7) by a pipe (6) including all electrical leads from the samples. The high pressure volume is closed from above by the mobile piston (8), transmitting the external force from a hydraulic press to the organic liquid, which fills up all empty space between the internal steel walls of (1) and the hydrogen container (4). All seals (9) are of conventional character, applying soft O-rings and Bridgman-type metallic rings. Pressure is measured by a manganine resistance gauge placed in the organic liquid, usually on the stopper (7). Higher pressures are attained by movement of the piston (8) downward. Because of the higher compressibility of gaseous hydrogen (3) than of the organic liquid, the vessel (4) moves down more quickly than the piston (8). Practically no pressure difference exists between gaseous hydrogen (3) and the transmitting organic liquid, as soon as the pressure supplied by the piston (8) exceeds the starting pressure of gaseous hydrogen.

The device shown in Fig. 4.1 is used to an upper pressure of 25 kbar. Besides the exhausted mechanical strength of the working steel vessel (1), a serious limiting factor was the piston (8) transmitting the external force to the organic liquid. The first limitation expressed itself in irreversible volume changes (extension) of the internal cylinder (1), while the second limitation was manifested by spectacular explosions of the piston (8) around the 20 kbar range. Both troubles were partially removed in the following way: The mechanical support of the working steel vessel (1) could be improved by a controlled penetration of this vessel into the supporting rings (2). This was accomplished by application of a second hydraulic force from below to the vessel (1), which was inverted to that presented in Fig. 4.1. A higher working pressure in the cylinder (3) was followed by a well-prescribed and controlled mechanical support of the vessel (1), realized by a known force exerted on the ending surface of (1) from below. The vessel (1) penetrated further into the supporting rings (2), followed by a mechanical indicator. It was found that the explosions of the working pistons (8) started mostly from the regions surrounding the upper surfaces; therefore the last 1–2 cm of these pistons were supported by an additional steel ring ($\phi \sim 4$ cm, thickness ~ 1.5 cm) thermally attached to the piston. These two improvements allowed an extension of the pressure range to 30 kbar [4.7] and in some experiments even to 32 kbar of gaseous hydrogen.

Another system of sealing and filling of the hydrogen vessel was described schematically elsewhere [4.32]. Although this device was developed and proved in our laboratory and successful measurements were carried out [4.32–36], it is less efficient than that presented in Fig. 4.1 and is therefore not recommended.

4.1.3 Devices Without Pressure Transmitting Media

As mentioned above, the separating piston (Fig. 4.1) is not completely protected from the destructive action of gaseous hydrogen on the steel elements. Therefore the best solution would be to avoid any contact between steel and hydrogen. We accomplished this in the apparatus presented in Fig. 4.1 by placing a tube of beryllium bronze (wall thickness ~ 1–2 mm) in the working cylinder of the steel vessel (1). The disadvantage of this protection is a reduction of the mechanical strength and, consequently, a lower pressure range. This is caused mainly by the extrusion of the lower end of the protecting tube, due to the action of the sealing rings of the steel stopper (7) in Fig. 4.1. The upper limit of pressure reached so far in such protected steel cylinders is 12–15 kbar.

Recently we introduced a further improvement by replacing the internal working steel cylinder in Fig. 4.1 by an equivalent element of beryllium bronze which is, in the conditions so far realized, completely inert to the action of gaseous hydrogen. Later the pressure transmitting liquid medium is no longer necessary; thus one can work with pure hydrogen in the complete absence of any foreign volatile component. This simplifies the construction of the high pressure device, making the hydrogen working vessel (4) in Fig. 4.1 superfluous. The first version of such a device was realized some years ago for x-ray investigations *in situ* [4.7, 37]. It could work to pressures of about 700 bar, whereby the limiting element was the transparent epoxy resin container for the x-rays, and a flowing or crashing took place in the upper pressure limit indicated. The recently improved version is presented in Fig. 4.2 [4.38].

The main parts of the above device are made of beryllium bronze. The sample container (1) consists of metallic beryllium which forms a double cone

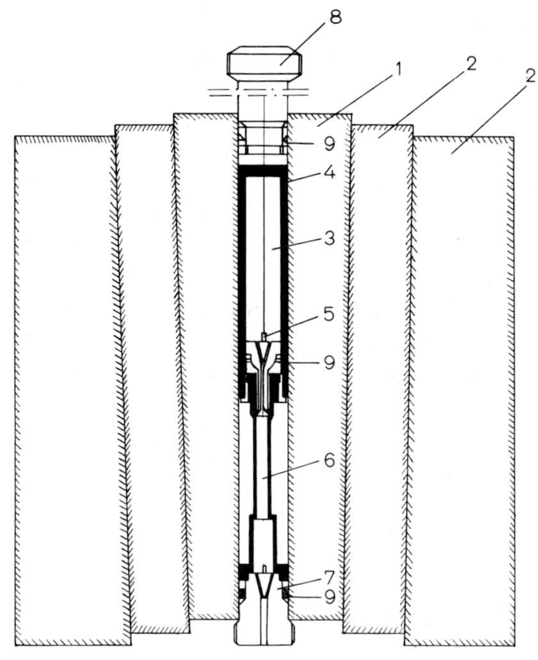

Fig. 4.1. Scheme of the high pressure device used up to 25 kbar of gaseous hydrogen

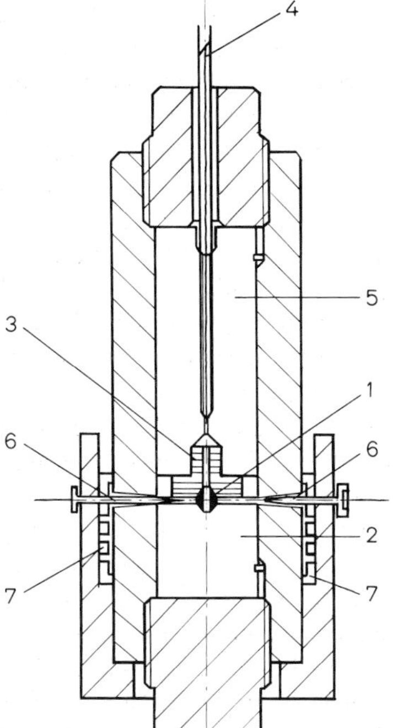

Fig. 4.2. Scheme of the high pressure device for roentgenographic investigations up to 10 kbar of gaseous hydrogen

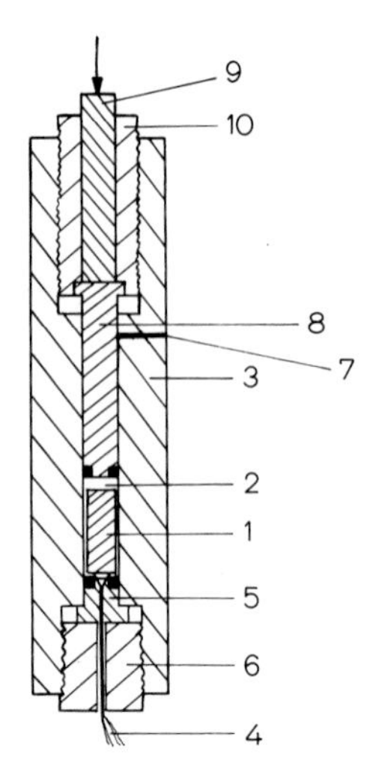

Fig. 4.3. Scheme of the manostatic device for low temperature investigations up to 12 kbar of gaseous hydrogen

ended cylinder, placed in the corresponding conical holes of the bottom (2) and the small mobile (3) pistons. High pressure tubing (4) fastened to the upper piston (5) connects the sample container (1) with the gas compressor, supplying gaseous hydrogen. The pressure is measured inside the compressor by a manganine gauge. The x-ray camera consists of the collimators (6) and the multiple film cassette (7) coaxial to the sample container. Suitable slits allow registration of the diffracted beams on both film parts in the full angle range. A commercially available x-ray unit could be used here.

For low temperature investigations, especially for superconductivity measurements, a manostatic device (Fig. 4.3) was developed [4.39]. The samples (1) are placed in the hydrogen working volume (2) which is included in the cylinder of a beryllium bronze vessel (3). The electrical leads from the samples (4) go through the closing stopper (5) supported by the screw (6). The gas inlet (7) supplies the working volume by gaseous hydrogen of pressures around 1.5 kbar from a separate multiplier, the mobile piston (8) being in an upper position and guaranteeing the seal. Higher pressures are achieved by moving the piston (8) downwards by exerting an external force on the steel cylinder (9). The screw (10) makes the monostating possible by keeping the piston (8) in the prescribed position. The pressure was measured with a manganin coil. After loading at room temperature the whole device could be cooled to helium temperatures.

The highest hydrogen pressures in "dry conditions" (up to 15 kbar) were reached in the device presented in Fig. 4.4 [4.40, 43] used mainly for measurements of ab- and desorption isotherms. A beryllium brass vessel (1) is mechanically supported by two steel rings (2). The circulation of a thermostated liquid through the jacket (3) serves for the temperature control inside the working volume (4) which is closed from above by the mobile steel piston (5)

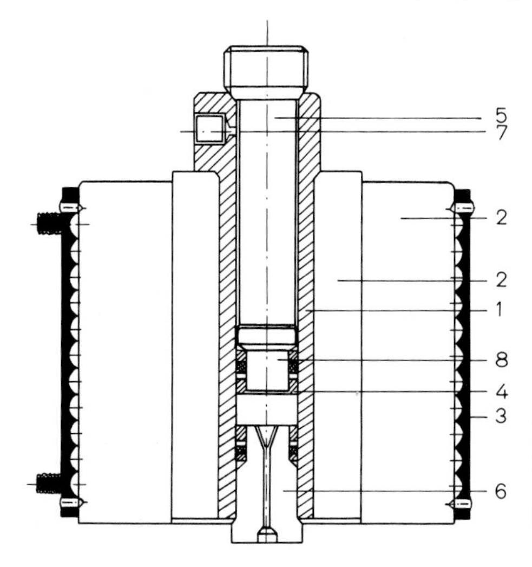

Fig. 4.4. Scheme of the high pressure device for determination of absorption and desorption isotherms up to 15 kbar of gaseous hydrogen

and from below by the fixed stopper (6). Gaseous hydrogen (starting pressures 1–2 kbar) is supplied through the capillary inlet (7). The end of the steel piston (5) is made of beryllium bronze (8). All seal are of conventional character. Great care must be given to the preparation of all surfaces to avoid leakage of the gaseous hydrogen. The movement of the piston can be followed by a precise mechanical indicator with a resolution of 10^{-2} mm.

In the pressure range between several bar and 1 kbar, a simple beryllium bronze container was used with dimensions similar to those given in Fig. 4.1. Pure hydrogen was supplied from a multiplier by a beryllium bronze capillary. A purifying assembly was eventually introduced between the multiplier and the hydrogen container. Pressure measurement was possible here by conventional Bourdon manometers.

The devices described above have all been used in recent investigations of metal-hydrogen systems in the high pressure range of gaseous hydrogen. We did not present any details of the earlier, less effective apparatus. All required information can be found in the references cited. The possible lines of future development will be discussed later after presentation of the results achieved with the devices described above.

4.2 Thermodynamic Properties

4.2.1 General Considerations

Because the vacancies available inside the metallic lattices are too small to be occupied by molecular hydrogen, it seems more convenient to express the condition of thermodynamic equilibrium between gaseous hydrogen and solid metals in terms of atomic hydrogen

$$\mu_{H(g)} = \mu_{H(m)}, \tag{4.2}$$

where $\mu_{H(g)}$ denotes the molar chemical potential of atomic hydrogen in the gaseous phase and $\mu_{H(m)}$ is the same quantity for the metallic phase. Besides very *high temperatures* or very low pressures, the majority of hydrogen atoms recombine at equilibrium conditions to molecular hydrogen. Therefore (4.2) can be rewritten in the form

$$\frac{1}{2}\mu_{H_2(g)} = \mu_{H(m)}, \tag{4.3}$$

where $\mu_{H_2(g)}$ denotes the molar chemical potential of molecular hydrogen. For high pressures the right-hand side of (4.3) can be given as

$$\mu_{H(m)} = \mu^0_{H(m)}(T) + RT\ln a_{H(m)} + V_{H(m)}(p - p_0), \tag{4.4}$$

where $\mu^0_{H(m)}(T)$ denotes the temperature-dependent standard molar chemical potential of hydrogen atoms in the metal, R, T are the gas constant and absolute temperature, $a_{H(m)}$ is the activity of hydrogen in the metallic phase, $V_{H(m)}$ is the partial molar volume of atomic hydrogen in the metallic phase and p, p_0 are pressure and standard pressure.

The last term in (4.4) is mostly neglected as being small beside the high pressure range. Rewriting the left-hand side of (4.3) in the conventional form

$$\tfrac{1}{2}\mu_{H_2(g)} = \tfrac{1}{2}(\mu^0_{H_2(g)}(T) + RT\ln p_{H_2} f_{H_2(g)}), \tag{4.5}$$

where p_{H_2} is the pressure of gaseous hydrogen, $f_{H_2(g)}$ is its fugacity coefficient, we can finally express the equilibrium condition in the form:

$$\ln p_{H_2} f_{H_2(g)} = \frac{2\mu^0_{H(m)}(T) - \mu^0_{H_2(g)}(T)}{RT} + \frac{2V_{H(m)}}{RT}(p - p_0) + 2\ln a_{H(m)}. \tag{4.6}$$

The main problem here is the activity of hydrogen in the metallic phase or more precisely the corresponding activity coefficient $f_{H(m)}$

$$a_{H(m)} = C_{H(m)} f_{H(m)}, \tag{4.7}$$

where $C_{H(m)}$ denotes the concentration of hydrogen in the metallic phase. The equilibrium condition (4.6) holds for any solid phase in contact with gaseous hydrogen. If more solid phases are present, the right-hand side can be written separately for each phase. The pressure p on the right-hand side of (4.6) can coincide with the pressure of gaseous hydrogen p_{H_2}, if dealing with pure hydrogen, or it can be higher if a mixture of gaseous components is present. Sievert's equation results from (4.6) in a simple way if the high pressure correction is neglected and ideal behavior in the gaseous and solid phases is assumed.

As the partial molar volume $V_{H(m)}$ of hydrogen in solid phases is usually positive, the increase of hydrogen activity by pressure—that is, the rise of the left-hand side of (4.6)—is partially compensated by the second term on the right-hand side. This causes a smaller increase of the hydrogen concentration in the solid phase due to the "extruding" action of the hydrostatic pressures. One can even imagine conditions in which the increase of the term $\ln p_{H_2} f_{H_2(g)}$ will be completely compensated by the change of the term $2V_{H(m)}/RT(p-p_0)$ or that the increment of the last term may even be higher than that corresponding to the previous one. The last situation would mean a decrease of hydrogen concentration in the solid phase when increasing the pressure of gaseous hydrogen.

4.2.2 Palladium and Palladium Alloys

Metallic palladium exhibits a unique response to variable thermodynamic activity of gaseous hydrogen. Within more than ten orders of magnitude of hydrogen activity, the change of concentration of this component in palladium can be followed [4.41]. In the high pressure region this manifested itself in the simplest way by a change of the electrical resistance to pressures up to 20 kbar of gaseous hydrogen at room temperature [4.31, 32, 42]. Two aspects here evoke special interest: The existence of a limiting concentration of hydrogen in palladium is one of the questions to be answered and, secondly, the appearance of superconductivity in the highly concentrated hydride phase makes high pressure research even more attractive. Therefore of great importance is the determination of the hydrogen concentration in this metal in in situ conditions for high content in palladium. This was recently realized, applying the device described above and presented in Fig. 4.4 [4.43]. A 20 g sample of powdered metallic palladium placed in a small copper container inside the working volume of the high pressure device was used for this purpose. The absorption of hydrogen was followed by the controlled displacement of the piston at isobaric and isothermal conditions. The results achieved at 25, 45, and 65 °C are presented in Fig. 4.5. The curves are valid for both the absorption and desorption procedures, thus exhibiting no hysteresis. The starting pressure was always chosen between 1.5 and 2 kbar, and the composition of this initial state was taken over from previous data [4.14, 44]. As can be seen from Fig. 4.5, in no case did the concentration of hydrogen exceed the full stoichiometry ($n_H/n_{Pd}=1$) for the octahedral vacancies in the palladium lattice. Let us remark that the data of Fig. 4.5 do not fulfill (4.1) even if the hydrogen pressure is replaced by its fugacity and if the term proportional to the partial volume of

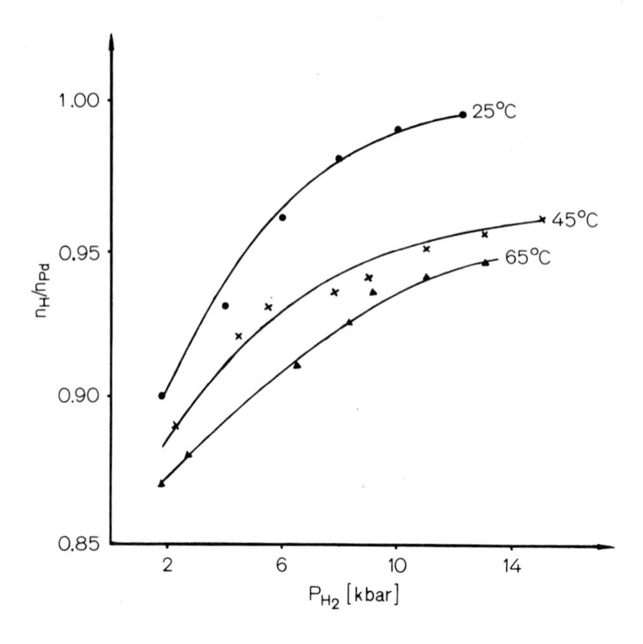

Fig. 4.5. Absorption isotherms of hydrogen in palladium in the high concentration range of the hydride

hydrogen in the metallic phase is added. To make this more clear, let us express the concentration c and the activity coefficient of (4.7) by an expression of an analytical character like in *Lacher*'s theory [4.45], which is found as mostly successful in the thermodynamics of the Pd–H system [4.46, 47]

$$C_{\mathrm{H(m)}} = \frac{\dfrac{n_{\mathrm{H}}}{n_{\mathrm{Pd}}}}{\left(\dfrac{n_{\mathrm{H}}}{n_{\mathrm{Pd}}}\right)_{\max} - \dfrac{n_{\mathrm{H}}}{n_{\mathrm{Pd}}}}, \tag{4.8}$$

$$f_{\mathrm{H(m)}} = \exp\left(B\frac{n_{\mathrm{H}}}{n_{\mathrm{Pd}}}\right), \tag{4.9}$$

where B is a constant. In terms of (4.8) and (4.9), (4.6) can be presented in the form

$$\ln p_{\mathrm{H_2}} + \ln f_{\mathrm{H_2(g)}} = A + 2\ln \frac{\dfrac{n_{\mathrm{H}}}{n_{\mathrm{Pd}}}}{\left(\dfrac{n_{\mathrm{H}}}{n_{\mathrm{Pd}}}\right)_{\max} - \dfrac{n_{\mathrm{H}}}{n_{\mathrm{Pd}}}} + \frac{2V_{\mathrm{H(m)}}}{RT}(p - p_0) + 2B\frac{n_{\mathrm{H}}}{n_{\mathrm{Pd}}}, \tag{4.10}$$

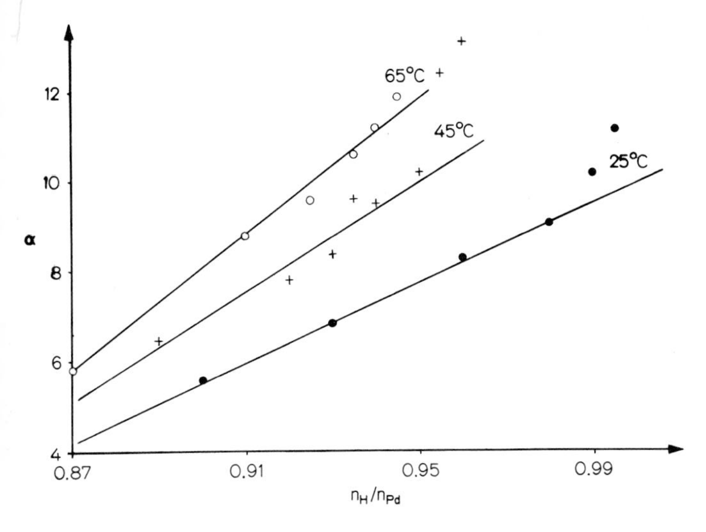

Fig. 4.6. Comparison of (4.10) with the experimental data (Fig. 4.5); $(n_H/n_{Pd})_{max} = 1.1$

where A is an abbreviation if the first term on the right-hand side of (4.6). It can be shown numerically that (4.10) reduces to (4.1) if $(n_H/n_{Pd})_{max}$ is replaced by 1 and if pressures between 200 and 1500 atm of gaseous hydrogen are considered at 25 °C. Thus the simplicity of (4.1) seems to be caused only by a compensation of contrary contributions within a suitable pressure range.

In order to prove the validity of (4.10) relative to the experimental data given in Fig. 4.5, let us plot the quantity

$$\alpha = \ln p_{H_2} f_{H_2(g)} - 2\ln \frac{\dfrac{n_H}{n_{Pd}}}{\left(\dfrac{n_H}{n_{Pd}}\right)_{max} - \dfrac{n_H}{n_{Pd}}} - \frac{2V_{H(m)}}{RT}(p_{H_2} - p_0) \tag{4.11}$$

as a function of n_H/n_{Pd}. In terms of (4.10) this should lead to a straight line. Figure 4.6 presents the curves for all three temperatures, whereby for $(n_H/n_{Pd})_{max}$ the value 1.1 was taken. In the calculations the data of hydrogen fugacity from [4.12] and for $V_{H(m)}$ the value of 0.46 cm^3/mole H [4.41] were introduced. Smaller values of this parameter give poorer agreement with (4.10). As can be seen from Fig. 4.6, only the highest concentrations do not fulfill the requirement of (4.10). Let us remark that experimental points and not smoothed values were taken over from Fig. 4.5 for the calculations presented in Fig. 4.6; therefore all experimental errors are included. Any more detailed quantitative interpretation of the absorption data in high hydrogen concentrations of palladium hydride would require an improvement of the experimental accuracy.

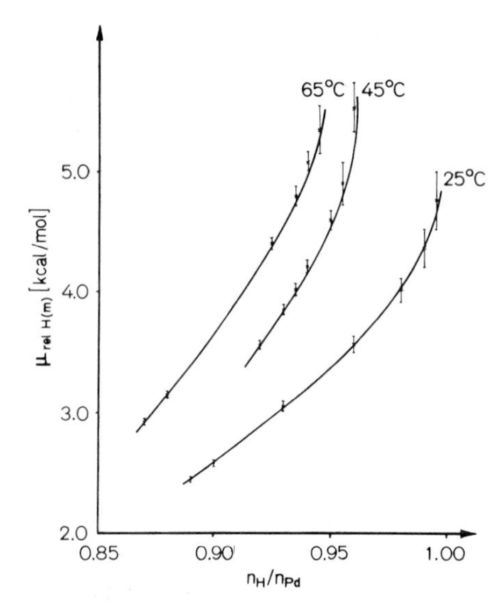

Fig. 4.7. Relative partial molar free energy of hydrogen in the metallic phase as a function of concentration (4.12)

Independent on the interpretation of the data presented in Fig. 4.5, the following relative partial molar quantities can be calculated. The relative partial molar free energy of hydrogen in the metallic phase $\mu_{\text{rel H(m)}}$ is

$$\mu_{\text{rel H(m)}} = \mu_{\text{H(m)}} - \tfrac{1}{2}\mu^0_{\text{H}_2\text{(g)}} = \tfrac{1}{2}RT\ln p_{\text{H}_2} f_{\text{H}_2\text{(g)}} . \tag{4.12}$$

The relative partial molar enthalpy of hydrogen in the metallic phase $h_{\text{rel H(m)}}$ is

$$h_{\text{rel H(m)}} = h_{\text{H(m)}} - \tfrac{1}{2}h^0_{\text{H}_2\text{(g)}} = \frac{R}{2}\frac{\partial \ln p_{\text{H}_2} f_{\text{H}_2\text{(g)}}}{\partial\left(\dfrac{1}{T}\right)} . \tag{4.13}$$

The relative partial molar entropy of hydrogen in the metallic phase $S_{\text{rel H(m)}}$ is

$$S_{\text{rel H(m)}} = S_{\text{H(m)}} - \tfrac{1}{2}S^0_{\text{H}_2\text{(g)}} = \frac{(h_{\text{H(m)}} - \tfrac{1}{2}h^0_{\text{H}_2\text{(g)}}) - (\mu_{\text{H(m)}} - \tfrac{1}{2}\mu^0_{\text{H}_2\text{(g)}})}{T} . \tag{4.14}$$

The above quantities present the difference between the actual state of hydrogen in the metallic phase and its value for the standard state of the gas (unit fugacity). The values calculated numerically or determined graphically are given in Figs. 4.7–9.

The characteristic feature of all three curves is the strong concentration dependence when approaching stoichiometry. This fact reflects the large pressure dependence of the hydrogen concentrations in the same range. The

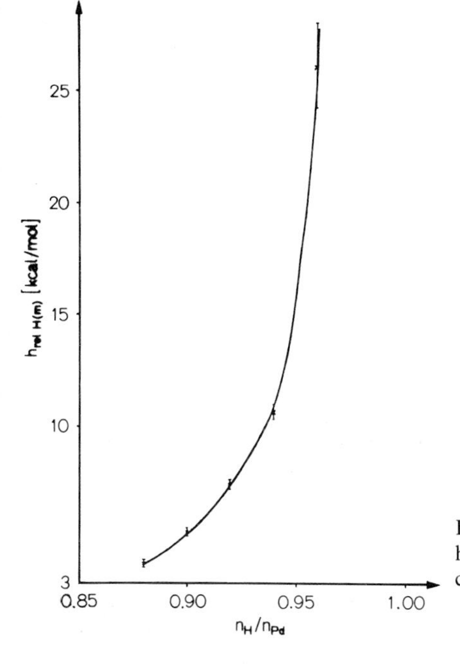

Fig. 4.8. Relative partial molar enthalpy of hydrogen in the metallic phase as a function of concentration (4.13)

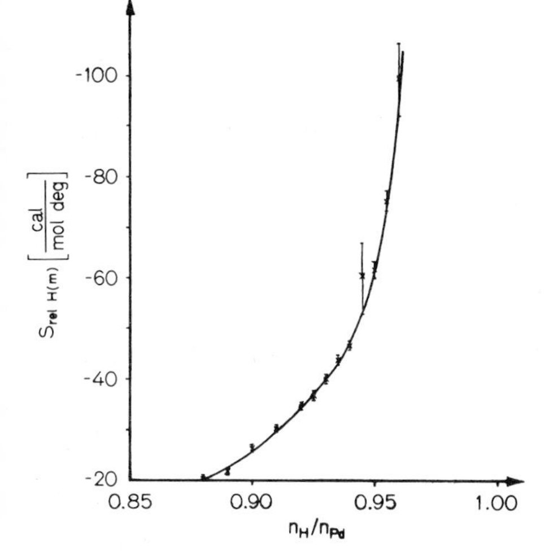

Fig. 4.9. Relative partial molar entropy of hydrogen in the metallic phase as a function of concentration (4.14)

values for the entropy are practically temperature independent. Plotting the relative enthalpies and entropies as a function of the concentration (4.8) with the parameter $(n_H/n_{Pd})_{max}=1$, a linear relationship is found in the atomic ratio range $0.88 \leqq n_H/n_{Pd} \leqq 0.94$.

The absorption of hydrogen in $Pd_{1-x}Au_x$ alloys $(0.057 \leqq x \leqq 0.684)$ was followed at 25 °C in the pressure range from 25 bar to 10 kbar [4.48]. Small samples of the alloys investigated (10–100 mg) were kept in contact with gaseous hydrogen at constant pressure and temperature conditions for time intervals sufficient to approach stationarity. The pressure vessel shown in Fig. 4.3 was used for this purpose. Before mass spectrometric analysis was carried out, the total device was quickly cooled to liquid nitrogen temperature. The results obtained could be satisfactorily presented by the relation

$$\frac{n_{\mathrm{H}}}{n_{\mathrm{Me}}}=a+b\log p_{\mathrm{H}_2} f_{\mathrm{H}2(\mathrm{g})}\,. \tag{4.15}$$

This equation is a generalization of (4.1), whereby the pressure of gaseous hydrogen is replaced by its fugacity. In terms of (4.10), the second and third terms on the right-hand side should be more or less constant if (3.14) has to be derived.

Interesting observations were made in the Pd–Rh–H system [4.49, 50]. Alloys with 20 and 30% of rhodium were reaching stoichiometry ($n_{\mathrm{H}}/n_{\mathrm{Me}}=1$) at much lower pressures than those expected for pure palladium. For alloys with 40% of rhodium, $n_{\mathrm{H}}/n_{\mathrm{Pd}}$ ratios of about 1.60 were observed. This has not been found in other alloys of palladium with components inert with respect to hydride formation (Ag, Au, Cu, Pt). Therefore, one can conclude that rhodium itself should form a hydride phase, and some estimations concerning the conditions required were given [4.50]. Unfortunately we were not successful in realizing this expectation. On the other hand, for alloys of 20 and 40% rhodium (see Table II in [4.50]) the concentrations of hydrogen achieved after exposure to 23 kbar of gaseous hydrogen were lower than those found at about 5 kbar. In addition to other considerations [4.50], the negative influence of the pure hydrostatic pressure may be responsible for this behavior, due to the positive values of the partial volume of hydrogen. This effect was discussed in details in Section 4.2.1.

In ordered and disordered $\mathrm{Pd_3Fe}$ alloys a considerable difference of hydrogen absorption was found [4.51]. The ordered alloys absorbed at the same pressures 5–10 times more hydrogen than the disordered one. This difference could not be explained in terms of a rigid band approach, where a parallel course is to be expected between the density of states and the hydrogen absorption capability. As the ordered alloy exhibits an approximately 30% lower γ-coefficient of the electronic specific heat [4.52, 53] than the disordered phase, a reduction of the hydrogen absorption should be expected, when long range order appears. The observed opposite behavior shows clearly that the density of states at the Fermi level is not the decisive criterion for the hydrogen solubility of palladium alloys. The explanation of the results observed may be the following [4.51]: In the ordered phase more "palladiumlike" interstitials are present—that is octahedral vacancies are surrounded by palladium particles. If we assume that such places are preferentially occupied by hydrogen particles, the difference observed can be understood. But this means that the nearest-neighbor environment is more essential for the hydrogen absorption

Table 4.1. Thermodynamic formation functions of nickel hydride and deuteride

System	Equilibrium pressure at 25 °C (atm)	Equilibrium fugacity at 25 °C (atm)	Standard-free energy of formation [kcal/mole H_2]	Standard enthalpy of formation [kcal/mole H_2]	Standard entropy of formation [cal/mole H_2 deg.]
$NiH_{0.5}$	3400 ± 70	$29{,}700 \pm 550$	5.64 ± 0.02	-2.10 ± 0.14	-25.4 ± 0.3
$NiD_{0.5}$	4010 ± 90	$46{,}600 \pm 1200$	5.86 ± 0.03	-2.10 ± 0.14	-26.2 ± 0.1

ability than the collective behavior, as represented for instance by the rigid band approach. Such a conclusion is equivalent to a new aspect to be taken into account in the theory of the Pd–H system.

In Pd–Pt alloys exposed to hydrogen pressures from 1 to 24 kbar at 25 °C [4.34], a considerable reduction of hydrogen solubility was observed when adding platinum to palladium. It is remarkable that platinum itself should be active to hydrogen, due to its electron vacancies in the *d*-band, and in reality no significance was found for a formation of platinum hydride even to 30 kbar of gaseous hydrogen. But it is interesting to remark [4.54] that a 70% Pt alloy systematically raised its electrical resistance in the pressure range 20–30 kbar of gaseous hydrogen, which could be caused by an uptake of hydrogen. Therefore it seems not improbable that a platinum hydride could be prepared at higher hydrogen activities than 30 kbar and perhaps in another temperature range than applied so far.

4.2.3 Nickel and Nickel Alloys

As it was already mentioned in Section 4.1.1, the first thermodynamic information about nickel hydride was achieved by determining the equilibrium between the decomposing nickel hydride and gaseous hydrogen [4.24, 25]. Similar measurements were later carried out with nickel deuteride [4.55]. Combined with calorimetry [4.56], all three principal formation functions—free energy, enthalpy and entropy—could be calculated [4.25, 55]. The results are summarized in Table 4.1.

No isotope effect has so far been noted in the enthalpy of formation [4.56], but the decomposition pressure and consequently the standard free energy of formation is higher for the deuteride than for the hydride phase [4.55]. The formation pressures of both phases exhibit a similar difference [4.29, 55]. This is the common feature of metal-hydrogen systems.

Absorption and desorption isotherms were recently determined by applying the high pressure device presented in Fig. 4.4. The curves obtained at temperatures of 25 and 65 °C are reproduced in Fig. 4.10.

According to the above results, nickel hydride is formed at 25 °C above 6 kbar of gaseous hydrogen, leading to a nearly stoichiometric hydride phase.

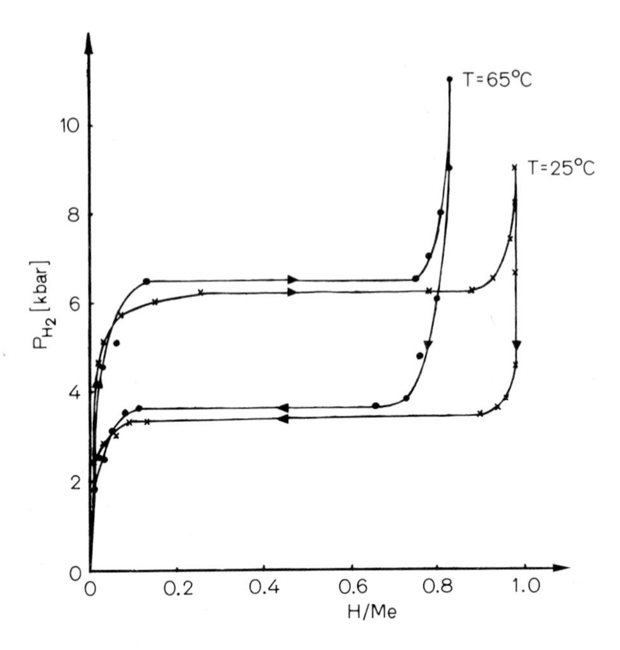

Fig. 4.10. Absorption and desorption isotherms of hydrogen in metallic nickel

Contrary to this, the decomposition to nickel takes place at around 3.4 kbar. Both data are in accordance with previous results [4.7, 8, 24, 25, 27]. Slightly higher values are observed for formation and decomposition pressures at 65 °C. In both temperatures a clear hysteresis exists, amounting in about 3 kbar difference between the formation and decomposition pressures. As this difference is nearly the same for both temperatures, one can conclude that a considerable distance to the critical region still exists. This problem was recently studied and the critical parameters (T_k, p_k) for the nickel hydride formation were estimated as $T_k > 623$ K and $p_k > 16.3$ kbar of gaseous hydrogen [4.57]. As a criterion for the critical region, the disappearance of the hysteresis was taken into account, whereby the formation and decomposition pressures were found from the course of the electrical resistance during the increase and decrease of the pressure of gaseous hydrogen. In our experiments the formation pressure of this phase could be followed in a unique way by the electrical resistance of the sample investigated [4.27, 29]—but we were not able to do it in respect to the pressure of decomposition. Therefore it seems probable that the exact critical parameters may be different from those mentioned above [4.57].

Systematic investigations of nickel alloys were carried out in the systems Ni–Cu–H, Ni–Fe–H, Ni–Mn–H, Ni–Co–H. Investigations to determine the absorption and desorption isotherms of the first alloys are now under way. An example of the results achieved so far is presented in Fig. 4.11 for an alloy $Ni_{0.9}Cu_{0.1}$ [4.58]. As compared with pure nickel (Fig. 4.10), much smaller hydrogen concentrations are obtained and the hysteresis behavior clearly indicated a closer approach to the critical region than in the Ni–H system.

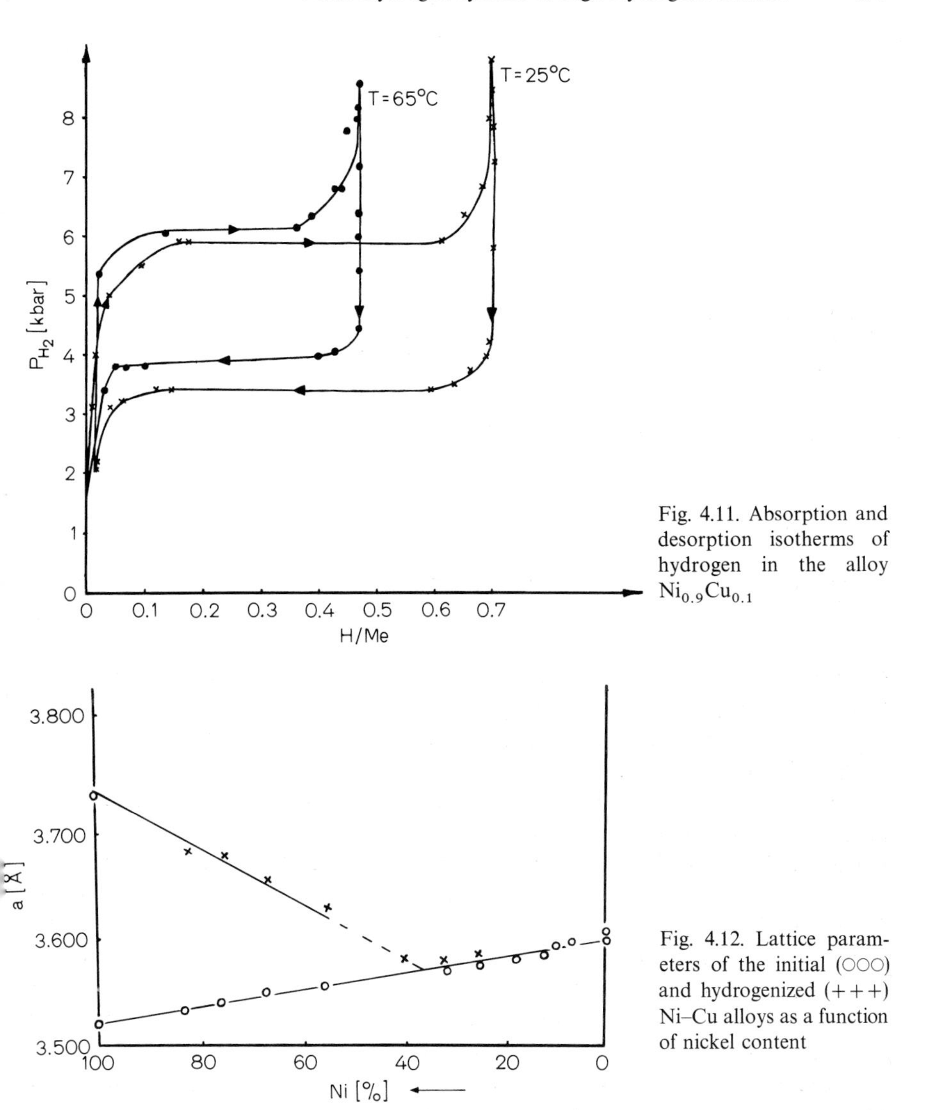

Fig. 4.11. Absorption and desorption isotherms of hydrogen in the alloy $Ni_{0.9}Cu_{0.1}$

Fig. 4.12. Lattice parameters of the initial (○○○) and hydrogenized (+++) Ni–Cu alloys as a function of nickel content

Another question is the closing of the two-phase region as shown in Fig. 4.12 for room temperature [4.59]. As can be seen, alloys with nickel contents above 40 at. % exhibit phase separation in the presence of hydrogen, whereby alloys with higher copper contents are always in the one-phase region. Similar curves were established at 25 °C for Ni-Fe-H [4.60] and Ni–Mn–H [4.61] systems, whereby the formation and decomposition pressures were determined from the course of the electrical resistance, when the pressure of gaseous hydrogen was incrementally increased and decreased. A remarkable difference exists between both systems. While the addition of iron to nickel shifts the

Table 4.2. Decomposition pressure and free energy of formation of hydrides in Ni–Fe alloys

at. % Fe	Decomposition pressure [kbar]	Free energy of formation [kcal/mole H_2]
0	3.40 ± 0.07	5.64 ± 0.02
1.20	3.50 ± 0.12	5.70 ± 0.04
2.11	4.00 ± 0.21	5.85 ± 0.10
4.22	4.80 ± 0.30	6.20 ± 0.10
9.6	9.70 ± 0.20	7.65 ± 0.05
16.25	15.70 ± 0.35	9.00 ± 0.09
21.4	17.20 ± 0.35	9.65 ± 0.07
25.8	19.60 ± 0.40	10.50 ± 0.09
30.2	21.30 ± 0.45	11.35 ± 0.10
36.5	23.10 ± 0.50	12.00 ± 0.12

formation pressure of hydrogen to higher values, an opposite effect is observed when adding manganese to nickel. This behavior makes it evident that the probable formation pressure for pure hydride will be much higher than that characteristic for pure nickel, but, conversely, manganese hydride can be expected to be formed at lower pressures.

The formation pressures for the hydrides in Ni–Fe alloys were determined by both electrical resistance and thermopower measurements [4.60, 62]. Following the procedure applied for nickel hydride and deuteride [4.25, 55], the standard free energies of formation were calculated [4.60] and are summarized together with the decomposition pressures in Table 4.2. The uptake of hydrogen in Ni–Fe alloys could be observed to 87.6% of iron, if the measurements were extended to 30 kbar at 25 °C. Similar investigations with deuterium led, generally, to a shifting of all formation and decomposition pressures to higher values [4.63] that coincide with the common results in metal-hydrogen systems. At room temperatures an active response of pure iron to gaseous hydrogen can be expected at pressures not lower than 35 kbar, which is beyond our present experimental facilities.

Formation of hydrides at 25 °C, manifested by discontinuous changes of the electrical resistance, was observed up to 40% of Co [4.63] in Ni–Co–H and Ni–Co–D systems. At this concentration, a significant hysteresis between the formation and decomposition pressures was still found. Therefore no estimation could be given for the closing of the phase separation gap. Comparison of the same atomic percentages of Fe and Co in the nickel matrix showed that formation pressures were always lower than in the Ni–Fe–H system. Thus one can conclude that pure cobalt is more easily transferred to the hydride phase than pure iron. This corresponds to the sequence of solubilities of hydrogen in pure metals at low pressures of gaseous hydrogen.

Because of the discouraging experience with nickel hydride in the determination of the decomposition pressure followed by resistance changes [4.27] and the good results obtained with thermopower measurements [4.64], the same method was applied to Ni–Pd alloys [4.65]. The formation pressures of the hydride phases are distributed continuously between the values characteristic for both pure components and were determined by discontinuities of the thermopower, measured in situ in high pressure conditions. No further thermodynamic investigations of this system have been carried out.

4.2.4 Other Systems

The Cr–H System

Chromium hydride was, like nickel hydride, prepared first by an electrochemical method [4.66]. Its thermodynamic characteristics were limited to low temperature heat capacity measurements [4.67] and preliminary results of the decomposition enthalpy [4.68]. The free energy of formation could be estimated [4.67] from both quantities. A direct determination of this quantity was difficult because the electrochemical conditions of formation [4.66] are even more unrealistic to any equilibrium measurement than in the case of nickel hydride [4.57, 58]. At normal pressure and temperature conditions, metallic chromium absorbs negligible amounts of gaseous hydrogen. An extrapolation to concentrations 10^{-1}–10^{0} in atomic ratios n_{H}/n_{Cr} would require enormously high activities of hydrogen. But as chromium undergoes a reconstructive transition during the hydride formation, one could eventually expect more realistic pressures sufficient for these purposes. Therefore, both the decomposition and formation of chromium hydride seemed to be possible at high pressures of gaseous hydrogen only. The simplest idea was to use the device applied previously for the determination of the equilibrium between nickel hydride and gaseous hydrogen [4.7, 25]. But this failed because chromium hydride did not decompose at room temperature with a reasonable velocity. The lowest acceptable temperature was 150 °C and here a new device had to be developed, where considerable troubles with the sealing of hydrogen had to be overcome. A 10 g sample of chromium hydride prepared electrochemically was placed in a brass cell with a minimal dead volume. The decomposition of the hydride created a high pressure of gaseous hydrogen, which was measured by a manganine gauge at room temperature. The reaction vessel and the manganine cell were connected by a steel capillary. The details are described in [4.69]. After several experiments with a large percentage of failures, the equilibrium pressure above solid chromium hydride at 150 °C was determined as equal to 3160 ± 150 atm [4.69, 70].

About three weeks was necessary to reach equilibrium; the pressure constancy during the last 5 days was taken as the criterion of stationarity. A plateau of the decomposition pressure was found in the concentration range $0.2 < n_{H}/n_{Cr} < 0.6$. Higher concentrations could not be achieved because the high pressure conditions were obtained by the partial decomposition of the hydride phase created. The free energy of formation was calculated from the

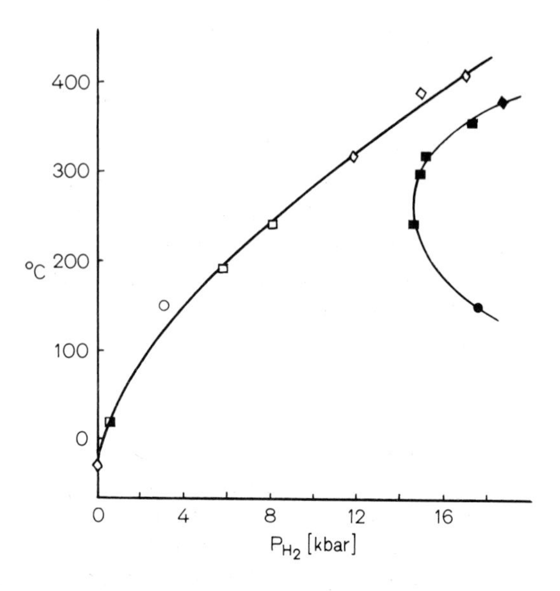

Fig. 4.13. Formation (■●) and decomposition (□◇○) curves for chromium hydride in a P_{H_2}, T plot [4.75]

above value of the decomposition pressure as equal to 4130 ± 80 cal/mole H_2 at standard conditions (25 °C and 1 atm) [4.69]. This corresponds to a decomposition pressure of chromium hydride at 25 °C as equal to 710 ± 90 atm H_2. Combining the above value of the free energy of formation with the recently determined enthalpy of formation [4.71] results in -25.9 ± 2.5 cal/mole K as the corresponding value of the entropy.

As could be expected, the formation of chromium hydride takes place at a much higher hydrogen pressure than does the decomposition. Thus at 150 °C a pressure range of 17,400–18,200 atm was found as necessary, if the large increase of the electrical resistance in a time interval of 30 min is chosen as the criterion of the phase transition [4.72]. For chromium deuteride the equivalent formation pressure was established as lying between 18,100 and 18,500 atm of gaseous deuterium at 150 °C [4.73]. The above results were summarized in a recent report [4.74]. Of course the formation of both phases was confirmed by x-ray and mass-spectrometric investigations.

An extended study of the formation and decomposition conditions of chromium hydride was carried out recently [4.75]. Changes of the electrical resistance of electrolytical chromium samples ($0.4 \times 1.5 \times 5.0$ mm^3) served as a criterion of the phase transitions. Unfortunately, as in the previous case [4.57], no details concerning the high pressure apparatus were given. Both isothermal and isobaric observations were performed in the temperature range 200 to 400 °C (accuracy ± 5 °C) and the pressure range 7 to 20 kbar (accuracy ± 0.5 kbar) of gaseous hydrogen.

Figure 4.13 presents the results obtained. Both curves represent the beginning of the formation and decomposition processes of chromium hydride. The curve characterizing the decomposition can be described by the analytical relation [4.75]

$$T = 0.84\,P^2 + 38.7\,P - 5.8\,, \tag{4.16}$$

where P denotes the hydrogen pressure in kbar and T the temperature in °C. Measurable desorption rates of hydrogen were observed even at -30 °C, contrary to chromium hydride obtained by the electrochemical method. Because no stationarity was achieved, the true values may differ from the reported data as can be seen, for example, in Fig. 4.13 for the 150 °C value of the decomposition pressure. The stationary value [4.69, 70] seems to be lower than that obtained under nonstationary conditions [4.75].

The Mn–H System

Nickel and chromium exhibit a lower solubility of hydrogen than does manganese. Therefore it seemed astonishing that the last metal did not form a hydride phase during the cathodic codeposition or cathodic saturation with hydrogen. The treatment of metallic manganese with gaseous hydrogen at room temperature up to 22 kbar gave no evidence of a hydride formation [4.76]. These results led us to a systematic investigation of Ni–Mn alloys of different compositions [4.61, 76]. As discussed above, the course of the miscibility gap clearly indicates a possible formation of a manganese hydride at pressures of gaseous hydrogen lower than those characteristic for nickel hydride. On the assumption that the main problem in this hydride formation consisted of kinetic barriers due to surface contamination, a higher temperature was chosen. Finally a successful synthesis was carried out at about 300 °C and a pressure of gaseous hydrogen of 14 kbar [4.76]. A hexagonal phase (h.c.p.) was formed, manifested by a discontinuous change of the electrical resistance of spectrographically standardized Johnson-Matthey samples ($0.5 \times 1 \times 15$ mm^3). Five months later an independent result was published [4.77] in which the formation of manganese hydride was carried out at 18 kbar of gaseous hydrogen and a temperature of 350 °C. In both papers the composition of $MnH_{0.8-0.82}$ was reported as the maximal concentration of hydrogen found. As could be expected from experience with nickel hydride [4.8, 26, 27], the above formation pressures for manganese hydride are higher than the minimal values observed, if stationarity is achieved at the given temperature/pressure conditions. Table 4.3 summarizes the results of the minimal formation pressures of the h.c.p. manganese hydride observed at different temperatures [4.61].

Table 4.3. The formation pressures of the h.c.p. manganese hydride, observed for various temperatures

T [K]	448 ± 10	500 ± 10	577 ± 10	673 ± 10	729 ± 10
P_{H_2} [kbar]	5.6 ± 0.3	6.0 ± 0.3	7.6 ± 0.3	9.6 ± 0.3	10.4 ± 0.3

The values given were determined from the discontinuities of the electrical resistance accompanying the formation process. A similar method was found inadequate for the determination of the decomposition pressures because uncertain and nonreproducible data resulted from this procedure. Therefore the complete thermodynamic characteristics of the manganese hydride remain an open question.

4.3 Electronic, Structural, and Transport Properties

4.3.1 Electrical Resistance

The introduction of interstitial hydrogen particles into a metallic lattice causes several consequences in respect to its electrical resistance. New scattering centra are created in metallic hydrides when interstitial vacancies are occupied by hydrogen. This leads to an increase of the residual resistance which can manifest itself even at room temperatures. In many cases this is the most important contribution to be considered in the overall changes observed. Approaching stoichiometry, a new ordered phase is formed, causing a decrease of the low temperature resistivity. Later the hydrogen introduced can influence the band structure of the electrons. In the transition metals discussed in this chapter, of special importance is the filling up of the d-band vacancies which reduces the s–d scattering. In addition to the reasons mentioned, the effective electrical resistance measured may include changes due to creation of cracks, microcrystallites and other mechanical damage. Such effects may often occur during phase transitions especially when lattice reconstruction takes place. Contrary to the previous changes—which exhibit reversible character that seems to be a unique function of the hydrogen concentration—the last effects are irreversible and usually not reproducible. Because all metal-hydrogen systems investigated so far in the high pressure region exhibit metallic conduction, the conventional four-probe potential technique was used for measurements of the electrical resistance in the presence of active gaseous hydrogen. Usually several samples—sometimes more than ten—could be measured simultaneously. In fact the electrical resistance, the first property we investigated in situ systematically, possess the greatest experimental simplicity. Mostly thin foils or wires (10^0–$10^2\,\mu$ thickness) with dimensions of several mm [about $1 \times (5$–$15)\,\mathrm{mm}^2$] were used. If necessary the surface of the samples was specially activated in order to accelerate the absorption process. As referred to above, in many cases the changes of the electrical resistance indicated phase transitions caused by gaseous hydrogen. As a typical example, Fig. 4.14 presents the course of the relative electrical resistance of two Ni–Mn alloys as a function of hydrogen pressure [4.61].

The alloy with 9.3 at. % Mn exhibits small changes of the electrical resistance up to about 5.5 kbar. This range is followed by a rapid increase of

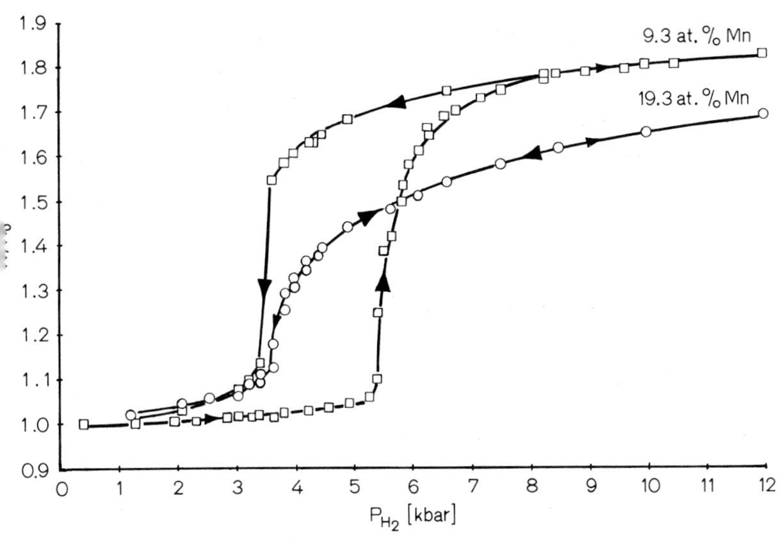

Fig. 4.14. Relative electrical resistance of Ni–Mn alloys as a function of the hydrogen pressure. R_0 = resistance at normal pressure in a neutral environment, R = resistance at pressure P_{H_2} of gaseous hydrogen

about 70% of the initial value, reaching a plateau around 8 kbar. Decreasing the pressure, one observes a clear hysteresis, whereby the discontinuous decrease of the resistance, characteristic for the decomposition of the hydride, takes place at hydrogen pressures around 3.2 kbar. Contrary to this the alloy with 19.3 at. % Mn shows no hysteresis and the changes involved are smoother than in the previous alloy. As discussed in Section 4.2.4, the last alloy lies outside the miscibility gap of the system considered. Similar curves were observed in other metals forming hydride phases [4.26, 27, 29, 34, 48–50, 54, 63, 72, 74]. In many cases, irreversible mechanical changes of the samples investigated contribute to the changes of the electrical resistance.

Even when no phase transition is to be expected, the changes of the electrical resistance can be treated as indicative of an uptake of hydrogen. Of course one has to distinguish between the influence of pure hydrostatic pressure and the contribution due to hydrogen. In doubtful cases an independent run in an inert gas or liquid medium in the same pressure range must be performed. The electrical resistance still remains the only property to be followed by *in situ* measurements in the highest pressure range available. Figure 4.15 presents such an example for a sequence of Ni–Fe alloys [4.60] where a clear response to hydrogen activity first starts above 20 kbar.

The alloy with the highest nickel content (34.1 at. % Ni) behaves similar to pure nickel—that is a small resistance increase is observed up to a certain pressure where a considerable increase of the electrical conductance takes place. This alloy is the last one, in the sequence represented on Fig. 4.15, which exhibits the fcc structure. The alloy with 71% of iron comes in contact with the

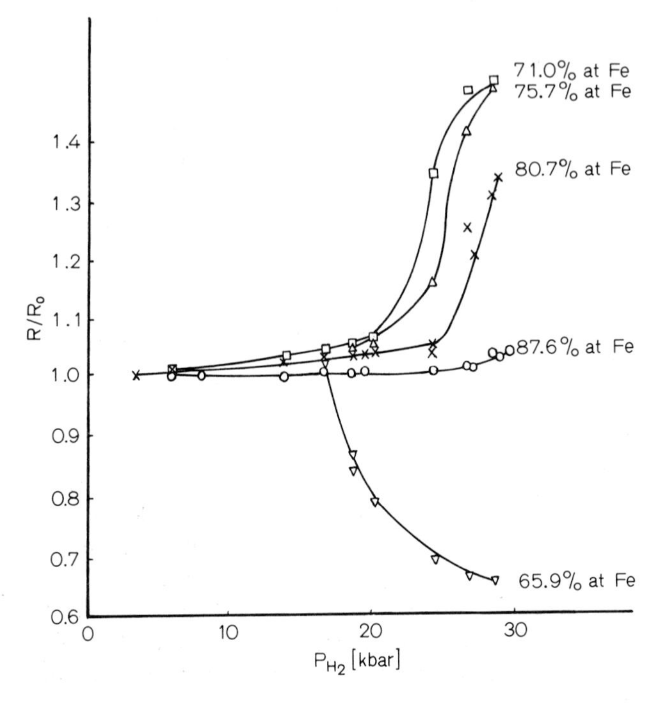

Fig. 4.15. Relative electrical resistance of Ni–Fe alloys as a function of the hydrogen pressure

bcc alloys of higher iron content. Simultaneously with this structure change, an opposite course of the electrical resistance is found. The discontinuities of the resistance changes are shifted to higher hydrogen pressure when increasing the iron concentration. From Fig. 4.15 it is obvious that hydrogen activities much higher than available now are required to reach a pressure-independent resistance.

For Ni–Fe alloys with 0.23 and 0.50 at. % iron it could be shown that the changes of the electrical resistances observed during the hydride formation are mainly caused by the increase of the residual resistivity. Contrary to this, in alloys with 4.22 and 9.6 at. % of iron the phononic contribution was decisive [4.63]. Because the alloys presented in Fig. 4.15 could not be taken from the pressure vessel without considerable hydrogen loss, an investigation at low temperatures was not possible; therefore a more detailed discussion of these curves is beyond the scope of this chapter.

An impressive example of the effectiveness of electrical conductance measurements as an indicator of hydrogen absorption is given in Fig. 4.16. The partially ordered Pd_3Fe alloy absorbs hydrogen at much lower pressures than the same alloy in the disordered state. For instance the relative resistance value of two is reached in the first case at pressures around 100 atm whereas the second alloy approaches the same relative value above 10 kbar of gaseous hydrogen. The explanation of this difference was given in Section 4.2.2. One can expect that residual resistance changes are mainly responsible for the above behavior.

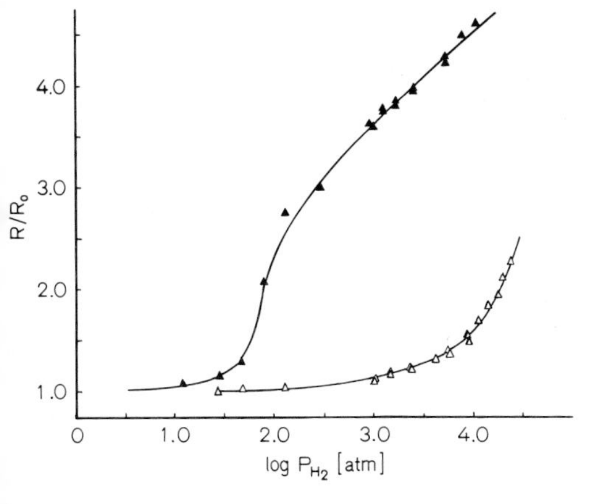

Fig. 4.16. Relative electrical resistance of ordered (▲▲▲) and disordered (△△△) Pd_3Fe alloys as a function of logarithm of the hydrogen pressure

In pure palladium the formation of the hydride phase is accompanied by a large increase of the electrical resistance, which at 25 °C rises above 80% of its initial value [4.31, 46, 48]. A maximum is reached at about 8 atm of gaseous hydrogen [4.31, 48]. Further increase of the hydrogen pressure causes a continuous decrease of the resistance to the range of 15 to 17 kbar, depending on the purity of the sample and the value of the hydrogen-free metall [4.31, 32, 42, 48]. Similar behavior can be achieved by intensive cathodic saturation of metallic palladium [4.78].

A systematic study of the electrical resistance as a function of temperature and hydrogen concentration was carried out in Pd–H, Pd–Ag–H and Pd–Au–H systems [4.31, 48, 79], in which previous results with low hydrogen contents [4.80, 81] were considerably extended by treating samples in high pressures of gaseous hydrogen. An example of the results achieved is presented in Fig. 4.17 [4.31]. The maximum of the electrical resistance is shifted to higher pressures of gaseous hydrogen when increasing the concentration of silver which simultaneously becomes shallower. The interpretation of the results achieved was based on the following model: According to Matthiessen's rule, the resistivity of the metal with hydrogen $\varrho_{\text{Me-H}}$ was divided into the residual ($\varrho_{\text{Me-H,O}}$) and the temperature-dependent part ($\varrho_{\text{Me-H,T}}$)

$$\varrho_{\text{Me-H}} = \varrho_{\text{Me-H,O}} + \varrho_{\text{Me-H,T}} . \tag{4.17}$$

Furthermore the low-temperature part could be split into metallic (disordered) ($\varrho_{\text{Me,O}}$) and hydrogen ($\varrho_{\text{H,O}}$) contributions and moreover each term could be treated as a sum of *s–s* and *s–d* scatterings:

$$\varrho_{\text{Me-H}} = \varrho^{sd}_{\text{Me,O}} + \varrho^{ss}_{\text{Me,O}} + \varrho^{sd}_{\text{H,O}} + \varrho^{ss}_{\text{H,O}} + \varrho^{sd}_{\text{Me-H,T}} + \varrho^{ss}_{\text{Me-H,T}} . \tag{4.18}$$

The filling up of the *d*-band was assumed at 0.6 atomic ratio $n_{\text{H}} + n_{\text{Me}}/n_{\text{Pd}}$ where n_{H}, n_{Me}, and n_{Pd} denote the number of particles of hydrogen, silver, or gold and palladium.

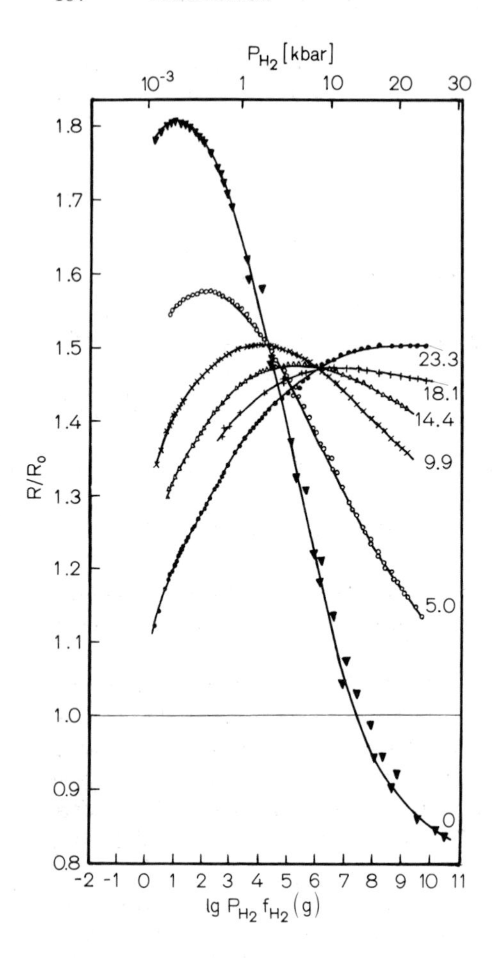

Fig. 4.17. Relative electrical resistance of Pd and $Pd_{1-x}Ag_x$ alloys as a function of logarithms of the hydrogen fugacity. The numbers indicate the values of x

An essential feature is the possible occupation of both octahedral and tetrahedral interstitial positions by hydrogen particles, whereas at low temperatures, tetrahedral positions only are filled up by hydrogen in the concentration range from zero to the maximal value of the d-band filling. At high temperatures the same concentration range is preferred by the hydrogen particles in the octahedral positions only. In the pure hydride phase, that is, for concentrations above the d-band filling, both positions can be partially occupied with different probabilities for the low and high temperature modifications. From the above assumptions the following relation results for the residual resistivity due to hydrogen $\varrho_{H,O}$:

$$\varrho_{H,O} = \varrho_{H,O}^{ss} + \varrho_{H,O}^{sd} = A \left\{ \left(\frac{n_H}{n_{Pd}}\right)^{oct} \left[1 - \left(\frac{n_H}{n_{Pd}}\right)^{oct}\right] + \left(\frac{n_H}{n_{Pd}}\right)^{tet} \left[1 - \frac{1}{2}\left(\frac{n_H}{n_{Pd}}\right)^{tet}\right] \right\}, \tag{4.19}$$

where the constant A includes the scattering matrix element and the indices "oct" and "tet" denote the octahedral and tetrahedral interstitials. Of course the identity holds

$$\frac{n_H}{n_{Pd}} = \left(\frac{n_H}{n_{Pd}}\right)^{oct} + \left(\frac{n_H}{n_{Pd}}\right)^{tet}. \tag{4.20}$$

A rich variety of experimental facts can be explained, at least qualitatively, by this simple model. Let us mention here some of these results.

a) The appearance of maxima in the room temperature resistance of the $Pd_{1-x}Ag_xH$ systems ($x=0$ to 0.266) at nearly the same atomic ratio.

b) The higher residual resistivity of the low temperature form than for the high temperature modification.

c) The existence of a low temperature anomaly which occurs because of the different resistivities of both forms of the hydride phase.

d) The disappearance of the low temperature anomaly at higher silver contents.

e) In pure palladium the resistance anomaly decreases for hydrogen contents more distant from the optimum value of 0.75 in the atomic ratio n_H/n_{Pd}.

A definitive check of the model presented would require extensive neutronographic investigations at low temperatures and high hydrogen contents in order to establish the existence of tetrahedral occupancy by hydrogen particles which is an essential consequence of this model. One early report supported and suggested for the first time this point of view [4.82]. More recent papers disagree with this conclusion [4.83, 84] and, therefore, more definitive results are required.

Of special importance is the investigation of the low temperature anomaly, discovered previously by heat capacity [4.85] and electrical resistance measurements [4.86] and extended later to other properties [4.87]. A palladium hydride prepared in high pressure conditions exhibited no anomaly [4.88] (Fig. 4.18). When desorbing hydrogen from the sample with the highest hydrogen concentration ($n_H/n_{Pd}=0.89$) in a controlled way, the electrical resistance anomaly appears and again becomes of negligible value in the low concentration range. These observations led later in an unexpected way to the discovery of superconductivity in concentrated palladium hydride samples [4.89, 90]. The curves presented in Fig. 4.18 and later measurements [4.48, 79] can be explained by the model described above.

The alloy system Pd–Pt was investigated systematically in the pressure range from 10–30,000 atm at 25 °C in which 15 different alloys ranging from 2.79 to 70 at. % of Pt were taken into account [4.34, 54]. Similar to the Pd–Ag system, the addition of the second metal to palladium causes a shift of the maxima of the electrical resistance to higher hydrogen pressures. An interesting relation, presented in Fig. 4.19, is fulfilled here. The positions of the maxima, expressed by the corresponding hydrogen fugacity, are a linear function of the

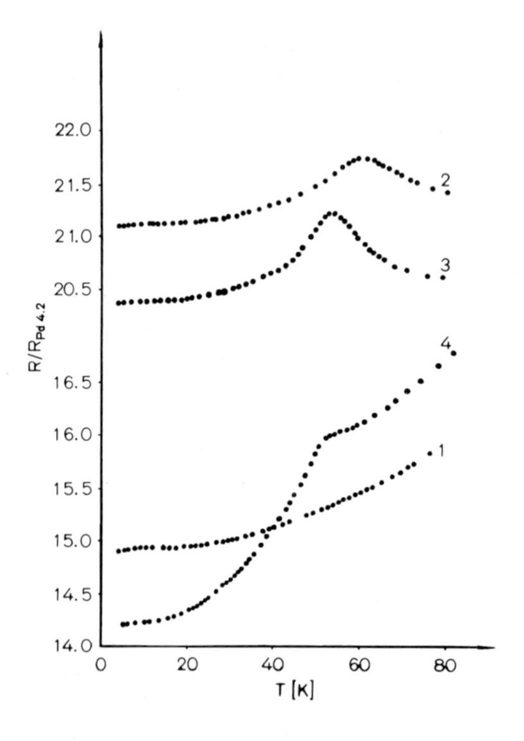

Fig. 4.18. Relative electrical resistance of a palladium hydride sample as a function of temperature. $R_{Pd\,4.2}$ = resistance of the pure sample at 4.2 K (without hydrogen). n_H/n_{Pd} = 0.89 (1); 0.74 (2); 0.67 (3); 0.58 (4)

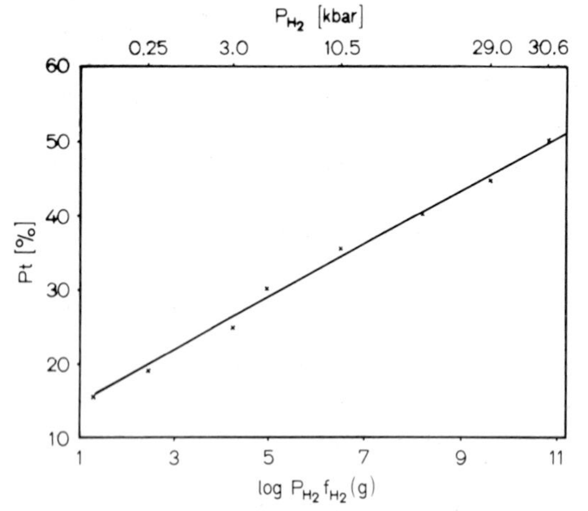

Fig. 4.19. Composition of the Pd–Pt alloys as a function of the position of the electrical resistance maxima expressed by the corresponding logarithm of the hydrogen fugacity at 25 °C

platinum concentration of the alloys. Because this result is equivalent to a proportionality to the chemical potential of gaseous hydrogen, a thermodynamic significance of the maxima observed seems to be very probable. The extrapolation for an alloy of 55 % of platinum shows that a pressure of nearly 39 kbar would be necessary if the maximum of the electrical resistance should be observed in this alloy.

The formation of nickel hydride is preceded by a small (few percent) increase of the electrical resistance [4.26, 27, 60] which is probably due to the residual contribution only. The hydride phase itself exhibits an electrical conductance 15% larger as compared with pure nickel [4.27, 60]. This is in accord with previous measurements of electrochemically prepared samples at normal conditions [4.91, 92] and at unidimensional pressures up to 14 kbar [4.93]. The partially filling of the *d*-band by hydrogen electrons as well as the disappearance of the ferromagnetic resistance anomaly [4.94] can be responsible for the behavior observed. But a more quantitative treatment of this concept does not support this observation [4.91], especially when the electronic heat capacity is taken into account [4.95]. Electronic energy band calculations confirm this point of view [4.96]. The decomposition of nickel hydride is followed by irreversible changes of the electrical resistance which never reaches the initial value of the starting metal. This is due to mechanical damage accompanying the recovery of the crystal lattice during the hydrogen loss.

4.3.2 Superconductivity

Because superconductivity in metal-hydrogen systems is treated separately in Chapter 6 by *Stritzker* and *Wühl*, we shall restrict ourselves to the application of high pressure technique only. One of the unforeseen indirect consequences of the high pressure research with gaseous hydrogen was the discovery of superconductivity in palladium hydride by *Skośkiewicz* in 1972 [4.89]. It resulted during the continued investigation of the low temperature anomaly in the electrical resistance of palladium hydride and hydrides of Pd–Ni alloys [4.90]. As shown in Fig. 4.18, in high concentrated palladium hydride samples the low temperature electrical resistance anomaly disappears [4.88]. We can say now that if the hydride with the hydrogen content $n_H/n_{Pd}=0.89$ could be cooled 1 K lower, the superconductivity of palladium hydride would have been known already in 1968!

The first investigations of the superconductivity of this phase were carried out on samples saturated by cathodic hydrogen [4.89, 90]. Later, preparations in high pressure conditions were performed, either by removing the hydrogenized samples from the pressure vessel at lower temperatures [4.97] or by making use of the *manostatic* device, presented in Fig. 4.3, and allowing the measurements in helium temperatures under in situ conditions [4.39]. Having the *manostatic* device at room temperature with pressures of about 11 kbar of gaseous hydrogen and about 12.5 kbar of gaseous deuterium and closing at these conditions, the cooling down to helium temperatures resulted in a solidification of hydrogen which exhibited a hydrostatic pressure of about 3 kbar as estimated from P, V, T, data [4.39]. Thus the superconductivity was measured here under the indicated hydrostatic pressure transmitted by solid hydrogen. On the other hand, one could open the *manostatic* device at liquid nitrogen temperature and release the hydrogen pressure. Assuming the same

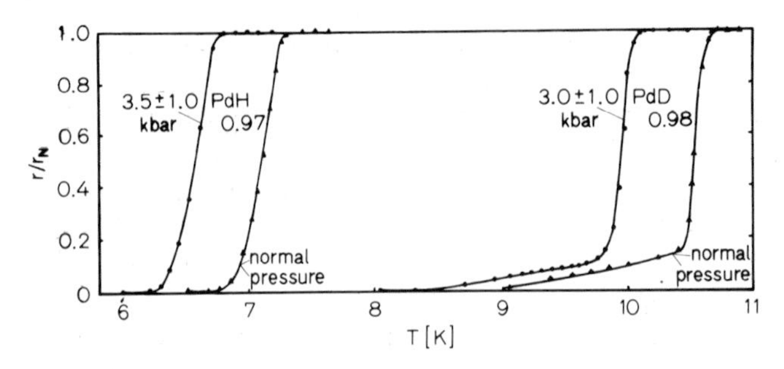

Fig. 4.20. Effect of hydrostatic pressure on superconductive transitions of palladium hydride and deuteride. r = resistance at the given temperature; r_N = residual resistance of the sample in the normal state

hydrogen concentrations of the samples investigated, both before and after pressure release, one was able to remeasure the transition temperatures and in this way determine the pressure effect on the superconductive transitions. The results are shown in Fig. 4.20 [4.39].

As can be seen, the hydrostatic pressure shifts the transition temperatures to lower values. The difference between the critical temperatures and the normal pressure values can serve for the calculation of the pressure coefficient $d \ln T_c/dp$ which equals -0.02 kbar^{-1}. This value is higher than that obtained either by using helium as the pressure transmitting medium [4.98] or by performing the measurement under unidimensional pressure [4.99]. This difference may be caused by the inaccurate estimation of the pressure at helium temperatures [4.39]. In Fig. 4.20 the isotope effect of the superconductivity is clearly indicated. At the same $n_{H(D)}/n_{Pd}$ ratios, the transition temperatures for the deuterides are higher by 1 to 2°.

In comparison with other preparation methods, the high pressure technique has the advantage of leading to uniform hydrogenized samples, thus affording the opportunity for a unique comparison of measurements carried out in different laboratories. For instance the results obtained by *Schirber* at pressures up to 5 kbar of gaseous hydrogen [4.100] (Fig. 4.1) in a device applied earlier to gaseous helium [4.101] disagree considerably when combining the absorption isotherm for 25 °C (Fig. 4.5) with the dependence of critical temperatures as a function of pressure given in [4.39] (Fig. 4.4). One simply comes to the conclusion that *Schirber*'s critical temperatures correspond to palladium hydride samples of higher concentrations than those resulting from equilibrium values at ~24 °C for the pressures of hydrogen indicated. The reason of this discrepancy is obvious if one takes into account that *Schirber* cooled his samples slowly to near 77 K at high hydrogen pressures. During this process, further uncontrolled uptake of hydrogen occurred and therefore the results presented in his figure cannot be compared with equilibrium values. Samples hydrogenized at high pressure conditions exhibit a much narrower transition to the superconducting state than those prepared by other methods [4.97] which is the simple consequence of the above-mentioned uniformity of the concentration profile.

Recently the high pressure technique was applied to charging thin palladium films [4.102]. Films with thicknesses of about 2000 Å seemed to exhibit behavior similar to that of bulk samples in respect to the sharpness of the transition to the superconducting state; this is contrary to films of 320 Å thickness which, when exposed to 1.5 kbar of gaseous hydrogen, remained in the normal state down to temperatures near 1 K. Unfortunately no direct determinations of the hydrogen concentrations were performed. The same charging technique was used for preparation of samples for tunneling experiments [4.102].

Recent investigations [4.103] of new palladium alloys led to the appearance of superconductivity at lower minimal concentrations of hydrogen than in previous measurements in pure palladium. Such a positive effect was stated by adding Cu, Ag, or B to palladium. Some Pd–Ni alloys showed the same property [4.65].

Let us mention finally that the first enhancement of superconductivity by the addition of hydrogen was found by *Satterthwaite* and co-workers [4.104, 105] in the hydride and deuteride of composition Th_4H_{15} prepared at higher gaseous pressures and exhibiting a higher transition temperature than pure thorium. Now other examples are known in which the introduction of hydrogen leads to higher transition temperatures than those known for the same metals without hydrogen [4.106].

4.3.3 Thermopower

The in situ measurement of the thermoelectric power requires the creation of a temperature gradient in the sample investigated. This was done by heating one end of the sample with a manganin heater; the other end was kept at the temperature of the high pressure vessel [4.64, 107]. The absolute thermoelectric power of palladium hydride at room temperature is shown in Fig. 4.21 at pressures up to 15 kbar of gaseous hydrogen [4.64]. A linear relationship similar to the previous measurements of the relative electrical resistance in the same pressure range [4.31, 32, 34, 42, 48] exists between the thermopower and the logarithm of the hydrogen fugacity. Therefore a linear dependence between the absolute thermopower and the relative electrical resistance can be checked [4.107]. The course presented in Fig. 4.21 is clear evidence of a further uptake of hydrogen by palladium hydride in the pressure range indicated. This behavior is completely different from the thermopower and electrical resistance measurements in nickel hydride [4.64, 107, 108]. The stationary formation of this phase at 25 °C manifests itself by a discontinuous jump of the absolute thermoelectric power from about $-17\,\mu V/deg$, characteristic for the α phase, to slightly positive values for the β phase.

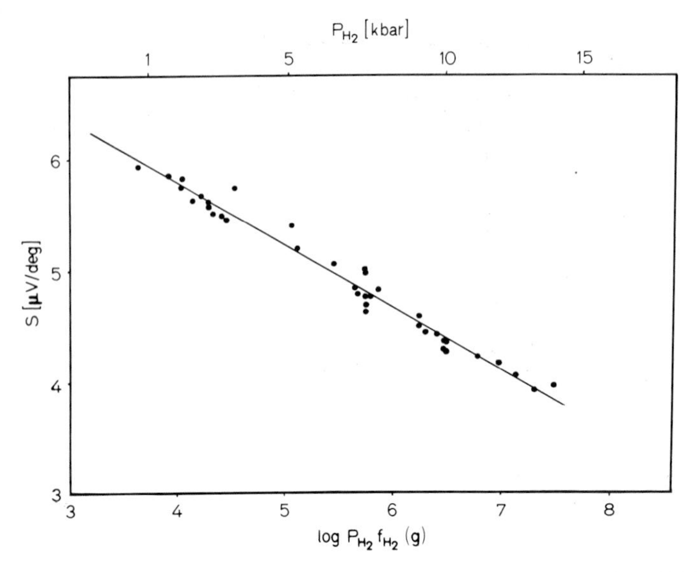

Fig. 4.21. Absolute thermoelectric power of palladium hydride at about 300 K as a function of the hydrogen pressure

Further increase of the hydrogen pressure is practically ineffective in respect to this property. The electrical resistance [4.60, 36, 109] shows similar insensitivity in the same pressure range. The difference between the Pd–H and Ni–H systems can be explained if one takes into account the absorption curves presented in Figs. 4.5 and 4.10. Nickel goes over to a nearly stoichiometric hydride phase ($n_H/n_{Ni} \sim 1$), quickly exhausting the available octahedral interstitial vacancies [4.110]. This fact makes plausible the course of the electrical resistance and thermopower at higher hydrogen pressures where no significant changes of the hydrogen content are taking place, and therefore the course observed is a pure hydrostatic effect only. Furthermore, measurements on samples with concentrations lower than those of stoichiometric hydrogen are probably representative for two-phase mixtures of the saturated α phase and the stoichiometric hydride. Such a simple concept is effective in interpretation of the thermopower for the Ni–H system as a function of the mean hydrogen concentration [4.64, 107, 111, 112] as opposed to the rigid band approximation which would require a uniform distribution of the hydrogen in the nickel lattice. The same concept explains the experiments on the electrical resistance which also make the existence of a two-phase mixture evident. Let us remark that the absolute thermopower of stoichiometric nickel hydride is very close to that of metallic copper, which would support the concept of the similarity of the electronic structure of both components.

Contrary to this, the palladium lattice absorbs hydrogen from the saturated α phase (about $n_H/n_{Pd} \sim 0.01$ at 25 °C [4.46]) to full stoichiometry ($n_H/n_{Pd} = 1$) with a uniform distribution of the hydrogen particles in the metallic matrix.

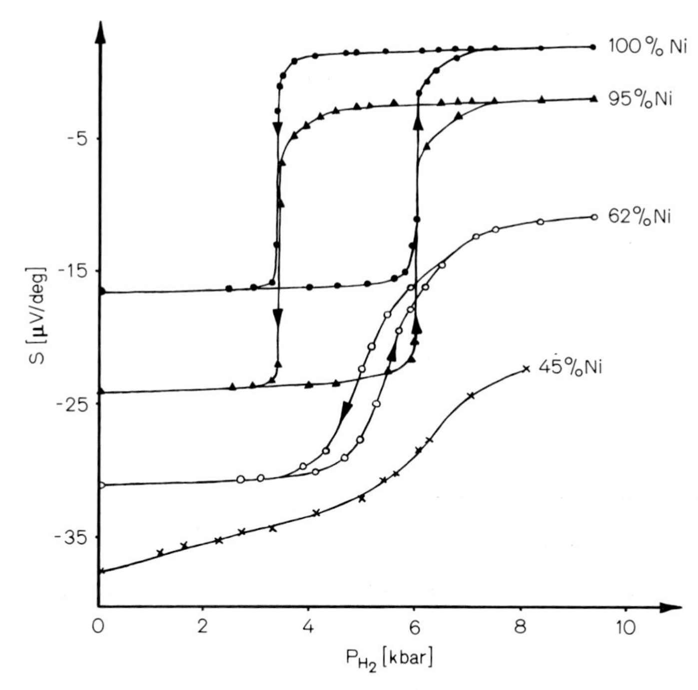

Fig. 4.22. Absolute thermoelectric power of Ni and Ni–Cu alloys as a function of the hydrogen pressure at 30 °C

Such a distribution makes the passing of the electrical resistance through a maximum [4.31–34, 42] and a further continuous decrease plausible. In the same way the curve in Fig. 4.21 can be understood. Previous explanation of the resistance maxima by co-conduction of the absorbed hydrogen [4.17] seems to be out of date, as a simple approach to an ordered stoichiometric hydride phase is sufficient for an interpretation.

In Ni–Cu alloys, the course of the absolute thermopower as a function of hydrogen pressure fulfills the expectation from pure nickel and the phase characteristics (see Sect. 4.2.3). The results obtained at 30 °C for nickel and three alloys are presented in Fig. 4.22 [4.108]. A clear hysteresis behavior is shown for 100, 95, and 62 at. % of nickel. The alloy with 45 % of nickel traces the same values with both increasing and decreasing hydrogen pressures. Thus the closing of the miscibility gap occurs at a higher nickel content than that shown in Fig. 4.12, which is simply due to the difference in temperature (Fig. 4.12 corresponds to a lower temperature than does Fig. 4.22).

An improved device for thermopower investigations which allowed the simultaneous measurement of eight samples with both thermopower and electrical resistance determinations was used in the Ni–Fe–H alloy systems [4.62, 63]. Besides pure nickel, ten alloys in the concentration range of 4 to 75 at. % Fe were investigated. Good agreement was found with previous resistance measurements [4.60, 63] concerning the transition pressures to hydride phases. An anomaly was found for two invar alloys (61.4 and 65.9% Fe) which were the only examples with large pure hydrostatic changes of the thermoelectric power. No influence of hydrogen on pure iron up to 30 kbar was confirmed by these observations. It seems too early to carry out a detailed discussion as the hydrogen concentration, and the magnetic characteristics of the samples investigated are as yet unknown.

4.3.4 Magnetic Measurements

More involved are measurements of the magnetic moment in situ conditions. The main difficulty is the limited working volume available in the high pressure device presented in Fig. 4.1. The even smaller working volumes of the existing apparatus without the liquid transmitting media make it, also, unsuitable for this purpose (Figs. 4.2–4). To avoid the complexity of an external magnetic generator, a simple "pull-out" method was recently chosen; the geometrical changes of the sample during hydride formation and decomposition have no effect [4.113]. The controlled movement of the sample inside an induction coil was carried out by a nonmagnetic mechanism, based on the thermal expansion and contraction of an electrically heated [4.114] wire. The magnetic field was supplied by a permanent magnet, which led to strengths of about 2000 Oe at the location of the investigated sample. The signal induced in the induction coil results from the total magnetic moment of the measured material.

As the first application of the device developed, the formation of nickel hydride was followed by the magnetic moment measurement in in situ conditions [4.113]. Figure 4.23 presents the results achieved in nonstationary conditions on fresh 4–8 μm thick nickel foils. Up to the numbered points on curve I, each measurement was performed after 30 min of constant pressure. Under such conditions the first reduction of the magnetic moment appears above 8 kbar of gaseous hydrogen. A quick decrease of the magnetization was observed at 9550 bar; the time behavior is shown on a separate time scale. Measurements on the sample, after previous complete decomposition of the hydride phase (curve II), led to an earlier reduction of the magnetic moment (below 7 kbar) which is in accord with the well-known phenomenon of "sample training" observed previously in both electrochemical and high pressure preparations of nickel hydride [4.8]. These results agree with magnetic moment determinations of electrochemically prepared nickel hydride samples [4.94, 115]. More recently the alloy Ni–Fe with 9.6 at. % of Fe was treated in a similar way, leading to a considerable reduction of the magnetization at higher pressures than pure nickel, which is in accord with earlier measurements of the electrical resistance and the thermopower in the same pressure range [4.60, 62, 63]. One can expect that the device developed [4.113, 114] will be very useful in further investigations, especially at the highest pressure range available.

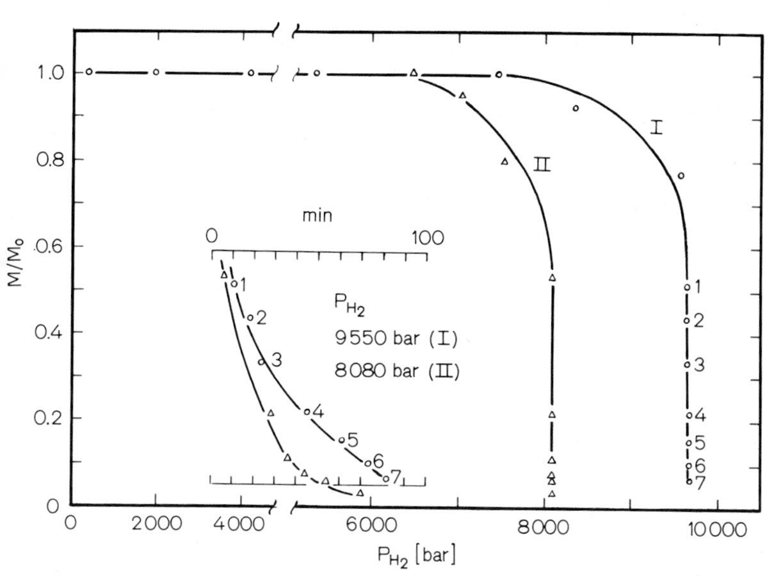

Fig. 4.23. Relative magnetic moment of nickel foils as a function of the hydrogen pressure (for details see text). M_0 = magnetic moment of the hydrogen-free sample

The Curie temperatures of Ni–H and of the alloys Ni-5 at. % Fe–H, Ni–10 at. % Fe–H and Ni-15 at. % Fe–H were determined in the temperature range 350–530 °C and pressure range 0–23 kbar of gaseous hydrogen by applying a differential transformer method [4.57, 116]. The temperature dependence of the initial magnetic permeability measured at a constant hydrogen pressure served for the determination of the Curie points. In the Ni–H system, the increase of the hydrogen pressure lowers the transition temperatures from the value of the pure metal (627 K) to about 551 K at 12.5 ± 0.5 kbar of gaseous hydrogen. Figure 4.24 presents the course of the Curie temperatures as a function of hydrogen pressures for Ni, Ni-10 Fe, and Ni-15 Fe [4.116].

In contrast to pure nickel (1) and the alloy Ni-5 at. % Fe (not shown on Fig. 4.24), both alloys with higher iron content 10 and 15 at. % exhibit Curie temperatures which do not intersect the phase boundary between the α and β phases—that is, they run above the critical regions. This means that the ferromagnetic ordering can be extended to the hydride phases. For the first time ferromagnetism can be expected for 3d-metals with a high hydrogen content.

The Hall effect was investigated in Pd–H [4.35], Ni–H [4.36] and recently Ni–Cu–H systems. Because all curves with concentration variables given in these papers are out of date now, we shall not reproduce them. Let us remark only that in the Ni–H system, Hall coefficients exhibit a course similar to the relative electrical resistance; in Pd–H, a rapid decrease was found in the hydride range. This would correspond to an increase of the effective number of free electrons per palladium atom when one uses a formula for the one-band model.

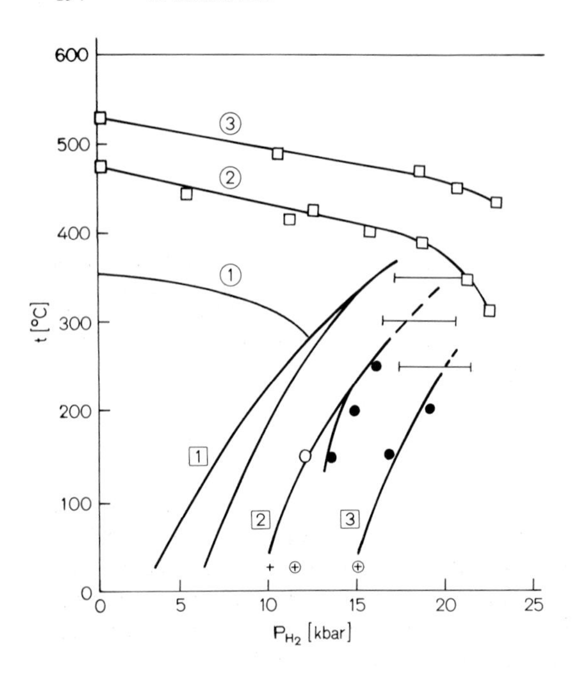

Fig. 4.24. Curie temperatures □, formation and decomposition curves ○ (electrical resistance courses) in P_{H_L}, t variables for Ni (1), Ni-10 at.% Fe (2) and Ni-15 at.% Fe (3)
+, ⊕ values from [4.60, 63]

4.3.5 X-Ray Investigations

The majority of the x-ray investigations have been performed at normal pressure on samples taken out of the high pressure vessel at low temperatures [4.7, 8, 27, 34, 59–61, 76]. The lattice parameter determinations are carried out under in situ conditions for the phase characteristics of special importance. Furthermore, based on the established general dependence between the unit cell increase and hydrogen content of fcc metal-hydrogen systems [4.117], in situ measurement of the lattice expansion could serve for the determination of the hydrogen concentration in such cases. Devices with liquid transmitting media should be avoided because such a solution would be too complex for the purpose of x-ray investigations. Two main problems arise here: a) The stationary supply of high pressure gaseous hydrogen for time consuming experiments. b) The choice of the transmitting material for the x-rays with sufficient mechanical strength. The registration of the x-rays could be carried out on a multiple film cassette using Mo radiation. Figure 4.2 presents the most recent device, which is based on earlier models [4.7, 37], used in our laboratory [4.38]. No mobile piston is used here for the pressure increase because the high pressure hydrogen gas is supplied by a special compressor. The sample container was made of beryllium, and the metals investigated could be used in the form of foils, wires, or powders. The upper limit of the pressure (9 kbar) was determined mainly by the mechanical strength of the beryllium element. Figure 4.25 summarizes the results of preliminary investigations on a wire sample of a

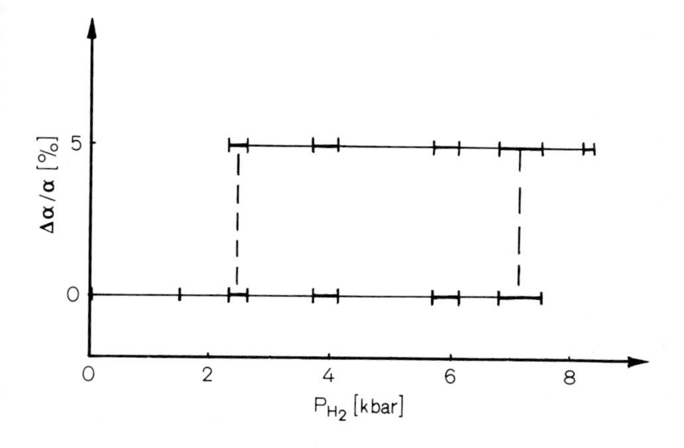

Fig. 4.25. Relative change of the lattice parameter of a Pd–Rh (50 at. %) alloy as a function of the hydrogen pressure

Pd–Rh alloy (50 at. % Rh) at room temperature [4.38]. The pressure indicated was obtained starting from lower values only. Around 3 kbar the characteristic lines of the hydride phase (lattice parameter about 5% larger) appear; they coexisted with the initial (α phase) lines to about 7 kbar. In the last measurement (above 8 kbar) only the characteristic lines of the hydride phase existed. Each point was taken after stabilization for at least one day at the pressure indicated. The scatter of the pressure values is due to temperature changes in the laboratory. The results shown in Fig. 4.25 exceed the pressure range by more than two orders of magnitude as compared with earlier results [4.118].

4.3.6 Thermal Conductivity

No data on thermal conductivity, measured at normal conditions, are known for the hydrides treated in this chapter. A method of measurement of this property was recently proposed for in situ conditions [4.48, 119]. It is of a comparative nature because an absolute procedure was not suitable for our purposes. An axial heat flow takes place simultaneously in two cylindrical metallic rods—one with a known heat conductivity and the other of the material to be investigated. An axial temperature gradient is created by heating one end of the rod with a manganin heater and keeping the other end in a good thermal contact with the bulk of the apparatus. A $1.5 \times 0.1\ \text{cm}^2$ platinum rod was taken as reference metal as it was found inert in the high pressure of gaseous hydrogen [4.34].

A palladium rod of the same dimensions served as the sample to be investigated. The measuring device was placed in the hydrogen container of the high pressure apparatus presented in Fig. 4.1. The method was tested in measurements of palladium and manganine—in respect to platinum - at pressures up to 20 kbar; helium was used as the pressure transmitting medium for palladium and hydrogen for manganine. In both cases no active influence of the gaseous medium on the rods investigated could be expected. Independent of the large differences in the heat conductivity of both metals and gases, the extrapolations to normal pressures coincide with the data from literature within the error range, which reaches about 10%, mainly due to the inaccuracy of the geometrical parameters. The pressure coefficients of the heat conductivity equal -1.8 and $+3.0 \times 10^{-6}$ bar^{-1} for palladium and manganine, respectively, at 25 °C.

Palladium was investigated in gaseous hydrogen in the pressure range 0.6 to 25 kbar at 25 °C. A continuous increase in heat conductivity was found from 0.48 W cm^{-1} K^{-1} at 0.6 kbar to 0.75 W cm^{-1} K^{-1} at 25 kbar; the changes of the lattice parameter due to the hydrogen absorption were taken into account.

The value for the pure palladium (0.72 W cm^{-1} K^{-1}) is reached at about the same hydrogen activity as is the case for the electrical conductivity [4.31, 32, 42]. Large deviations form the Wiedemann-Franz law were found.

4.3.7 Diffusion

The formation kinetics of nickel hydride was followed by the electrical resistance measurements under in situ conditions [4.27, 28]. Using nickel foils of different thickness, the apparent diffusion coefficient could be determined [4.28] to be of the order of 10^{-12} cm^2 s^{-1} at 25 °C. But evidently the kinetics observed did not fulfill quantitatively the analytical requirements of pure diffusion in the thickness range of 2 to 20 μm, probably due to surface barriers. Reasonable agreement was found with earlier formation kinetics followed during the electrochemical process [4.120].

A relaxation method was applied for the determination of the diffusion coefficients of hydrogen in palladium hydride [4.121]. In the pressure range 40 bar to 23 kbar, small rapid changes of the hydrogen pressure were introduced and the electrical resistance was registered as a function of time at a constant pressure. Assuming for each step a linear relation between the concentration and the average electrical resistance, the relaxation time measured could be simply related to the intrinsic diffusion coefficient through a suitable solution of Fick's law. Figure 4.26 presents the results obtained. As can be seen, the intrinsic diffusion coefficient reduces its value nearly by two orders of magnitude in the pressure range indicated[2]. Knowledge of the hydrogen concentration and the character of the occupied sites is required for a detailed discussion of the above data. As the results presented in Fig. 4.5 show, a change of the diffusion mechanism in the higher pressure range, as compared to normal conditions, seems to be unavoidable. This point of view is supported by the observed change of the activation energy as a function of the hydrogen pressure [4.121].

The curve presented in Fig. 4.26 includes two effects: a) The change of the intrinsic diffusion coefficient caused by the increase of the hydrogen con-

[2] The diffusion coefficient used here contains the thermodynamic factor $\partial\mu/\partial c$.

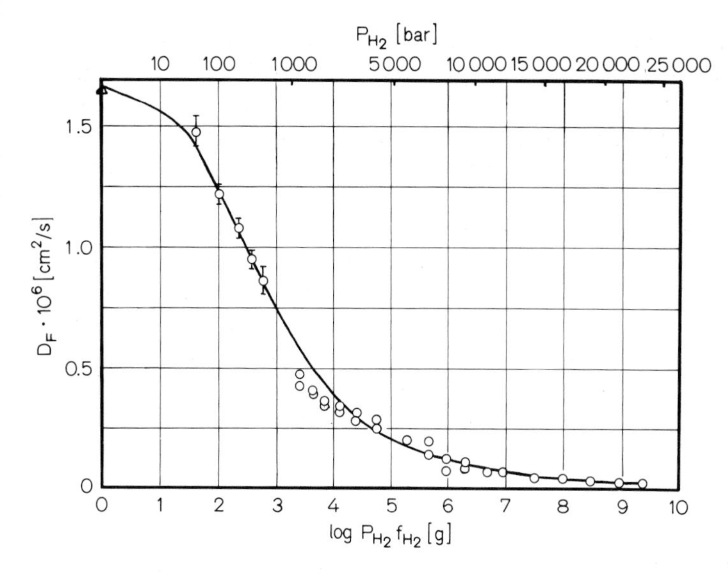

Fig. 4.26. Intrinsic diffusion coefficient (D_F) of hydrogen in palladium hydride as a function of the logarithm of the hydrogen fugacity

centration. The course shown on Fig. 4.26 is—from this point of view—due to the reduction of free interstitials with increasing hydrogen activity in the gaseous phase. b) The change of the intrinsic diffusion coefficient caused by the increase of the pure hydrostatic pressure. The normal trend of this effect is the same as the previous one but it is probably less significant because the activation volume of the diffusing component is usually lower than its partial volume. Because this quantity is smaller than 1 cm^3/mole H for palladium hydride [4.41], the expected change of the diffusion coefficient due to the hydrostatic pressure is below the experimental resolution.

4.4 Conclusions and Perspectives

As this chapter has shown, several physicochemical properties can be now quantitatively investigated under in situ conditions of high pressure of gaseous hydrogen. This extends our knowledge to heretofore unavailable ranges of hydrogen activity and has already led to the discovery of new hydride phases and unexpected behavior (superconductivity of palladium hydride, probable magnetic ordering in *d*-metal hydrides). It is desirable to enrich the in situ investigations by new methods. Electrochemical and calorimetric measurements are recent developments in our laboratory. As shown above, the extension to higher hydrogen pressures and higher temperatures makes the achievement of new interesting results quite probable for several metallic systems. Therefore it is our intention to follow these lines in future development. Finally, let us remark that known hydrides prepared in high pressure conditions (Ni, Cr, Mn) are relatively inexpensive and rich in natural resource materials; therefore, they may be useful for storage purposes and other applications if the high pressure technology becomes more common than it is now.

References

4.1 D.P.Schumacher: Phys. Rev. **126**, 1679 (1962)
4.2 D.R.Stephens, E.M.Lilley: J. Appl. Phys. **39**, 177 (1968)
4.3 St.M.Filipek, Koji Wakamori, Akira Sawaoka, Shiuroku Saito: Jap. J. Appl. Phys. (in press)
4.4 B.J.Alder, R.H.Christian: Disc. Faraday Soc. **22**, 44 (1956)
4.5 B.Baranowski, M.Śmiałowski: J. Phys. Chem. Sol. **12**, 206 (1959)
4.6 B.Baranowski: Plat. Met. Rev. **16**, 10 (1972)
4.7 B.Baranowski: Ber. Bunsenges. Physik. Chem. **76**, 714 (1972)
4.8 B.Baranowski: „Hydride und Deuteride des Nickels und von Kupfer-Nickel-Legierungen", in *Festkörperchemie*, ed. by V.Boldyrev, K.Meyer (VEB Deutscher Verlag für Grundstoffindustrie, Leipzig 1973) pp. 364–383
4.9 P.W.Bridgman: *The Physics of High Pressure* (G.Bell and Sons, London 1952) Chap. I, pp. 1–29
4.10 E.H.Amagat: Ann. Chim. Phys. (6) **29**, 1 (1893)
4.11 P.W.Bridgman: Proc. Am. Acad. Arts Sci. **59**, 173 (1924)
4.12 A.Michels, W.de Graaf, T.Wassernaar, J.M.Levelt, P.Louverse: Physica **25**, 25 (1959) W.de Graaf: Compressibility Isotherms and Thermodynamic Functions of Hydrogen and Deuterium. Conclusions Regarding the Intermolecular Field. Thesis, Amsterdam (1960)
4.13 D.S.Cyklis, W.A.Maslennikova, S.D.Gavrilov, A.N.Jegorov, G.W.Timofiejeva: Dokl. Akad. Nauk SSSR **220**, 1384 (1975)
4.14 P.S.Perminov, A.A.Orlow, A.N.Frumkin: Dokl. Akad. Nauk SSSR **84**, 749 (1952)
4.15 H.O.von Samson-Himmelstjerna: Z. Anorg. Allg. Chem. **186**, 337 (1930)
4.16 A.J.Fedorova, A.N.Frumkin: Ż. Fiz. Chim. **27**, 247 (1953)
4.17 P.L.Levine, K.E.Weale: Trans. Faraday Soc. **56**, 357 (1960)
4.18 B.Baranowski, M.Śmiałowski: Bull. Polon. Acad. Sci. **7**, 663 (1959)
4.19 B.Baranowski: Naturwissenschaften **46**, 666 (1959)
4.20 B.Baranowski: Bull. Polon. Acad. Sci. **7**, 891 (1959)
4.21 B.Baranowski: Bull. Polon. Acad. Sci. **7**, 907 (1959)
4.22 B.Baranowski: Roczn. Chemii **38**, 1019 (1964)
4.23 B.Baranowski: Bull. Polon. Acad. Sci. **10**, 451 (1962)
4.24 B.Baranowski, K.Bocheńska: Roczn. Chemii **38**, 1419 (1964)
4.25 B.Baranowski, K.Bocheńska: Z. Physik. Chem. (N.F.) **45**, 140 (1965)
4.26 B.Baranowski, R.Wiśniewski: Bull. Acad. Polon. Sci. **14**, 273 (1966)
4.27 B.Baranowski, K.Bocheńska, S.Majchrzak: Roczn. Chemii **41**, 2071 (1967)
4.28 B.Baranowski, K.Bocheńska: "Penetration of High Pressure Hydrogen into Nickel", in *Atomic Transport in Solids and Liquids*, ed. by A.Lodding, T.Lagerwall (Verlag der Zeitschr. für Naturforschung, Tübingen 1971) pp. 360–362
4.29 A.Stroka, A.Freilich: Roczn. Chemii **44**, 235 (1970)
4.30 B.Baranowski, W.Bujnowski: Roczn. Chemii **44**, 2271 (1970)
4.31 A.W.Szafrański, B.Baranowski: Phys. Stat. Sol. (a) **9**, 435 (1972)
4.32 R.Wiśniewski: Rev. Sci. Instr. **41**, 464 (1970)
4.33 B.Baranowski, R.Wiśniewski: Phys. Stat. Sol. **35**, 593 (1969)

4.34 B. Baranowski, F. A. Lewis, S. Majchrzak, R. Wiśniewski: J. Chem. Soc. Faraday Trans. I **68**, 653 (1972)
4.35 R. Wiśniewski, A. J. Rostocki: Phys. Rev. B **3**, 251 (1971)
4.36 R. Wiśniewski, A. J. Rostocki: Phys. Rev. B **4**, 4330 (1971)
4.37 S. Majchrzak, B. Baranowski, W. Bujnowski, M. Krukowski: Roczn. Chemii **46**, 1173 (1972)
4.38 S. Majchrzak: Roczn. Chemii **51**, 1549 (1977)
4.39 T. Skośkiewicz, A. W. Szafrański, W. Bujnowski, B. Baranowski: J. Phys. C: Solid State Phys. **7**, 2670 (1974)
4.40 B. Baranowski, M. Tkacz, W. Bujnowski: High Temperature—High Pressure **8**, 656 (1976)
4.41 B. Baranowski, T. Skośkiewicz, A. W. Szafrański: Fiz. Nisk. Temp. **1**, 616 (1975)
4.42 B. Baranowski, R. Wiśniewski: J. Phys. Chem. Sol. **29**, 1275 (1968)
4.43 M. Tkacz, B. Baranowski: Roczn. Chemii **50**, 2159 (1976)
4.44 E. Wicke, G. H. Nernst: Ber. Bunsenges. Physik. Chem. **68**, 224 (1964)
4.45 J. R. Lacher: Proc. Roy. Soc. London **161** A, 525 (1937)
4.46 F. A. Lewis: *The Palladium Hydrogen System* (Academic Press, New York, London 1967)
4.47 M. J. B. Evans, D. H. Everett: J. Less-Common Metals **49**, 123 (1976)
4.48 A. Szafrański: Electrical and thermal conductivity of some alloys Pd + H, Pd + Ag + H and Pd + Au + H (in Polish). Thesis, Warsaw (1976)
4.49 T. B. Flanagan, B. Baranowski, S. Majchrzak: J. Phys. Chem. **74**, 4299 (1970)
4.50 B. Baranowski, S. Majchrzak, T. B. Flanagan: J. Phys. Chem. **77**, 35 (1973)
4.51 T. B. Flanagan, S. Majchrzak, B. Baranowski: Phil. Mag. **25**, 257 (1972)
4.52 C. A. Beckman, W. E. Wallace, R. S. Craig: Phil. Mag. **27**, 1249 (1973)
4.53 P. Merker, G. Wolf, B. Baranowski: Phys. Stat. Sol. (a) **26**, 167 (1974)
4.54 W. D. Mc Fall: Studies of Phase Relationship in Palladium Alloy/Hydrogen Systems. Thesis. Belfast (1974)
4.55 A. Stroka: Differences in the formation and decomposition of nickel hydride and deuteride (in Polish). Thesis, Warszaw (1970)
4.56 I. Czarnota, B. Baranowski: Bull. Acad. Polon. Sci. **14**, 191 (1966)
4.57 E. G. Poniatowskij, W. E. Antonov, I. T. Belash: Dokl. Akad. Nauk SSSR **229**, 391 (1976)
4.58 B. Baranowski, M. Tkacz, W. Bujnowski: Roczn. Chemii **49**, 437 (1975)
4.59 B. Baranowski, S. Majchrzak: Roczń. Chemii **42**, 1137 (1968)
4.60 B. Baranowski, S. Filipek: Roczn. Chemii **47**, 2165 (1973)
4.61 M. Krukowski, B. Baranowski: J. Less-Common Metals **49**, 385 (1976)
4.62 S. Filipek, B. Baranowski: Roczn. Chemii **49**, 1149 (1975)
4.63 S. Filipek: The Systems Ni–Fe–H(D) and Ni–Co–H(D) in High Pressures of Gaseous Hydrogen (in Polish). Thesis, Warszaw (1977)
4.64 T. Skośkiewicz: Phys. Stat. Sol. (a) **6**, 29 (1971)
4.65 T. Skośkiewicz: Phys. Stat. Sol. (a) **48** (1978)
4.66 C. A. Snayely, D. A. Vangham: J. Am. Chem. Soc. **71**, 313 (1949)
4.67 G. Wolf: Z. Physik. Chem. **246**, 403 (1971)
4.68 A. Sieverts, A. Gatta: Z. Anorg. Chem. **172**, 1 (1928)
4.69 B. Baranowski, K. Bojarski: Roczn. Chemii **46**, 1403 (1972)
4.70 B. Baranowski, K. Bojarski: Roczn. Chemii **45**, 499 (1971)
4.71 S. Randzio, K. Bojarski: Roczn. Chemii **48**, 1375 (1974)
4.72 B. Baranowski, K. Bojarski: Roczn. Chemii **46**, 525 (1972)
4.73 B. Baranowski, M. Tkacz: Roczn. Chemii **48**, 713 (1974)
4.74 B. Baranowski, K. Bojarski, M. Tkacz: Cr–H and Cr–D Systems in the High Pressure Region, in Proc. IV. Intern. Conference on High Pressure, Kyoto, 1974 (The Physico-Chemical Society of Japan, Kyoto 1975)
4.75 E. G. Poniatowskij, I. T. Belash: Dokl. Akad. Nauk SSSR **229**, 1171 (1976)
4.76 M. Krukowski, B. Baranowski: Roczn. Chemii **49**, 1183 (1975)
4.77 E. G. Poniatowskij, I. T. Belash: Dokł. Akad. Nauk SSSR **224**, 607 (1975)
4.78 R. J. Smith, D. A. Otterson: J. Phys. Chem. Sol. **31**, 187 (1970)
4.79 A. W. Szafrański: Phys. Stat. Sol. (a) **19**, 459 (1973)
4.80 G. Rosenhall: Ann. Phys. (5) **24**, 297 (1935)

4.81 A.W.Carson, F.A.Lewis, W.H.Schurter: Trans. Faraday Soc. **63**, 1453 (1967)
4.82 G.A.Ferguson, A.I.Schindler, T.Tanaka, P.Morita: Phys. Rev. **137**, A 483 (1965)
4.83 I.R.Ertin, V.A.Somenkov, Ya.S.Umanskij, A.A.Chertkov, S.Sh.Shil'stein: Sov. Phys.-Solid State **15**, 1840 (1974)
4.84 M.H.Mueller, J.Fober, H.E.Flotow, D.G.Westlake: Bull. Am. Phys. Soc. **20**, 421 (1975)
4.85 D.M.Nace, J.G.Aston: J. Am. Chem. Soc. **79**, 3623 (1957)
4.86 A.J.Schindler, R.J.Smith, E.W.Kammer: Proc. X. Intern. Congr. Refrigeration **1**, 74 (1959)
4.87 J.K.Jacobs, F.D.Manchester: J. Less-Common Metals **49**, 67 (1976)
4.88 T.Skośkiewicz, B.Baranowski: Phys. Stat. Sol. **30**, K 33 (1968)
4.89 T.Skośkiewicz: Phys. Stat. Sol. (a) **11**, K 123 (1972)
4.90 T.Skośkiewicz: Phys. Stat. Sol. (b) **59**, 329 (1973)
4.91 B.Baranowski: Acta Met. **12**, 322 (1964)
4.92 H.J.Bauer: Z. Physik **177**, 1 (1964)
4.93 B.Baranowski, A.Freilich: Roczn. Chemii **42**, 1983 (1968)
4.94 H.J.Bauer, E.Schmidbauer: Z. Physik **164**, 367 (1961)
4.95 G.Wolf, B.Baranowski: J. Phys. Chem. Sol. **32**, 1649 (1971)
4.96 A.C.Switendick: Ber. Bunsenges. Physik. Chem. **76**, 6 (1972)
4.97 T.Skośkiewicz, A.W.Szafrański, B.Baranowski: Phys. Stat. Sol. (b) **59**, K 135 (1973)
4.98 J.E.Schirber: Phys. Lett. **46** A, 285 (1973)
4.99 W.Buckel, A.Eichler, B.Stritzker: Z. Physik **263**, 1 (1973)
4.100 J.E.Schirber: Phys. Lett. **45** A, 141 (1973)
4.101 J.E.Schirber: Cryogenics **10**, 418 (1970)
4.102 J.Igalson, L.Śniadower, A.J.Pindor, T.Skośkiewicz, K.Blüthner, F.Dettmann: Solid State Commun. **17**, 309 (1975)
4.103 A.W.Szafrański, T.Skośkiewicz, B.Baranowski: Phys. Stat. Sol. (a) **37**, K 163 (1976)
4.104 C.B.Satterthwaite, J.L.Toepke: Phys. Rev. Lett. **25**, 741 (1970)
4.105 C.B.Satterthwaite, D.T.Peterson: J. Less-Common Metals **26**, 361 (1972)
4.106 C.G.Robbins, M.Ishikawa, A.Treyrand, J.Muller: Solid State Commun. **17**, 903 (1975)
4.107 T.Skośkiewicz: Electrical resistance and thermopower in systems Ni–H, Ni–Cu–H, and Pd–H (in Polish). Thesis, Warsaw (1969)
4.108 B.Baranowski, M.Tkacz: Europhys. Conf. Abstr. **1** A, 108 (1975)
4.109 R.Wiśniewski: Phys. Stat. Sol. **5**, K 31 (1971)
4.110 E.O.Wollan, J.W.Cable, W.C.Koehler: J. Phys. Chem. Sol. **24**, 1141 (1963)
4.111 B.Baranowski: Phys. Stat. Sol. **7**, K 141 (1964)
4.112 B.Baranowski, T.Skośkiewicz: Acta Phys. Polon. **33**, 349 (1968)
4.113 H.J.Bauer, B.Baranowski: Phys. Stat. Sol. (a) **40**, K 35 (1977)
4.114 H.J.Bauer: J. Phys. E **10**, 332 (1977)
4.115 G.K.Wertheim, D.N.E.Buchanan: Phys. Lett. **21**, 255 (1966)
4.116 E.G.Poniatowskij, W.E. Antonov, I.T.Belash: Dokl. Akad. Nauk SSSR **230**, 649 (1976)
4.117 B.Baranowski, S.Majchrzak, T.B.Flanagan: J. Phys. F **1**, 258 (1971)
4.118 A.J.Maeland, T.R.P.Gibb: J. Phys. Chem. **65**, 1270 (1961)
4.119 A.W.Szafrański, B.Baranowski: J. Phys. E **8**, 823 (1975)
4.120 A.Mituya, K.Sekine, G.Toda: J. Res. Inst. Cat. **15**, 21 (1967)
4.121 M.Kuballa, B.Baranowski: Ber. Bunsenges. Physik. Chem. **78**, 335 (1974)

5. Hydrogen Storage in Metals

R. Wiswall

With 16 Figures

5.1 Background

5.1.1 The Need of Storage

Hydrogen storage is an important topic chiefly because of its relevance to the energy economy of the future. Since hydrogen is a versatile fuel, and can be readily generated from and converted to other forms of energy, to store hydrogen is to store energy. There are many sources of energy which can be fully utilized only if the product can be stored. Indeed, one needs not only to store but also to package energy and to be able to release it at a later time and a distant place—a place not necessarily linked with wires to the source.

A case in point is the power generated by nuclear reactors. Economy of scale is conspicuous here; they must be built in very large units. At present such reactors supply only the base load of electric grids, but as their numbers increase, the time will come when their output could be used, if storage methods were available, to take over certain tasks now reserved to fossil fuels; even automotive propulsion is a possibility. This statement will apply with particular force to thermonuclear reactors when they are developed. They will be extremely large, and by the time they are ready to make an important contribution to the world's energy needs—probably A.D. 2010 at the earliest—fossil fuel reserves will be well on their way to depletion.

All intermittent energy sources are, of course, obvious beneficiaries of an energy storage technology. Solar, tidal, and wind energy become far more valuable if partial storage is possible. Also in this class belongs the output of electric generating plants at times of low demand. If this "off-peak" power could be stored and released a few hours later, at the time of high demand, the amount of generating equipment required could be substantially reduced. The probable mechanism of load leveling would be to electrolyze water, store the hydrogen, and later feed it to a fuel cell stack. The latter's electric product would, after conversion from dc to ac, be supplied to the grid. Variations that have been suggested include 1) storing oxygen as well as hydrogen and using it rather than air as the cathode fuel to raise the fuel cell voltage; 2) electrolyzing aqueous hydrochloric acid instead of water, storing hydrogen and chlorine, and causing them to recombine in a suitable cell—possibly the same cell as that used for electrolysis [5.1]; 3) using the hydrogen as a fuel for a high temperature, high efficiency turbine; and 4) the combustion of hydrogen with oxygen in magnetohydrodynamic generators. One feature all the alternatives have in common is the need for hydrogen storage.

There are, of course, other ways of storing and packaging electrical energy than by the production of hydrogen, and one of them may eventually prove more cost effective. Meanwhile, however, the hydrogen path appears promising enough to justify its further exploration. Furthermore, one must remember that there are sources of hydrogen fuel other than water electrolysis. The earth's most abundant fuel, coal, is convertible to hydrogen by the familiar water-gas and CO-shift reactions. Part of its fuel value is lost in the process, but the hydrogen product has several important advantages over coal. It is fluid; it is usable in internal combustion engines with better than the efficiency of gasoline, and with much less pollution; and it is the ideal fuel for fuel cells, in which efficiencies of over 50% can be expected. The storage problem will be equally important, whether one is dealing with hydrogen made from nuclear (or solar) electricity, or from coal.

Finally, mention should be made of certain nonfuel uses of hydrogen in which improved storage methods could play a part. Quite large amounts of the gas are used to hydrogenate oils of animal and vegetable origin, in the food industry; it is important to have a reserve supply of hydrogen on hand. Another large use is for the cooling of electrical generating equipment, and here too a reservoir is necessary. These two uses may be small compared to projections of hydrogen fuel production in a "hydrogen economy", but they are at least actual and contemporary.

5.1.2 Methods of Storage

A brief consideration of the existing technology of hydrogen storage shows it to be inadequate for many of the applications just mentioned. Storage for automotive use is in particular need of development. No matter how fine an automotive fuel hydrogen may be, it will never find extensive use if it has to be supplied as a compressed gas or in the liquid form. Gas cylinders are far too heavy and bulky, and to liquefy hydrogen requires a power input of 14,450 kJ/kg [5.2], or 6.96 kcal/mole. Since this is mechanical work, its heat equivalent is at least 2.5 times as large, or 17.4 kcal/mole. But the heat of combustion of one mole of H_2, to liquid H_2O, is 68.32 kcal/mole, which is only four times this quantity. Other published figures for the power required for liquefaction are somewhat smaller [5.3], but even if these are accepted, the use of liquid H_2 as a fuel is energetically very wasteful. Compression also is rather inefficient in this respect. To compress an ideal gas isothermally from 1 atmosphere to 140 atmospheres (a common tank pressure) requires a minimum work of $RT \ln 140$, or 2.925 kcal at 298 K. Compressor inefficiencies will raise this to at least 4 kcal. This, if we again apply a factor of 2.5 to obtain the equivalent heat, corresponds to 10 kcal, which also is an appreciable fraction of the potential fuel value of the gas.

For stationary applications, low-pressure and medium-pressure gas storage is a possibility, either in underground caverns or in atmospheric-pressure telescoping gas holders such as are used in city gas systems for manufactured gas. Considerations of availability for the former, and of economics and esthetics for the latter, provide incentive to seek alternatives here also.

Certain chemical compounds can be considered as storage media or carriers of hydrogen fuel. They include NH_3, N_2H_4, and CH_3OH. For all of them, hydrogen is a major raw material in their synthesis, and its fuel value is recovered in large part when they are oxidized. They differ, however, in the types of application for which they would be suitable. Ammonia is a poor fuel, but it can be cracked to yield a N_2–H_2 mixture which might be useful for fuel-cell or other purposes. Methanol has its advocates as a motor fuel, but is probably less satisfactory in fuel cells. Hydrazine has a higher energy content than the others but is toxic and expensive to make. All require rather large installations for their manufacture. (As will shortly be seen, a useful feature of metal hydrides is that they can be prepared as well on a small as on a large scale.)

A more truly reversible class of systems for hydrogen storage involves the hydrogenation of unsaturated hydrocarbons. An example is the reaction

$$C_6H_6 + 3H_2 \rightleftarrows c\,C_6H_{12}\,.$$

Experimental work has been done on this system [5.4] but its feasibility has not yet been thoroughly demonstrated. This cycle also suffers from the requirement of a relatively elaborate setup.

5.1.3 Other Reviews

The general subject matter of this chapter has recently been treated, from slightly different points of view, by *Hoffman* et al. [5.5] and *Garg* and *Mc Claine* [5.6].

5.2 Hydrides as Storage Media

5.2.1 Reversible and Irreversible Systems

Metal hydrides can serve as hydrogen sources either through chemical reaction or by thermal decomposition. The hydrolysis of calcium hydride is an example of the former; this reaction has been proposed for the provision of hydrogen for fuel cells, for the inflation of balloons, and for automotive use. Lithium hydride can serve a similar purpose. Such a system is inherently expensive, however. The hydrolysis reaction is highly irreversible. If the process is to be cyclic, the $Ca(OH)_2$ or LiOH which is produced must be converted back to the

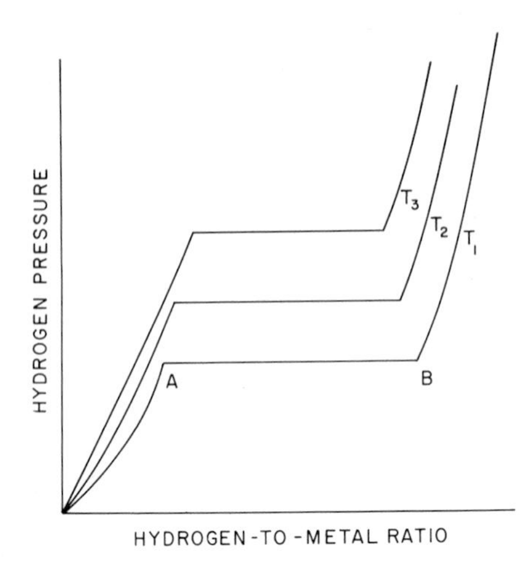

Fig. 5.1. Pressure vs composition isotherms in a typical hydrogen-metal system

corresponding hydride [5.7, 8]. This consumes a great deal of energy, whether it is done by high-temperature reactions or electrochemical processes. If, on the other hand, one is able so to arrange things that hydrogen is obtained by a thermal decomposition reaction, the metallic product can (usually) be reconverted to hydride very simply, by direct combination with hydrogen. In this chapter, we shall be almost entirely concerned with those hydrogen-metal systems that show approximately reversible behavior.

5.2.2 Pressure-Temperature-Composition (P–T–C) Relations

The first example of reversible hydriding was that of palladium, investigated by Thomas Graham 111 years ago. Since his time, hundreds of other metal-hydrogen systems have been shown to possess most of the important features of $Pd–H_2$. Much hydrogen can be easily absorbed and later, by a slight change of conditions, recovered, and (which is most important for storage purposes) most of the recovered gas is furnished at approximately constant pressure. This is due to the happy circumstance that we are not really dealing with a solubility phenomenon but with a reversible chemical reaction. In a storage reservoir where dissolution of gas in solid was the mechanism, the pressure would decrease more or less rapidly as gas was withdrawn. If, however, two solid phases and a gas phase maintain an equilibrium, the gas pressure will be independent of the relative amounts of the phases as long as any of each remains. It is true that, in some cases, rather large amounts of hydrogen can dissolve interstitially in a metal phase without causing changes, other than expansion, in the relative positions of the metal atoms. More often, however, the addition of hydrogen to metal soon results in its becoming saturated with respect to the formation of a hydrogen-rich "hydride" phase. Figure 5.1

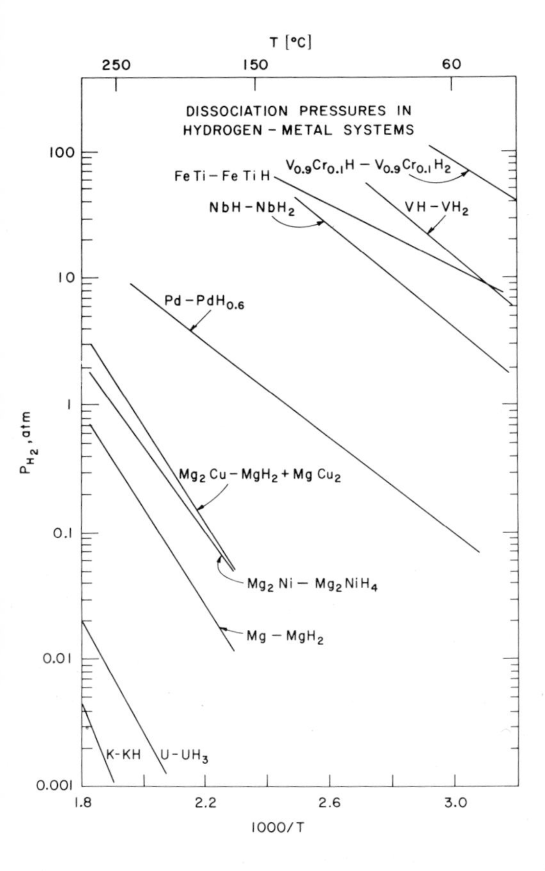

Fig. 5.2. Dissociation pressures in hydrogen-metal systems

illustrates the behavior of a typical, if slightly idealized, metal-hydrogen system. Here we plot equilibrium pressure against composition; either can be regarded as the independent variable. In the first section, from the origin to point A, a true solution exists, and the pressure-composition dependence is approximately described by the Sieverts' law parabola, $p = k(H/M)^2$. (This is simply Henry's law for a dissociating solute.) Between A and B, there coexist the saturated solution, of composition $(H/M)_A$, and the hydride phase, of composition $(H/M)_B$. The value of the latter may or may not be an integer. Beyond B, a steep increase in pressure is necessary to alter the composition of the hydride phase. The curves labeled T_2 and T_3 show the effect on the pressure-composition relation of raising the temperature. It is convenient, and customary, to characterize the pressure-temperature relationship of a hydride system by plotting the logarithm of the "plateau" pressure against the reciprocal of the absolute temperature; such graphs are usually linear. Figure 5.2 gives data of this sort for a number of hydride systems of possible storage interest.

It is worthwhile at this point to distinguish two uses of the word "solubility" in the hydride literature. One simply refers to the total amount of hydrogen

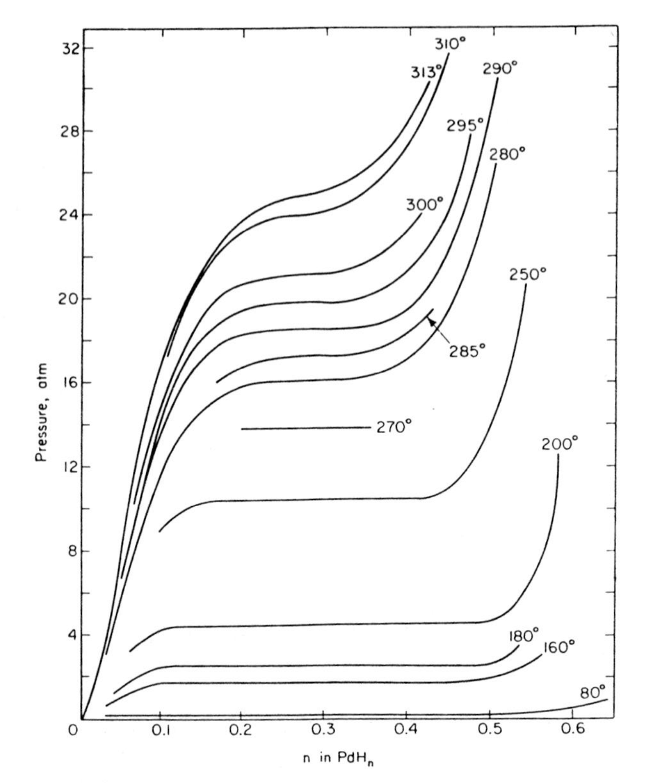

Fig. 5.3. The palladium-hydrogen system [Ref. 5.9, p. 322], courtesy Wiley-Interscience

contained in the solid phase when the pressure is specified. This is sometimes termed the overall solubility and is especially appropriate for systems containing a single solid phase. The other is the terminal or limiting solubility. This refers to the single-phase saturation concentration, above which a second solid phase begins to appear, e.g., the abscissa value corresponding to point *A* in Fig. 5.1.

5.2.3 Complications of Real Systems, Hysteresis

Plateaus are seldom truly horizontal, especially in alloy systems. Even with single elements, plateaus tend to tilt as the temperature is raised, and to become narrower. Sometimes they disappear altogether and are replaced by a point of inflection. That is, above a definite critical temperature, two phases are no longer to be distinguished. Figure 5.3 shows part of the pressure-temperature-composition data for Pd–H_2, which illustrate this phenomenon.

An apparent departure from the horizontal can also arise as an experimental artifact. *Libowitz* [5.10] showed that the existence of a temperature gradient across a hydride sample during addition or removal of hydrogen can result in a

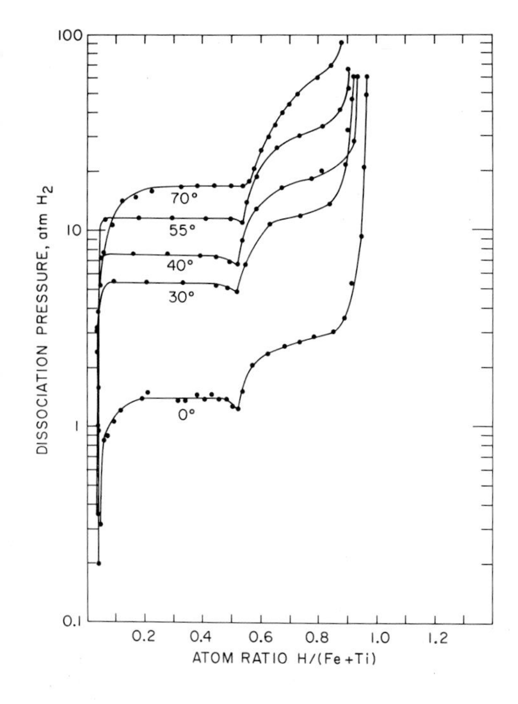

Fig. 5.4. The system, TiFe–H_2

sloping plateau. This would apply as much to a practical reservoir as to a research experiment.

It often happens that two or more hydride phases can exist in a given system. These show up on the pressure-composition isotherm as successive plateau regions. Both may be well defined, or it may happen that one is hardly more than a strong inflection. Figure 5.4 gives examples.

The most important omission from Fig. 5.1 is the hysteresis phenomenon. The experimental points from which such curves are derived can be obtained either by adding a measured quantity of hydrogen gas to a closed system, at constant temperature, and waiting for the pressure to equilibrate, or by reversing the procedure, and removing aliquots of hydrogen. The latter process almost invariably results in lower plateau pressures, unless the temperature is above about 400 °C. In Fig. 5.5, absorption and desorption isotherms are compared for a typical low-temperature system, VH–VH_2 [5.11]. It is to be noted that this phenomenon is not due simply to a failure to wait long enough for equilibrium to be reached. Very long waiting does not bring the curves closer together, and both absorption and desorption points are quite reproducible, even as between different laboratories. Much has been written on the subject of hydride hysteresis. Useful discussions will be found in [5.12–20]. While quantitative treatments and predictions are still lacking, most authorities relate the effect to the abrupt increase of molar volume on going from a lower to a higher hydride, and the obstacle which this imposes to the growth of the higher in a matrix of the lower. This is consistent with the observation that

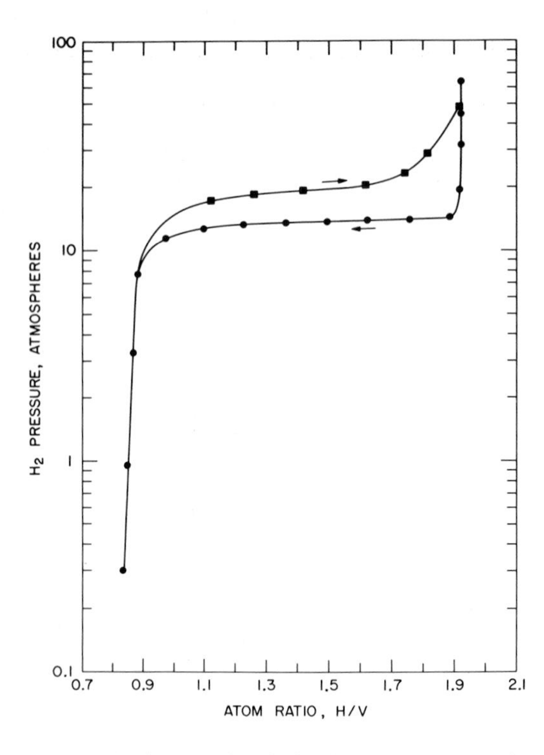

Fig. 5.5. Hysteresis in hydrogen absorption by commercial vanadium, at 45 °C

hysteresis is usually limited to the parts of the isotherms where two phases coexist.

Some hydride workers consider that the desorption curves are the more likely to represent the "true" equilibrium situation than the absorption curves [5.18], and should be taken as standard. Desorption points are certainly easier to measure; constant pressure is more quickly obtained than on absorption, results are more reproducible, and plateaus often are more nearly horizontal. Also, it is sometimes found [5.18] that the shape of the desorption isotherm is less sensitive to variations in the method of preparing the metal sample.

The magnitude of the hysteresis effect varies widely. The largest known to the writer is that for a U–Nb alloy, reported by *Katz* and *Gulbransen* [5.21], in which absorption pressures were fifty times those of desorption. At the other extreme, an unusually small low-temperature effect was recently reported by *Bechman* et al. [5.22] for systems $GdCo_3$–H_2 and $GdFe_3$–H_2. Here the plateau pressures differed by <5% at 150–175 °C. Hysteresis magnitudes for systems of possible storage interest will be presented below (Sect. 5.3).

The temperature dependence of hysteresis depends on the system and on the temperature. Below 100 °C, it is often found that the ratio of absorption to desorption pressures is approximately constant (see, for example, the results of *Kuijpers* and *Van Mal* on $LaNi_5$ [5.23]). With these systems, that is, and in a limited low-temperature region, the difference between absorption and desorption plateaus increases with temperature.

Table 5.1. Thermodynamics of representative hydrides: Standard enthalpies and entropies of formation[a, b]

Initial and final compositions	ΔH_f^0 [kcal mole^{-1} H_2]	ΔS_f^0 [cal deg^{-1} mole^{-1} H_2]
Li–LiH	−43.3	−32.8
Na–NaH	−27.0	
Mg–MgH_2	−17.8	−32.3
$Mg_2NiH_{0.3}$ − $Mg_2NiH_{4.2}$	−15.4	−29.0
Ca–CaH_2	−41.7	−30.4
Al–AlH_3[d]	− 2.7	−30.9
La–LaH_2[e]	−49.6	
$LaNi_5$–$LaNi_5H_6$[f]	− 7.6	−26
Ti–$TiH_{1.97}$	−29.9	−30.0
$TiFeH_{0.1}$ − $TiFeH_{1.0}$[g]	− 6.7	−25.0
Zr–ZrH_2	−39.0	
V–$VH_{0.2}$[h]	−15.3	−18.8
$VH_{0.9}$ − $VH_{2.0}$[i]	− 9.6	−34.0
U–UH_3	−20.2	−28.9

[a] Values taken from [5.12] unless otherwise noted.
[b] Values are given for 25 °C unless otherwise noted.
[c] [5.24].
[d] [5.25].
[e] At about 800 °C.
[f] [5.26].
[g] [5.27].
[h] [5.28].
[i] [5.29].

We have gone into some detail on this subject because of its obvious importance in practical systems. A further understanding of the causes of hysteresis, and its reduction in practice, are high-priority goals of future research.

5.2.4 Thermodynamics of Hydrides

Detailed consideration of hydride thermodynamics is beyond the scope of this chapter; good treatments of the subject are already available elsewhere [5.12]. For storage purposes, a few important points need to be remembered. In all practical systems, the hydrogen-metal reaction is exothermic. The heat evolved ranges from about 4 to 50 kcal per mole of H_2. Table 5.1 gives some representative values of this quantity. Most of the numbers were derived from P–T–C desorption data taken in plateau regions, using the *Van't Hoff*-type equation,

$$\frac{d(\ln p)}{d(1/T)} = \frac{\Delta \mathrm{H}}{R}.$$

A number of approximations and assumptions enter into the application of this equation, and the results must be accepted with some reserve. (For a critique of the procedure, see an article by *Flanagan* and *Lynch* [5.30].) Calorimetry yields more accurate heats of reaction, but unfortunately not many metal-hydrogen systems have been so dealt with. Where they have been, reasonable agreement with the *Van't Hoff* results is usually found, and the latter should at least serve engineering needs.

Table 5.1 also includes data on entropies of formation of hydride phases. These, it will be noted, show some variation, but much less than the heats. A typical entropy has the rather large negative value of 30 cal. deg^{-1} (mole $H_2)^{-1}$; most of this is, of course, simply attributable to the disappearance of a mole of gas during the reaction. The entropy term can have a rather large destabilizing effect on hydrides with a relatively low heat of formation. An interesting example is AlH_3, for which are reported the following, at 25 °C: $\Delta G_{f,298'}$ +11.1 kcal/mole; $\Delta H_{f,298'}$ −2.7 kcal/mole [5.25]. It is thus unstable toward decomposition, but not explosively so because the decomposition is endothermic. In practice it does not dissociate at an appreciable rate below 100 °C.

Passing mention should be made here of those metals that do not form hydrides but will dissolve a little hydrogen endothermally. Iron is an example. Since the heat of solution is positive, the solubility increases with increasing temperature. It never becomes large enough to be of storage interest, although the phenomenon is of great metallurgical importance.

5.2.5 Alloy-Hydrogen Reactions

Let a hydride-forming metal A capable of the reaction

$$A + x/2H_2 \rightleftarrows AH_x \tag{5.1}$$

be alloyed with a non-hydride-forming metal B with which it forms a solid solution, and let the relative partial molar free energy of A in the alloy be $\overline{\Delta G_A}$. If the alloy reacts with hydrogen according to the equation

$$AB_n + x/2H_2 \rightarrow AH_x + nB\,, \tag{5.2}$$

the equilibrium hydrogen pressure P in the system will be given by

$$P = P' \exp\left(-\frac{2\overline{\Delta G_A}}{xRT}\right),$$

where P' is the pressure in equilibrium with pure A and AH_x. P is thus always higher than P', since the sign of $\overline{\Delta G_A}$ is always negative. The difference will be greater, the more the alloy system shows negative deviation from Raoult's law ideality. If A and B form an intermetallic compound, the effect may be much greater. Indeed, since many intermetallic compounds are far more exothermic than the corresponding hydrides, they will, for all practical purposes, not react with hydrogen at all. By the same token, the alloying metal can react violently with the pure hydride of A, liberating hydrogen. Thus, rare earth hydrides are decomposed by mercury [5.31]. Also, the effect has found practical use in the manufacture of foamed metals. Since zirconium forms a very exothermic compound with aluminum, its hydride decomposes when added to molten aluminum [5.12, 32].

The P–T–C relations in a system where a reaction of type (5.2) occurs are describable, ideally, by curves like that of Fig. 5.1. The alloy will in general dissolve a little interstitial hydrogen without change, up to point "A". Then the formation of the hydride phase takes place, at constant pressure, and finally the third branch of the curve describes the same process as in the original. The enthalpy change, like the free energy change, will have a smaller negative value for the alloy reaction than for that of pure metal, and consequently the slope of the $\log P$ vs $1/T$ isochores will be less.

It often happens that an alloy AB does not decompose in hydrogen but reacts to form a wholly new chemical species known as a ternary hydride,

$$AB + x/2H_2 \rightarrow ABH_x. \tag{5.3}$$

Most such compounds exist only in the solid state; their claim to chemical individuality rests on the combination of P–T–C data with x-ray evidence of characteristic structure. For these systems too, the curves of Fig. 5.1 apply. Generally, the plateau pressures are higher than for the corresponding reaction of type (5.1), sometimes by many orders of magnitude. Thus, in one of the pioneering researches in this field, *Libowitz* et al. [5.33] found that the dissociation pressure of $ZrNiH_2$ was about 10^{10} times as large as that of ZrH_2, at 250 °C. An inverse correlation between the stability of certain intermetallic compounds and the stability of their ternary hydrides was proposed by *Miedema* et al. [5.34].

More complex reactions than those of (5.2) and (5.3) can easily be imagined, and some have actually been found. For example, the type

$$2A_2B + 3x/2H_2 \rightarrow 3AH_x + AB_2 \tag{5.4}$$

occurs with the intermetallic compound Mg_2Cu. Also, if both A and B are hydride formers, further variations are clearly possible.

Alternative hydriding reactions are usually possible in systems where ternary hydrides form. For example, the alloy phase $LaNi_5$ takes up hydrogen to form a homogeneous ternary with six to seven hydrogen atoms [5.26, 35–37]. The thermodynamics of the system are such that it could as well have decomposed to form the binary hydride LaH_2 plus nickel. In fact, the free energy change for binary formation has a greater negative value than that for the ternary. That it does not happen, in this and in many other cases, is one of the many intriguing questions in metal hydride chemistry. From a practical point of view, the possibility of the gradual development of thermodynamically possible side reactions over the course of repeated hydriding-dehydriding cycles must be kept in mind.

5.2.6 Complex Hydrides

The ternary hydrides we have been discussing are to be distinguished from the so-called complex hydrides. These are salts of the boranate (borohydride, tetrahydroborate) and alanate ions, BH_4^- and AlH_4^-, and differ in many ways from alloy hydrides. As molecular compounds, they can exist as solids, liquids, or gases, and can dissolve without decomposition in appropriate solvents. They are nonconducting in the pure state, but dissociate in ionizing solvents. Some of them are unstable and decompose readily, but this process is usually irreversible and often complex. Similarly, they are not usually able to be synthesized by direct combination of the elements, but must be made from other hydrides by metathesis reactions in anhydrous solvents. A compound like $NaBH_4$ is to be thought of as the double hydride $NaH \cdot BH_3$ rather than the hydride of the nonexistent alloy NaB. In spite of their irreversibility, the complex hydrides are not to be excluded from consideration as hydrogen storage media. Some possibilities are discussed below (Sect. 5.3).

5.2.7 Kinetic Considerations

Much less research has been done on kinetic than on equilibrium aspects of hydrogen-metal systems, and the subject is in a rather unsatisfactory condition. The available results hardly lend themselves to generalizations, each worker's data being applicable only to a particular material under particular circumstances. There is more than one answer to each of the fundamental kinetic questions, viz.: What is the functional relationship between the extent of reaction and the rate? How are rate constants affected by temperature and pressure? What process will be rate limiting in a given situation? etc. We shall here briefly consider several factors which complicate hydrogen-metal kinetics, illustrating with occasional references to actual systems that have been reported.

A very important consideration is the state of the metal surface. All hydride-forming metals except palladium have an even greater affinity for oxygen than for hydrogen and are normally coated with an oxide film. Such films often provide a high degree of protection against attack by gaseous reactants, and must be removed or disrupted before hydride formation can proceed. Various methods are effective for this purpose. In the case of zirconium and titanium, surface oxide can be caused to go into solution in the bulk metal by a heat treatment in vacuo or under hydrogen. With other metals the oxide film is not very soluble, but its protective quality can be destroyed by cycling the sample between high and low temperatures in the presence of hydrogen [5.27]. In some cases it suffices to expose the metal to a high pressure of hydrogen at room temperature for a long period (the "induction period"). Once it has been hydrided, further increase in reactivity can be achieved by subjecting the material to successive cycles of hydriding and dehydriding. A good illustrative case, that of Th_2Al, is described, with graphs, by *Fast* [5.38].

When a metal, once it is activated, reacts with hydrogen, the reaction product will at first consist of a solution of hydrogen in the alpha phase of the metal. If, at the operating temperature, the limiting solubility of the hydride phase is very small, a film of that phase can form on the surface. In some cases, e.g., sodium, this film inhibits further reaction, but generally it has little protective quality. If the limiting solubility of the hydride is large, the reacted hydrogen can more easily diffuse away from the surface and further reaction will not be impeded. In such cases, the rate may be limited by the diffusion process, or by a slow step, such as dissociation of the hydrogen molecule, at the surface. Both equilibrium and kinetic data are consistent with dissociation being the necessary precursor to reaction; hydrogen enters the metal only as single atoms, or ions. It is known that not every collision of hydrogen with a clean metal surface results in reaction. *Reimann* [5.39] exposed a magnesium film that had been freshly prepared by evaporation to hydrogen at low pressures and determined that only about one in every 1.5×10^6 incident molecules was absorbed by the surface. When the gas was partially atomized, by contact with incandescent tungsten, the hydrogen uptake was "enormously faster". *Atkinson* et al. [5.40] recently reported much higher sticking probabilities for molecular hydrogen on evaporated films of rare earth metals.

Many of the same mechanistic considerations apply to the reverse process, the desorption of hydrogen from a metal solution or a hydride. Both the diffusion of hydrogen atoms and their recombination at the surface may be important slow steps. *Pryde* and *Tsong* [5.41] consider the latter to be limiting in the case of the outgassing of a tantalum wire; *Schoenfelder* and *Swisher* [5.42] concur, in the case of 3.8×10^{-2} cm titanium sheet. The relative importance of the two processes will depend on the dimensions of the sample, or the particle size in the case of a powder.

A surface film can inhibit the release of hydrogen from a hydride, presumably by hindering the recombination of hydrogen atoms. In such cases the observed pressure of hydrogen may be very much less than its equilibrium value. An example is the alloy hydride $TiFeH_{1.9}$, which has a dissociation pressure of around 5 atmospheres at room temperature. If it is cooled to liquid nitrogen temperatures, exposed to air, and warmed back up to room temperature it is found to retain its hydrogen indefinitely [5.27]. (This circumstance permits the taking of x-ray diffraction pictures without high-pressure apparatus.) A similar effect was recently reported by *Gualtieri* et al. [5.43], who found the release of hydrogen from the hydride of $LaNi_5$ to be inhibited by sulfur dioxide.

Still another slow step can be the transformation of a saturated solution of hydrogen in metal into a hydride phase. As mentioned above, resistance to this process is believed responsible for the hysteresis phenomenon, and *Boser* [5.44] suggested that it is rate limiting in the absorption of hydrogen by $LaNi_5$ and also in its desorption.

Practically, one would like to relate reaction rates with the extent of reaction, but there is little consistency in the literature. In a recent very careful research on uranium powder, *Condon* and *Larson* [5.45] found first-order kinetics, the hydriding rate being proportional to the mass of unreacted uranium. *Peterson* and *Westlake* [5.46] obtained a parabolic rate law on bulk thorium. *Gulbransen* and *Andrew* found a rather slight dependence of rate on extent of reaction for titanium [5.47] and zirconium [5.48]. For dehydriding, *Condon* and *Larson*'s kinetics were approximately zero order.

There is no better agreement on the effect of hydrogen pressure on reaction rate. *Condon* and *Larson* reported an absorption rate proportional to the square root of pressure. *Wicke* and *Otto* [5.17] gave the relation for bulk uranium in hydrogen below 100 °C as

$$\frac{dn}{dt} = \frac{ab\sqrt{p}}{b + a\sqrt{p}},$$

where n is moles of H_2 consumed, a and b are temperature-dependent constants, and p is the pressure. As the temperature was raised, however, the functional relationship changed, and by 250 °C there was a linear dependence of rate on pressure. With thorium, *Peterson* and *Westlake* [5.46] found a rate nearly independent of pressure at temperatures below 550 °C and pressures above 100 Torr.

It sometimes happens that one wants to hydride a metal at a temperature at which the product has an appreciable decomposition pressure. In such cases, it is appropriate to seek correlations between the rate and the quantity $p - p_0$ rather than p, where p_0 is the equilibrium absorption pressure. Where hysteresis exists, p_0 would be given by the upper curve. In order to get a reasonably fast uptake in practice, the applied pressure must often be much higher than p_0. With iron-titanium, for example, good rates are obtained only under 30 atmospheres at 25 °C; this is three times as high as the pressure of the higher hydride at that temperature [5.49].

The effect of increased temperature is, as one might expect, increased reaction rates. And yet there is a reported exception even to this: *Condon* and *Larson* [5.45] found a negative temperature coefficient for uranium and hydrogen. However, most hydride workers report the normal temperature effect. It must be remembered, of course, that the dissociation pressure p_0 increases with increasing temperature. In a constant-pressure system the quantity $p - p_0$, which is related to the driving force of the hydriding reaction, consequently decreases with increasing temperature. In such a system, therefore, the hydriding rate could go through a maximum and decrease to zero as p_0 approached p.

In situations where surface reactions are rate limiting, catalysts may be useful. The first direct synthesis of magnesium hydride involved the use of MgI_2 [5.50]. Other workers have used iodine [5.51], carbon tetrachloride [5.51], $HgCl_2$ [5.52], allyl iodide [5.53], and Mg_2Ni [5.24]. How most of these act remains unknown. The last named, however, probably owes its activity to its own ability to form a hydride, namely, Mg_2NiH_4. This then acts as what *Wicke* terms a "transference catalyst". It had been observed by him and his co-workers [5.18, 54, 55] that the hydrides of uranium, titanium, zirconium, thorium, and cerium catalyze the absorption of hydrogen by bulk palladium and that uranium and titanium hydrides catalyze the formation of tantalum hydride. A self-catalysis was also observed; when titanium was put in contact with titanium hydride powder, it reacted with hydrogen more readily. *Carstens* et al. observed catalysis by $LaNi_5$ [5.56].

When conditions are such that reaction at the gas-solid interface is fast, rate limitation by heat transfer can set in. This may well be the most important factor to consider in engineering—scale reactions, especially in dehydriding. The rate at which heat can be supplied to a reservoir will depend on the extent and arrangement of heat transfer surfaces, the temperature and flow rate of heat transfer fluid, and the thermal conductivity of the material. This last will be a function of the nature of the material, its particle size, and the gas pressure. Values of the order of $1\,\mathrm{Btu\,h^{-1}\,ft^{-1}\,{}^{\circ}F^{-1}}$ ($0.017\,\mathrm{W\,cm^{-1}\,{}^{\circ}C^{-1}}$) have been reported [5.57].

Additional kinetic data may be found in [5.56, 58].

5.2.8 Hydrogen Content

An important consideration in the choice of a hydride for hydrogen storage is the amount stored per unit volume and per unit weight of the medium. The first is often very large, amounting in one case to 2.5 times the density of liquid hydrogen. Table 5.2 gives densities for a number of hydrides, expressed as N_H, the number of atoms of hydrogen per cubic centimeter of hydride, times 10^{-22}. Also in this table are values for the corresponding weight percent of stored hydrogen. This is a quantity that is important to maximize in applications where the hydrogen reservoir must be mobile, as in automotive use.

5.2.9 Temperature and Enthalpy Criteria

A necessary condition for a hydride's usefulness for storage is that the hydrogen be recoverable under the conditions of use. That is, heat of the requisite quality and quantity must be available to supply the enthalpy of dissociation. The thermal qualities desired in a hydride will thus depend on the application. If the

Table 5.2. Hydrogen content of various media[a]

Medium	Density [g cm^{-3}]	Weight percent H	N_H, atoms H per cm^3, $\times 10^{-22}$
H_2, gas at 100 atm	8.2×10^{-5}	100	0.49
H_2, liquid	0.071	100	4.2
Water	1.0	11.2	6.7
Ammonia, liquid	0.6	17.8	6.5
LiH	0.8	12.7	5.3
NaH	1.4	4.2	2.3
MgH_2	1.4	7.6	6.7
Mg_2NiH_4	2.6	3.8	5.9
CaH_2	1.8	4.8	5.1
$CaH_2 + H_2O$	—	6.7[b]	—
AlH_3	1.48	10.1	8.9
$LiAlH_4$	0.91	10.6	5.74
CeH_3	5.5	2.1	7.0
TiH_2	3.8	4.0	9.0
$TiFeH_{1.93}$	5.47	1.8	6.0
$LaNi_5H_{6.7}$	8.25	1.5	7.58
VH_2	4.5	2.1	10.3

[a] Sources are standard tables [5.12] and the references in Table 5.1.
[b] Hydrogen evolved in the reaction $CaH_2 + H_2O \rightarrow CaO + 2H_2$.

application involves thermal storage, like the load-leveling scheme described on p. 230, a rather large enthalpy may be desirable. If, on the other hand, hydrogen is to be stored for later use as a combustion fuel, the use of waste heat is indicated for the dissociation process. Fortunately, every energy-producing device to which hydrogen is likely to be supplied as a fuel produces waste heat equal to at least half the heat of combustion. This can be used to dissociate the hydride if the operating temperature of the device is such that at that temperature the dissociation pressure of the hydride is at least 1 atm. One might thus select quite different hydrides to supply aqueous fuel cells, on the one hand, and internal combustion engines, on the other.

Equally important is the total amount of heat that can be transferred to the hydride reservoir, relative to the enthalpy of dissociation. The latter covers a wide range of values depending on the nature of the hydride; see Table 5.1 for the ΔH_f of some representative hydrides. The heat of combustion of hydrogen is -68.3 kcal/mole, if the product is liquid water, or -57.8 for a vapor product. Assume the upper value and suppose that 75% of it takes the form of waste heat. In an actual integrated system of engine plus hydride reservoir, perhaps one-third of this waste heat can be transferred to the hydride at a temperature equal to or higher than the temperature corresponding to 1 atm dissociation pressure. The fraction transferrable is quite dependent on that temperature, but it is clear that for most uses it will be best to focus attention on those hydrides with ΔH_f values with smaller (less negative) values than -20 kcal/mole.

5.3 Survey of Hydride Storage Systems

In this section we shall present a synopsis of most of the known solid metal hydrides with regard to their suitability for hydrogen storage. Hydride properties correlate rather well with the column of the periodic table of elements to which the parent metal belongs, so it will be convenient to group them accordingly in what follows.

5.3.1 Alkali Metals and Their Alloys

The hydrides of lithium, sodium, potassium, rubidium, and cesium are saltlike compounds in which hydrogen plays the part of an anion. All are relatively stable; their dissociation pressures reach 1 atm at 894, 421, 428, 364, and 389 °C respectively [5.12]. Lithium hydride is unusual in that it is stable to temperatures well above its melting point, 680 °C; no other binary hydride exists in the liquid state. For all of the hydrides in this group, the melting point of the parent metal is lower than the decomposition temperature of the hydride, a circumstance which could pose difficulties for a cyclical storage unit. The main obstacle to their use in such units, however, is the magnitude of the required operating temperatures. This effectively rules them out.

One alloy of lithium, LiAl, is of possible interest for hydrogen storage. It apparently reacts with hydrogen by simple decomposition:

$$LiAl + 1/2\,H_2 \rightleftarrows LiH + Al\,. \tag{5.5}$$

The free energy of formation of the alloy is such that the equilibrium pressure of hydrogen is much higher than in the Li–LiH system [5.59]. It reaches 1 atmosphere at about 580 °C (extrapolated) and the enthalpy change accompanying reaction (5.5) is 9.6 kcal, versus 12.6 for the hydriding of pure lithium, at 500 °C. The lowering of the decomposition temperature that is effected by the presence of aluminum is still insufficient to make the system practical for most applications. Note that the system LiH + Al is not to be confused with the complex compound $LiAlH_4$, which will be touched on below.

Two double hydrides of LiH with alkaline earth hydrides have been reported: $LiSrH_3$ and $LiBaH_3$ [5.60]. Structural information is available on these (they have the perovskite structure) but no dissociation pressures. From the preparation conditions, they are evidently stable to 300 °C or higher. An analogous double hydride with europium, $LiEuH_3$, has also been found [5.61]. Other double hydrides of alkali metals that have been reported are Li_3CdH_3 and Li_3CdH [5.62]; Li_4RhH_4 and Li_4RhH_5 [5.63]; a series involving LiH and Rh, Ir, Pd, and Pt [5.64]; K_2TcH_9 [5.65] and K_2ReH_9 [5.66]; and Li_2BeH_4 and Na_2BeH_4 [5.67]. The last two are said to decompose at 350 and 390 °C respectively.

No other ternary systems in this group have been examined carefully, although some qualitative statements about the ability of lithium hydride to react with silver [5.62], mercury [5.68], and other materials [5.69] have been published.

5.3.2 Alkaline Earth Metals and Their Alloys

Beryllium differs so much from the other IIA elements that it can scarcely be called an alkaline earth. The difference is especially marked in the case of its hydride, which is much less stable than those of its heavier homologs. It is, in fact, thermodynamically unstable relative to decomposition into its elements. It can be made only by indirect methods involving metal-organic compounds in organic solvents, and there is some question as to whether it has ever been obtained in truly solvent-free state. In spite of its high hydrogen content (15.6% by weight), both the high cost of beryllium and the difficulty of making the hydride rule out BeH_2 as a storage medium.

Magnesium hydride is more promising. It contains 7.65% hydrogen, dissociates at 287 °C, requires only 17.8 kcal for dissociation, and is inexpensive. Given a rather careful design of the integrated system, it might be possible to use MgH_2 as the source of hydrogen fuel for a combustion engine in which the exhaust was unusually hot and contained most of the waste heat, or for a high-temperature fuel cell such as those using solid oxides or molten carbonate as electrolyte. The reconstitution of the hydride ordinarily requires rather high temperatures and pressures, but the addition of catalysts such as Mg_2Ni permits the use of milder conditions. Problems of safety might arise when a reservoir was in the dehydrided condition, since the residue would be a high-area, flammable magnesium, and an accidental rupture could result in ignition. In the hydrided state it is much more stable. Such safety considerations, which apply to many other metal-hydrogen systems, probably pose no greater problem to the designer than minimizing gasoline fires in most present-day automobiles. Experiments on hydride safety are reported in [5.58, 70].

Magnesium forms many alloys which can react with hydrogen. Several of the systems appear interesting for storage purposes, and pressure-composition-temperature data have been obtained. One of the earliest was the magnesium-copper intermetallic compound Mg_2Cu. It has a rather complex orthorhombic structure C_b. Its reaction with hydrogen apparently proceeds as follows:

$$2MgCu + 3H_2 \rightarrow 3MgH_2 + MgCu_2 .$$

The equilibrium pressure reaches 1 atm at 239 °C, notably lower than that for MgH_2 [5.71]. The enthalpy change of the reaction is -17.4 kcal per mole H_2.

In the magnesium-nickel system there is also a compound in the 2:1 proportion, Mg_2Ni. Its structure is different (C_a, hexagonal) and its reaction with hydrogen differs. It proceeds

$$Mg_2Ni + H_2 \rightleftarrows Mg_2NiH_4 .$$

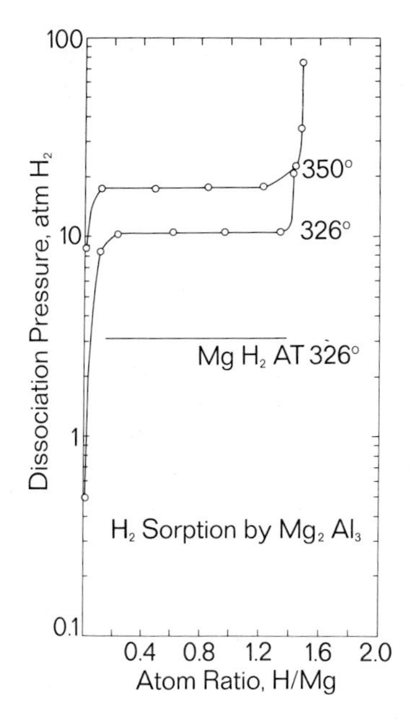

Fig. 5.6. Pressure vs composition isotherms for the system, Mg_2Al_3–H_2

Here too the equilibrium pressure is higher, reaching 1 atm at 253 °C [5.24]. The enthalpy change of the reaction is −15.4 kcal.

Both Mg_2Cu and Mg_2Ni are much more easily hydrided than Mg. In fact, small amounts of Mg_2Ni added to Mg will promote the hydriding of the latter. The lower dissociation temperatures and enthalpies of both make the design engineer's requirements for an integrated system less stringent than for Mg, but this is somewhat offset by the loss in hydrogen content. The values are 3.6 wt. % for Mg_2NiH_4 and 2.6 for ($MgCu_2 + 3MgH_2$).

In the magnesium-aluminum system there are several intermediate phases which react with hydrogen, and some of these are of practical interest [5.11]. At magnesium contents of 79 wt. % (80.7 at. %) and above, the dissociation pressures are hardly distinguishable from those of MgH_2. The γ phase alloy, however, of approximate composition Mg_5Al_4 gives pressures that are definitely higher, and for the β phase alloy, approximately Mg_2Al_3, the plateaus are higher still (Fig. 5.6). The nature of the reactions has not been definitely established, but in the product of hydriding Mg_2Al_3 there was x-ray diffraction evidence only for MgH_2 and Al, pointing to the reaction

$$Mg_2Al_3 + 2H_2 \rightarrow 2MgH_2 + 3Al.$$

This calls for a H/Mg ratio of 2 in the hydrided material, but as Fig. 5.6 shows, the maximum actually obtained was about 1.4. Rather higher ratios were found with Mg_5Al_4. These results are all of a preliminary nature, and subject to revision. The ease with which Al–Mg alloy can be hydrided is very dependent on the particle size, and reproducible results are harder to get than in most alloy systems. More work needs to be done.

A large number of other magnesium alloys have been screened by the Brookhaven group [5.11]. The results, while of considerable scientific interest, did not uncover any likely new candidates for storage. Among the alloys investigated were Mg_3Ag, $Mg_{17}Ba_2$, Mg_3Cd, Mg_9Ce, $Mg_2Cu_{0.5}Ni_{0.5}$, Mg_3Sb_2, MgSn, and MgZn. Other workers have studied the reaction of hydrogen with MgCe [5.72] and MgAu [Ref. 5.73, p. 7] but without results of practical importance.

The remaining three alkaline earth hydrides are much more stable than MgH_2. They are considered to be saline in nature, like the hydrides of the alkali metals, whereas MgH_2 probably has covalent bonding. The dissociation temperatures are as follows: CaH_2, 1074 °C; SrH_2, 992 °C; BaH_2, 943 °C [5.12]. These are so high as to effectively eliminate them as reversible storage media. By hydrolysis, CaH_2 can furnish large amounts of hydrogen on a one-time basis, but the cost limits the use of this reaction to rather special applications.

An alloy of calcium, $CaNi_5$, is of possible interest for hydrogen storage. It is said to take up hydrogen to approximately the composition $CaNi_5H_6$, with a room-temperature dissociation pressure variously reported as 15 atm [5.26] and 0.5 atm [5.74]. If part of the calcium is replaced by a rare-earth mixture (Mischmetal), the pressure increases [5.26]. Information has also been published on the hydriding of alloys of calcium with magnesium [5.75], aluminum (Ref. [5.73], p. 3), and lanthanum [5.76] and on complex hydrides of calcium with noble metals [5.64], but none of the products appears to be useful for storage.

5.3.3 Metals of Group III and Their Alloys

Aluminum, like beryllium, is very different from the heavier members of its group. Its hydride AlH_3 is unstable and must be made by reactions like

$$3LiAlH_4 + AlCl_3 \xrightarrow{(C_2H_5)_2O} 4AlH_3 + 3LiCl .$$

At one time it was thought to be of interest as a rocket fuel, and its chemistry was investigated rather thoroughly. Methods were developed for obtaining solvent-free material [5.77] and much information was obtained on its thermodynamic and crystallographic properties. If it could be made cheaply it would be an excellent storage material, since it contains 10% hydrogen and decomposes on heating to a little over 100 °C. Although theoretically unstable, it can be kept indefinitely at room temperature. The indirect syntheses are too costly, however, and the compound cannot be made directly from the elements even when hydrogen pressures of 4000 psi (2.8×10^7 Pa) at -196 °C, or 8200 psi

(5.7×10^7 Pa) at room temperature are used [5.11]. This lack of reaction is consistent with the thermodynamic data.

Aluminum forms alloys with many hydride-forming metals; their properties are discussed under those metals. As for its alloys with other metals, a number were tested at Brookhaven under 20 atm or more of hydrogen, but none reacted. They included alloys of the approximate formulas Al_5Fe_2, Al_5Co_2, Al_9Co_3, Al_3Ni, $AlNi_3$, AlCu, and AlAg [5.11].

Aluminum hydride is somewhat stabilized by combination with other hydrides. The resulting compounds, known as aluminohydrides or alanates, are examples of the saltlike complex hydrides mentioned on p. 212. They can be considered as constituted of various metal cations, and the appropriate number of anions AlH_4^-, so that typical formulas are $NaAlH_4$ and $Mg(AlH_4)_2$. They are quite unstable, and are made by such indirect, and expensive, reactions as

$$4LiH + AlCl_3 \xrightarrow{\text{solvent}} LiAlH_4 + 3LiCl\,.$$

It has been reported, however, that some of the alkali metal alanates can be made from the elements, under high hydrogen pressures [5.78–80].

The alanates give off hydrogen at readily accessible temperatures, but the reactions may be complex [5.81]. Thus, $NaAlH_4$ first decomposes to another complex hydride, Na_3AlH_6, which at a higher temperature itself decomposes to NaH,

$$NaAlH_4 \rightarrow 1/3Na_3AlH_6 + 2/3Al + H_2$$

$$Na_3AlH_6 \rightarrow 3NaH + Al + 3/2H_2\,.$$

Dymova et al. [5.82] reported equilibrium pressures for these reactions. For the first, they found 153.5 atm at 210 °C; for the second, 21.4 atm at the same temperature.

It is possible that further research on the alanates, both on their synthesis and decomposition reactions, would produce systems of interest for storage. On paper, the hydrogen content of $Mg(AlH_4)_2$ is attractive, but little is known of its properties. The analogous borohydrides, or salts of the anion BH_4^-, may also be worth a research effort. They have an even higher hydrogen content than the aluminohydrides, but on decomposition they may give off gaseous boron hydrides as well as hydrogen.

The Group III A metals beyond aluminum, namely, scandium, yttrium, and the rare earths, all form extremely stable dihydrides. That of yttrium is of interest for applications in which a solid of high hydrogen content is wanted at high temperatures (1200–1300 °C). Reversible storage, however, is out of the question. Most of this series also form a less stable trihydride. Only that of ytterbium has interesting P–T–C characteristics [5.83], but the amount of available hydrogen is small and of course the cost is prohibitive.

With the alloys of the rare earth metals we come to a large number of materials of great interest. A extensive series of papers from the Philips

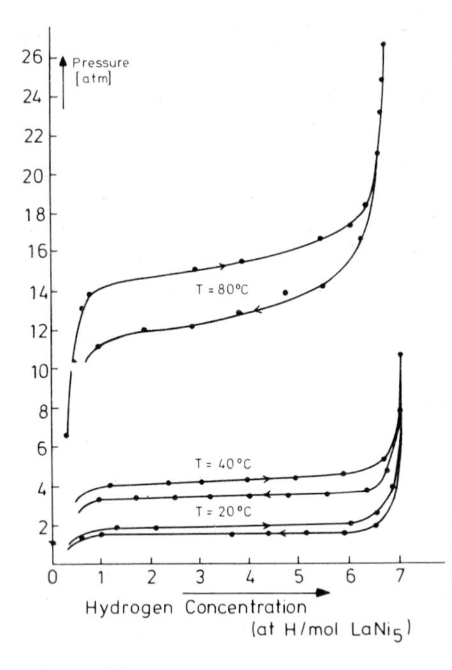

Fig. 5.7. Pressure vs composition isotherms for the system $LaNi_5$–H_2 ([5.23], courtesy Elsevier-Sequoia)

Research Laboratories, Eindhoven [5.26, 35–37, 84, 85] described the reaction with hydrogen of alloys of the formula AB_5, where A is a rare earth and B is a non-hydride-forming transition element, usually nickel or cobalt. Subsequently, other laboratories confirmed and extended these results [5.58, 86–89]. Pressure-temperature-composition information is available on systems of the following alloys with hydrogen: $LaNi_5$, $(La, Ce)Ni_5$, $PrNi_5$, $NdNi_5$, $SmNi_5$, $GdNi_5$, YNi_5, $(Y, Th)Ni_5$, $(La, Th)Ni_5$, $LaCo_5$, $PrCo_5$, $NdCo_5$, $CeCo_5$, $SmCo_5$, $GdCo_5$, $TbCo_5$; $LaNi_4M$, where M is Pd, Ag, Cu, Fe, Cr, or Co; $La_{0.8}M'_{0.2}Ni_5$, where M′ is Er, Y, Gd, Nd, Th, or Zr.

Space does not permit more than a brief summary of the results. The starting alloys are all structurally similar, having the hexagonal $CaCu_5$ structure. In their reactions with hydrogen they are alike in forming less stable hydrides than the parent rare earth metals, but the reactions differ very much in detail. The number of successive hydride phases in a system, the structures, the total amount of hydrogen that can be absorbed and the pressure-temperature relationships all vary. Recourse to the original literature is recommended for details, but we shall try to summarize the situation here with regard to storage requirements.

Lanthanum pentanickel, $LaNi_5$, may be taken as the prototype, and the rest can be viewed in comparison with it. Its P–T–C relationships are illustrated in Fig. 5.7. Notice that one $LaNi_5$ unit contains over six atoms (1.37 wt. %) of easily recoverable hydrogen, the room temperature dissociation pressure being around 2 atm. The extent of hysteresis is small, as low-temperature hydrides go.

An activated sample will rehydride under rather mild conditions; at 5 atm, the hydriding rate is quite adequate at 20 °C.

If a slightly different pressure-temperature relationship is desired, this is easily obtained. When cobalt replaces nickel in an AB_5 alloy, the new alloy forms a more stable (lower pressure) hydride. On progressing through the series $LaNi_5$, $LaNi_4Co$, $LaNi_3Co_2$, etc., to $LaCo_5$, a whole spectrum of hydrides of increasing stability becomes possible. If, on the other hand, a less stable hydride is wanted, it can be made by replacing more or less of the lanthanum in $LaNi_5$ by cerium. Alternatively, the lanthanum-nickel alloy can be made in non-stoichiometric proportions, instead of the 5 to 1 ratio. On varying x, in $LaNi_x$, from 4.9 to 5.5, the hydride plateau pressure changes from 2.75 to 9.2 atm [5.90].

Most of the AB_5 compounds take up rather less hydrogen than $LaNi_5$. The highest known content, however, was found for $LaCo_5$ when it was exposed to a pressure of 137.3 MPa (1334.7 atm); $LaCo_5H_9$ was obtained [5.86]. In the pressure range of practical interest, however, this alloy's isotherm does not have a suitable form for storage purposes.

A useful feature of certain AB_5 alloys is their relative immunity to deactivation by oxygen-containing gases. Most hydride-forming metals are very sensitive to traces of air or moisture. Lanthanum pentanickel, however, proved able to pick up hydrogen out of gas mixtures, as shown in Table 5.3. The components added were those that would be encountered in gas produced by the steam reforming of a hydrocarbon fuel. Of these, carbon monoxide was the most effective in inhibiting hydrogen uptake. When part of the nickel in the alloy was replaced by copper, the carbon monoxide poisoning effect was slightly reduced [5.87].

Rare earths and transition metals form alloys in other than the 1 to 5 proportion, and some of these have also been tested in hydrogen. They include $GdCo_3$, $TbCo_3$, $DyCo_3$, $HoCo_3$, and $ErFe_3$ [5.22]; $ErCo_3$, $DyCo_3$, and $HoCo_3$ [5.91]; a series of La–Ni alloys from La_3Ni to La_2Ni_7 [5.75, 92]; some alloys of yttrium with Mn, Fe, Co, and Ni [5.75]; and a series of Pr–Co alloys, from $PrCo_2$ to Pr_2Co_{17} [5.93]. In these last two series, the greater the A/B ratio, the greater the hydride stability. As a result, none of the materials tested appears to be superior to $LaNi_5$. Of the AB_3 compounds, several took up reasonable amounts of hydrogen. The N_H of $GdCo_3H_{4.6}$ was 6.2×10^{22} atoms H per cubic centimeter, and its upper plateau pressure was 3.69 atm at 150 °C [5.22]. All rare earths are expensive, and even a 1–5 alloy with nickel may be ruled out on this ground for most purposes. A possible economy could come from the use of Mischmetal, the commercial mixture of rare earths, which is much cheaper than lanthanum or cerium. The Mischmetal pentanickel alloy "$MmNi_5$" forms a hydride with a rather high dissociation pressure—about 10 to 11 atmospheres at room temperature [5.11]. Mischmetals of different origins give different pressures, because of the varying proportions of the several rare earths. With a given Mischmetal, the dissociation pressure could presumably be lowered appreciably by replacing part of the nickel with cobalt.

Table 5.3. Hydriding of various alloys in hydrogen-containing gas mixtures

Alloy	Composition of gas	Temp. [°C]	Pressure [atm]	Flow rate [l/min]	Total flow [l]	Product composition
$LaNi_5$	74 H_2 26 CO_2	25	27		0.6	$LaNi_5H_{6.1}$
$LaNi_5$	99 H_2 1 CO	25	27 – 17	0.4	116	(No H pickup)
$LaNi_5$	99.95 H_2 0.05 CO	25	33 – 14		8.0	$LaNi_5H_{5.7}$
$LaNi_5$	97.0 H_2 3.0 air	25	32 – 26	0.1	1.0	$LaNi_5H_{5.1}$
$LaNi_5$	72 H_2 28 CO_2 sat. H_2O	25	19 – 13		6.0	$LaNi_5H_{4.1}$
$LaNi_5$	79.3 H_2 20.3 CO_2 0.3 CH_4 700 ppm N_2 20 ppm CO sat. H_2O	25	12	0.17	4.00	$LaNi_5H_{6.4}$
$LaCu_4Ni$	100 H_2	22	39	static	—	$LaCu_4NiH_{4.97}$
$LaCu_4Ni$	72 H_2 24 CO_2 4 CO sat. H_2O	124	29 – 31	static	—	$LaCu_4NiH_{2.7}$

5.3.4 Titanium, Zirconium, Hafnium, and Their Alloys

As we move across the periodic table, metal hydrides increasingly take on the character of alloys. In Group IV, they still resemble chemical compounds in their highly exothermic character, however. The hydrides of titanium, zirconium, and hafnium are stable to very high temperatures. It is hard to give a meaningful dissociation temperature for them, since plateaus are narrow at high temperature, and over most of the composition range the pressure is a more or less steeply varying function of the hydrogen content of the solid phase. Detailed phase diagrams may be found in [5.12]. For our present purposes, it may be said that the three metals are unsuitable for reversible storage. Many of their alloys, however, react with hydrogen to give ternary hydrides, and some of these have attracted interest as storage media. The best known is the intermetallic compound TiFe, which will take up nearly two atoms of hydrogen. Its P–T–C behavior is illustrated in Fig. 5.4. The shapes of the isotherms suggest the existence of two successive hydride phases, and this is confirmed by x-ray evidence [5.27]. Figure 5.8 is a phase diagram of the system. From the temperature dependence of the dissociation pressure, the integral heat of formation of the dihydride is about 7 kcal (29 kJ).

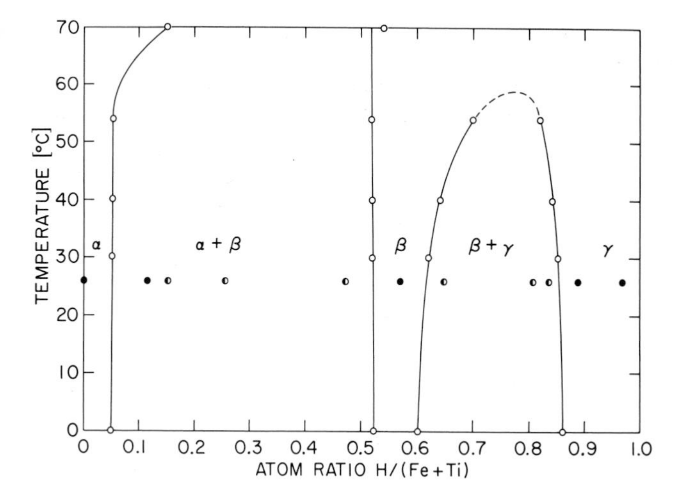

Fig. 5.8. Phase diagram for the system TiFe–H_2. X-ray evidence for one or two solid phases is indicated by ○ or ◑, respectively. Phase boundaries are derived from P–T–C data

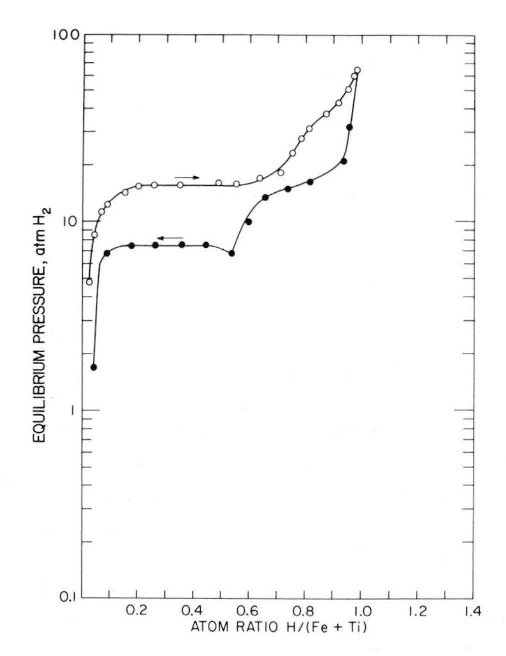

Fig. 5.9. Sorption hysteresis in the system TiFe–H_2, at 40 °C

There is a rather large hysteresis effect, as shown in Fig. 5.9. In a practical hydriding operation, a pressure much higher than the upper branch of the hysteresis loop is required. With an activated sample, rapid hydrogen uptake is obtained with an applied pressure of 30 atm at 20 °C. In $TiFeH_2$, the weight percent of hydrogen would be 1.91; but in an actual system, perhaps cycling between $TiFeH_{0.1}$ and $TiFeH_{1.8}$, this is reduced to 1.62%. Powdered TiFe

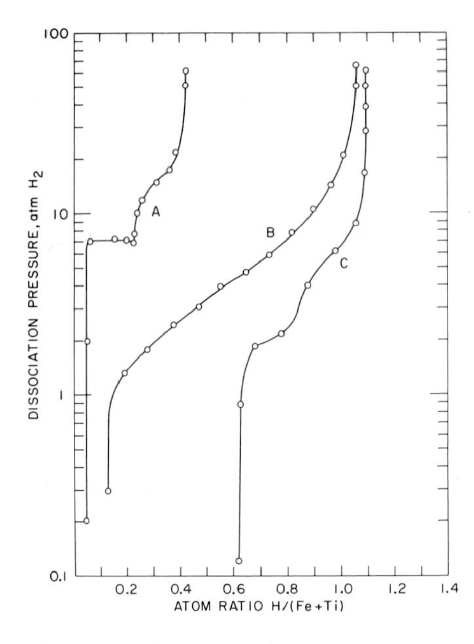

Fig. 5.10. Effect of composition on TiFe–H_2 isotherm, at 40°. (A) 39.5 Ti, 60.5 Fe; (B) 49.2 Ti, 50.5 Fe; (C) 63.2 Ti, 36.7 Fe (all in wt-%)

hydride is not pyrophoric, and rather extensive tests have shown it to be a safe material to handle [5.70]. Titanium-iron is, on the whole, probably the best hydrogen storage medium yet found for stationary applications where only low-grade waste heat is available. It has been the subject of development work which is described below in Section 5.4.2.

Many alloys that can be regarded as derivatives of TiFe have been tested in hydrogen and present a wide variety in their P–T–C behavior. The phase TiFe can vary from the integral proportion, and changing the amount of titanium by a few atomic percent changes the isotherm very appreciably, as Fig. 5.10 shows. Adding still more titanium results in a two-phase system, Ti and TiFe; the two phases hydride separately to give a mixture of products including TiH_2. On the iron-rich side, a mixture of TiFe and $TiFe_2$ is obtained. The latter phase does not react with hydrogen [5.27].

Part of the Fe in TiFe can be replaced by other transition metals of similar size, like Cr, Mn, Co, and Ni, and these substitutions result in changed behavior toward hydrogen. Figure 5.11 gives an example [5.94].

Titanium forms intermetallic compounds with other transition elements, and several of these have been found to react with hydrogen. TiCo, which is structurally similar to TiFe, also forms a ternary hydride $TiCoH_x$. One pressure plateau is found, extending from $x=0.16$ to $x=0.9$. The plateau pressure is 1 atm at 110 °C [5.95] or 130 °C [5.11]. In any case, it is much more stable than the corresponding $FeTiH_x$. $TiCo_2$ does not react with hydrogen at up to 150 atm [5.11].

In the Ti–Ni system, TiNi forms a ternary of the approximate compositon $TiNiH_{1.4}$. It is more stable than the hydride of TiCo, reaching 1 atm only at

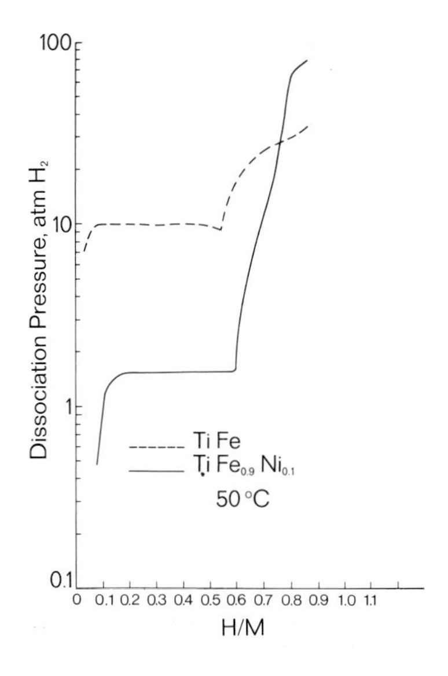

Fig. 5.11. Effect of nickel substitution on the system $TiFe–H_2$

200 °C. Ti_2Ni undergoes a decomposition in hydrogen, to form TiH_2 and $TiNiH_{1.4}$. $TiNi_3$ does not react with hydrogen. In the Ti–Cu system, all three intermediate phases—Ti_2Cu, TiCu, and $TiCu_2$—decompose to give TiH_2 and Cu [5.95]. The titanium-chromium system contains a single intermediate phase, approximately $TiCr_2$. This is interesting in that it forms what is perhaps the least stable reversible hydride yet known. Figure 5.12 illustrates its P–T–C behavior [5.94].

These systems have been singled out for mention because of their possible storage interest, although it is much less than that of TiFe. There is, in addition to the above references, quite an extensive literature on the behavior of titanium alloys in hydrogen. Space does not permit even a summary of these, especially since the systems studied are of little practical interest. The reader interested in alloy-hydrogen chemistry is referred particularly to the work of *McQuillan*, in such publications as reference [5.96]; to the results obtained at the Denver Research Institute, where a great many alloys were screened (though unfortunately at only 1 atm pressure) [5.73]; to the work of the Battelle Memorial Institute [5.97]; and to a number of other workers [5.98–100].

Zirconium alloys are of interest historically in providing one of the first examples of a ternary hydride, that of ZrNi [5.33]. The nickel exerts a strong destabilizing effect, and a plateau pressure of about 25 Torr is found at 200 °C. This is still too stable to be of practical interest, especially since the amount of available hydrogen is only about 1% by weight. The same is true for an interesting series of zirconium alloys investigated by *Pebler* and *Gulbransen*: ZrV_2, $ZrCr_2$, $ZrFe_2$, $ZrCo_2$, Zr_2Cu, Zr_2Ni, and $ZrMo_2$ [5.101]. As for hafnium, a few alloys have been hydrided, but none of the results is relevant to our present purpose [Ref. 5.73, p. 7].

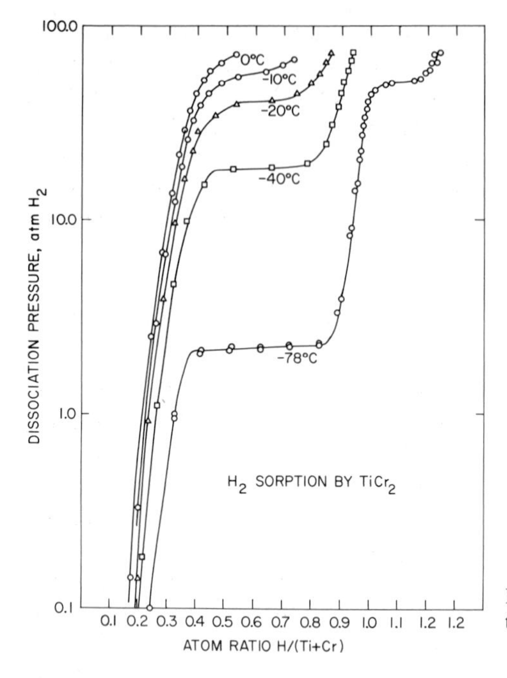

Fig. 5.12. Pressure vs composition isotherms for the system $TiCr_2$–H_2

5.3.5 Vanadium, Niobium, Tantalum, and Their Alloys

Each of these forms a relatively stable lower hydride of limiting composition $MH_{1.0}$. The two lighter elements also form much less stable dihydrides. The 1 atm dissociation temperature of VH_2 is 13 °C; that of NbH_2, 31 °C [5.29]. Figure 5.5 shows the P–C behavior of VH_2 at 45 °C. The plateau is reasonably wide, corresponding to about 1.9% available hydrogen, and the hysteresis is not excessive. The heat of the reaction

$$2VH_2 \rightarrow 2VH + H_2$$

is 9.6 kcal, that of the corresponding niobium reaction is the same. Large scale storage use of vanadium is ruled out by the high cost of the metal, but it could find use in applications that require only small quantities. One such is described below in Section 5.4.6.

An interesting feature of the V–H and Nb–H systems is the sensitivity of the dihydride stability to small amounts of metallic impurities in the starting metal. Thus, VH_2 made from commercial vanadium has twice the dissociation pressure of that from a zone-refined sample. The chief impurities in commercial vanadium are Ta, Mo, Si, and Fe, and of these, Si is the most effective in raising the VH_2 pressure. On adding 0.4 at. % of Si to V, the pressure is raised by 50%; while 1.66 at. % raises the pressure by a factor of 3.4 [5.102]. Additions of chromium also raise the VH_2 pressure; titanium additions, on the other hand, lower it. A spectrum of systems with different pressure-temperature characteristics thus becomes available here, as in the case of AB_5 alloys.

Little has been published on the effect of alloying on the properties of the lower hydrides of the Group V metals. *Kirschfeld* and *Sieverts* [5.103] found that iron diminished the hydrogen uptake. The Denver Research Institute [Ref. 5.73, p. 7] and the Battelle Memorial Institute [5.97] screened a number of alloys, but the storage value of the hydrides was either nil or impossible to ascertain from the published results. Two ternary hydrides of high hydrogen content, TiV_4H_8 and $TiV_{1.4}H_{4.6}$, were reported by *Gibb* [5.9], but they are rather stable [5.11].

5.3.6 Palladium, Uranium, Thorium, and Their Alloys

For completeness, we touch briefly on these three systems, which are of more scientific than practical interest. More research has been done on the reaction of hydrogen with palladium and its alloys than any other metal. A recent book on the subject summarizes the voluminous literature [5.104].

Uranium hydride UH_3 has also been investigated extensively, and has actually been used as a source of hydrogen for experimental purposes. It decomposes at a temperature, 430 °C, that is easily accessible in the laboratory, and the hydrogen evolved is very pure. For other than laboratory use, the temperature is too high and the hydrogen content too low. A few alloys of uranium have been hydrided, but no products of importance for storage purposes resulted [5.12] and [Ref. 5.73, p. 7].

Thorium forms a very stable dihydride that dissociates at 883 °C, and a higher hydride Th_4H_{15} that dissociates at 365 °C. Its alloys form a number of interesting ternary hydrides, detailed studies of which have recently been published [5.34a, 105]. Nickel, cobalt, and iron all result in less stable hydrides, just as they do with the rare earths, which thorium resembles in many ways. It forms such analogs of the lanthanide AB_5 compounds as $ThNi_5$ and $ThCo_5$, and the corresponding hydrides are likewise unstable. At 40 °C, the plateau pressure for $ThCo_5H_{4.6}$ is 20 atm [5.34a]. This, and some of the other thorium alloy hydrides, might find use for special purposes but the cost is too high and the hydrogen content too low for general use.

Hydrogen also reacts with alloys of thorium and aluminum [5.38, 97]; silver [Ref. 5.73, p. 7]; cobalt (equi-atomic) [Ref. 5.73, p. 21] and [5.106]; iron, manganese, and nickel (equi-atomic) [Ref. 5.73, p. 7]; titanium [5.107]; and zirconium [5.107]. The products are all quite stable.

5.4 Specific Applications

In the first three sections of this chapter, we considered the needs for hydrogen storage and reviewed the relevant properties of hydrides. We now examine how well these properties match the storage needs of particular systems, and review a number of actual or proposed applications.

5.4.1 Remote Generation of Electricity

The first kilogram-scale use of metal hydrides for reversible hydrogen storage was occasioned by a military requirement. Hydrogen was needed for a fuel-cell stack that would provide small amounts of power to unattended instruments at remote locations. Since the unit should function at temperatures as low as $-32\,^{\circ}C$, a fairly unstable hydride was indicated. Mischmetal pentanickel was selected and a reservoir containing 36.7 kg of the alloy, crushed to pass through 1/4-inch mesh, was constructed and tested as a joint effort of Brookhaven National Laboratory [5.11] and the U.S. Army Mobility Equipment Research and Development Center, Fort Belvoir, Virginia, in 1972. The container was made of stainless steel pipe of outside diameter 11.4 cm and length 106 cm. It contained about 0.6 kg available hydrogen. The heat of dissociation was supplied by conduction through the cylinder walls from the surroundings; this mode of heat supply proved adequate for the low rate of hydrogen evolution that was called for by the specification. Although the unit was tested successfully, it was never actually used in the field.

The production of hydrogen for fuel cells by the irreversible process of hydride hydrolysis has often been proposed. Recent articles by *Baker* et al. [5.108] are illustrative of the concept.

5.4.2 Electric Utility Load Leveling

It was mentioned in Section 5.1 that the production and storage of hydrogen constitutes a possible method of storing off-peak electric energy. Several American utilities have expressed interest in this concept [5.109], and a small research and development program was begun in 1973. In work that was jointly supported by Brookhaven National Laboratory and Public Service Electric and Gas Co. of New Jersey, a reservoir using the alloy TiFe was built. It was developed to be a component of a system which also included an electrolyzer, a compressor, and a fuel-cell stack. The heat of dissociation was supplied by warm water ($45\,^{\circ}C$) which flowed through tubes embedded in the granular hydride; in the hydriding half of the cycle the same tubes circulated cooling water at $17\,^{\circ}C$. Figures 5.13–5.15 illustrate the construction of the reservoir, which somewhat resembles a shell-and-tube heat exchanger. The hydrogen was drawn off or added through a porous central tube. Stainless steel was used throughout, since time did not permit testing of cheaper materials such as mild steel. The charge was 400 kg of TiFe; the amount of available hydrogen, 6.4 kg [5.49, 57].

Since the construction of this reservoir, an enlarged research and development program on the use of TiFe for load leveling has been undertaken. Its ultimate goal is the design of a storage unit capable of taking up or giving off hydrogen at a rate equivalent to 26 MW, electric (MWe). Among the problems being dealt with are the effects of repeated cycling on storage capacity and physical state of the alloy, the possible promotion of hydrogen embrittlement at

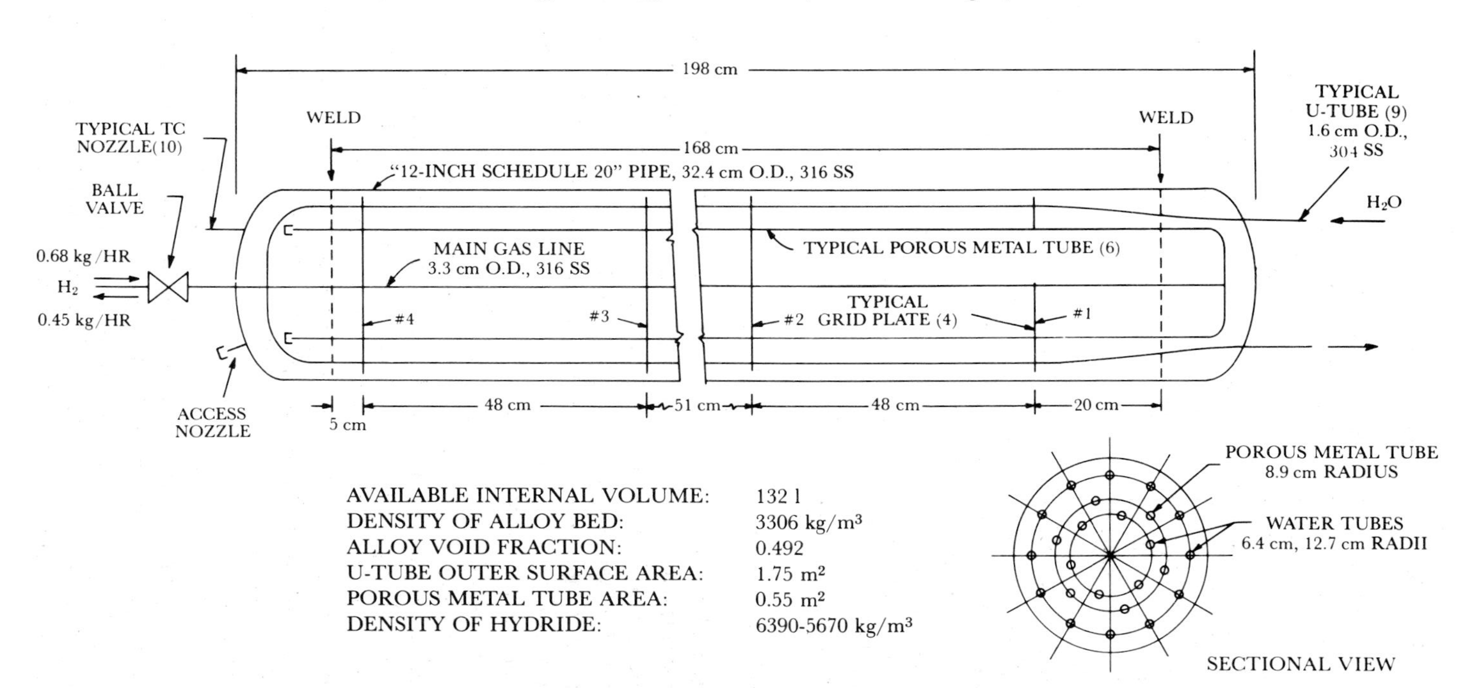

Fig. 5.13. Diagram of TiFe hydrogen reservoir

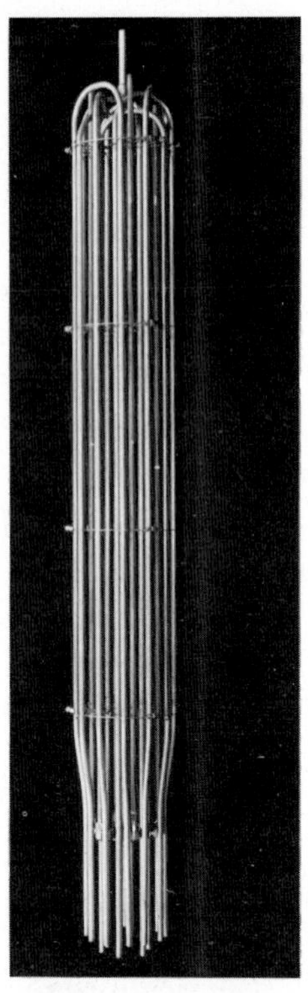

Fig. 5.14. Tube bundle before insertion into TiFe hydrogen reservoir

Fig. 5.15. TiFe hydrogen reservoir, completed

points of hydride-container contact, rates of heat and gas flow through the granular material, hazards in case of rupture, the bulging of containers which sometimes results from the expansion that attends hydride formation, the avoidance of oxygen contamination in large-scale alloy melting, and many others [5.110–112]. In a separate program at Jülich, *Pick* et al. obtained results on the same hydride [5.113]. The results of this work to date may be summarized by saying that none of the technical problems seem to present insuperable difficulties. Economic problems are another matter. At present there is some question as to whether a hydrogen-air cycle for peak shaving can be cost effective.

5.4.3 Automotive Propulsion

The use of hydride hydrogen as an automotive fuel was proposed as early as 1970, by *Meijer*, who considered its applicability to Stirling engines [5.114]. The subject was later discussed by other writers, but it was not until 1974 that actual vehicles appeared that were powered solely by hydrogen supplied from metal hydride reservoirs. The Billings Energy Corporation of Provo, Utah, and the Daimler-Benz Co. of Stuttgart, Germany, were the pioneers in this field, and each has built several such vehicles, see Fig. 5.16. Billings' latest and largest is a Winnebago 19-passenger bus powered by a modified 1976 Dodge 440 CID V-8 engine [5.115]. The hydrogen reservoir consists of 44 stainless steel tubes of 7.62 cm o.d. containing titanium iron hydride; the hot exhaust gases from the engine pass over them to supply the heat of dissociation. On recharge, which is effected by connecting the system to a supply of hydrogen at 500 psig (3447 kPa), the tubes are cooled with water. Some of the vehicle's characteristics are given in Table 5.4.

The most recent experimental vehicle from Stuttgart is a Mercedes-Benz van provided with 200 kg titanium iron hydride. Heat of dissociation is supplied by cooling water from the 44 kW engine. This is circulated through the hydride bed. A range of 130 km is reported, on a hydrogen usage of 48 m^3, or 3.96 kg, at a speed of 60 km/h. Total vehicle weight was not given. For recharge, cold water is circulated through the system while gaseous hydrogen at high pressure is supplied. Recharge is about 75% complete in the first ten minutes, and 100% complete in 45 min [5.116, 117].

The performance of these vehicles does not, of course, bear comparison with that of gasoline burners. However, the proper comparison to make is with other alternative modes of propulsion, and in this field the hydride systems look promising. In terms of extra weight and of total range, even these primitive models compare very well with electric vehicles powered by lead-acid batteries. The power density of such batteries is about 25 W-h per kilogram while the power density of a hydride, such as that of TiFe, containing 1.5 wt-% of available hydrogen is 500 W-h per kilogram, based on combustion to water vapor. If this number is halved, to allow for the weight of the container, and then divided by four, for combustion at 25% efficiency, the resulting figure of 62.5 is still more than twice the energy density of the battery.

Fig. 5.16a and b. Vehicles employing metal hydride storage. (a) by Billings Energy Research Corp.; (b) by Daimler-Benz Co.

There is a good possibility that substantial improvements will be made in automotive hydride systems. Magnesium hydride has more than three times the available hydrogen of $TiFeH_{1.9}$, and its use would permit a correspondingly longer range or lighter weight. Its enthalpy and temperature of dissociation are so high that it could probably be used as the sole source of hydrogen only for an engine especially designed to produce its waste heat in a form suitable for maximum transfer to a MgH_2 reservoir. Recently, however, *Buchner* et al. [5.117] proposed an interesting scheme in which two kinds of hydride are used alternately on the same vehicle. Hydrogen is first drawn from a $TiFeH_{1.9}$ reservoir, then from Mg_2NiH_4 heated by exhaust gas. The latter gradually cools off, because of an enthalpy deficit, and the pressure falls until it is necessary to switch back to the $TiFeH_{1.9}$ tank. As soon as hydrogen is no longer being drawn from the Mg_2NiH_4 it starts to heat up again, and soon the cycle begins again.

Table 5.4. Characteristics of the Billings hydrogen-powered bus[a]

Total vehicle weight loaded:	
a) Before conversion to hydrogen	5302 kg
b) After conversion	6785 kg
Metal hydride	$TiFeH_x$
Net weight of hydride	1016 kg
Total stored hydrogen	12.7 kg
Useable hydrogen at 50 mph (80.5 km/h) sustained road speed	7.7 kg
Range[b] at 50 mph (80.5 km/h) sustained road speed	121 km
Refueling times:	
for 80% of charge	15 min
for complete charge	1 h

[a] From [5.115].

[b] In these tests, only 1.25 wt.% of stored hydrogen, relative to the TiFe, was obtained. Higher loadings will extend the range.

Complicated problems of mass and heat transfer are encountered in the hydriding and dehydriding of large reservoirs, both stationary and mobile. Useful general analyses were published by *Yu* et al. [5.118] and by *Cummings* and *Powers* [5.119]. *Onischak* et al. [5.120] discussed heat-transfer problems encountered when hydrides are used to supply fuel cells. *Ron* [5.121] proposed the use of a porous metallic matrix to improve thermal conductivity.

Systems using the higher-temperature hydrides like MgH_2 and Mg_2NiH_4 will be rather restricted in their mode of operation. The sensible heat needed to bring them to the operating temperature range is not negligible; once heated, they should not be allowed to cool down until the charge is used up, for maximum efficiency.

This consideration makes them more appropriate for urban fleet vehicles than for use in private cars. Another factor favoring fleet use, at least initially, is the hydrogen supply problem. The provision of large quantities of hydrogen at the pressures necessary to recharge the present generation of hydrides will require special plants. The first such refueling centers are likely to be built only if there is a captive market.

When and where hydrogen actually comes into widespread use as an automotive fuel will depend on a variety of economic factors. It is possible that the first use of hydrogen engines will be in special applications where their environmental beneficence is a unique asset. One example might be the powering of fork-lift trucks used indoors; here, the economic comparison would be with batteries, rather than with gasoline. A review of the vehicular hydrogen storage problem, and a comparison with electric propulsion, was recently published by *Ecklund* and *Kester* [5.122].

5.4.4 Battery-Related Uses

Hydrogen electrodes, which are so useful in fuel cells, can also serve as components of secondary batteries if the hydrogen generated in the charging part of the cycle is somehow retained. Hydrides have been proposed for this purpose. In the researches of *Justi* and co-workers [5.123], a hydride-forming metal or alloy actually forms a part of the hydrogen electrode, and comes in contact with the electrolyte. Materials included Ti_2Ni–TiNi mixtures, and $LaNi_5$. Both were compatible with a 6N KOH electrolyte, and showed more or less reversible behavior. The use of $LaNi_5$ has also been studied by *Bronoël* et al. [5.124].

The hydride reservoir can also be kept separate from the electrolyte. In the design of *Earl* and *Dunlop* [5.125], a nickel-hydrogen cell is connected to a cylinder of $LaNi_5$. Tests of experimental models showed good performance. *Onischak* et al. [5.120] discussed some engineering problems encountered in such a system.

5.4.5 Nonfuel Applications

We now come to an interesting class of devices, in which the same quantity of hydrogen is repeatedly absorbed and desorbed without ever being consumed as a fuel. These fall basically into two classes: those in which the hydride functions as a gas compressor, and those in which the latent heat of hydride formation is made use of for energy storage.

Hydride Engines

One can in principle design a heat engine employing hydrogen as the working medium. The gas is heated, and then caused to expand through a turbine, at the low-pressure end of which the hydrogen is "condensed" by absorption in a hydride-forming metal. This is then heated to produce the high-pressure gas to which heat is supplied. Such a cycle has been proposed as a way of using the heat from a high-temperature nuclear reactor [5.126]. According to these authors, an especially efficient system will result if there is also available at the reactor site a source of cheap, low-grade heat. Hydride characteristics are such that this heat will suffice for the decomposition of the hydride into metal and high-pressure hydrogen. The particular hydride proposed in this case is $TiFeH_x$.

The hydride reservoir in this scheme may be regarded as a replacement for the gas compressor in the standard Brayton cycle—a compressor with no moving parts. Such a compressor also lends itself to refrigeration purposes, and *Van Mal* [5.127] has described the application of $LaNi_5$ hydride to the production of very low temperatures, around 20 K. An actual compressor capable of delivering $0.015\ g\ s^{-1}$ of hydrogen at 50 atm was built, and its operating characteristics were measured by *Boser* and *Lehrfeld* [5.128]. The medium was $LaNi_5H_6$. A gas circulator operating on the same principle was built by *Reilly* et al. [5.129], for use in an experiment requiring the circulation of hydrogen-tritium mixtures. A quantity of VH_2 was warmed and cooled in 2-min cycles. This resulted in pressure fluctuations in a closed system which was connected to one leg of a *U* tube containing mercury. The liquid transmitted the fluctuations to an experimental system on the far side, where, by means of a suitable arrangement of check valves, circulation of gas around a loop was effected.

Heat Storage

Any reversible reaction which is accompanied by a large enthalpy change can, in principle, be used for the storage of thermal energy. The number of reactions that, in practice, reverse in a convenient temperature or pressure range and are inexpensive, noncorrosive, etc., is quite limited, but hydrogen-metal systems are among the more promising candidates. *Alefeld* [5.130] examined the suitability of a number of chemical systems for storing off-peak power in thermal, rather than electric, form and carried out a cost analysis. This indicated that certain hydrides, as well as such other systems as $CaCl_2$–NH_3, would show an economic advantage over the storage of steam, pumped water, or compressed air.

Solar energy is another intermittent heat source for which hydride storage is particularly suitable. System designs incorporating hydrides were described by *Libowitz* and *Blank* [5.131], *Van Mal* [5.85], *Gruen* et al. [5.132], and *Cottingham* [5.133]. They vary in detail, but have in common the use of two hydrides of differing stability. Solar heat is used to decompose the more stable one, the hydrogen evolved being then stored on a metal which forms a hydride that can be decomposed at ground-water temperature. When heat is later desired, it is produced by allowing hydrogen to flow back from the second metal to the first. With somewhat more elaborate apparatus, air conditioning can be obtained. *Gruen* et al. [5.132] described a solar harvesting system which also incorporates an expansion engine and generator. Some pairs of hydrides that have been suggested are VH_2, $TiFeH_{1.7}$ [5.131]; $LaNi_5H_6$, $MnNi_5H_6$ [5.132]; and $VNbH_3$, $MnNi_5H_6$ [5.133]. A single hydride plus a mechanical compressor can be used for a heat pump; this concept was analyzed by *Wolf* [5.134].

Thermometers

Quite precise temperature readings can be obtained at some distance from the point of measurement by using a small quantity of a metal hydride as a probe. This is connected by a capillary tube to a room-temperature pressure gauge. By a suitable choice of hydride, pressure changes of as much as 100 Torr per degree can be obtained. Hysteresis must be avoided, which means that thermometers of this sort will generally be useful only above 400 °C. In scouting research by *Chen* et al. [5.135], titanium hydride was shown to be a useful probe material for the temperature range 525–605 °C.

Acknowledgement. This review was prepared under the auspices of the Division of Physical Research, U.S. Energy Research and Development Administration.

References

The references cover literature published up to the end of 1976. It is hoped that no major articles on the practical aspects of hydrogen storage have been overlooked. The field of alloy-hydrogen chemistry, however, was too large to cover exhaustively; consequently, Section 5.3 and the references to it are intended only to constitute an introduction to the literature. When a statement of fact about a particular system is not individually documented, the source is usually [5.12].

5.1 E.Gileadi, S.Srinivasan, F.J.Salzano, C.Braun, A.Beaufrere, S.Gottesfeld, L.J.Nuttall, A.B.LaConti: "An Electrochemically Regenerative Hydrogen-Chlorine Energy Storage System for Electric Utilities", Rpt. BNL-21820, Brookhaven National Laboratory (1976)
5.2 G.G.Haselden: *Cryogenic Fundamentals* (Academic Press, New York, London 1971) p. 23
5.3 R.Barron: *Cryogenic Systems* (McGraw-Hill, New York 1966) p. 10
5.4 L.G.O'Connell: "The Role of Energy Storage Power Systems in Transportation" Rpt. UCID 17274, Lawrence Livermore Laboratory (1976)
5.5 K.C.Hoffman, J.J.Reilly, F.J.Salzano, C.H.Waide, R.H.Wiswall, W.E.Winsche: International J. Hydrogen Energy **1**, 133 (1976)
5.6 S.C.Garg, A.W.McClaine: "Metal Hydrides for Energy Storage Applications" Rpt. TN 1395 [AD-A014174] Civil Engineering Lab., Port Hueneme, California (1975)
5.7 P.S.Rudman: Hydrogen Vehicular Fuel Storage as a Step in a Water Splitting Cycle in Proceedings, First World Hydrogen Energy Conference, Vol. 2 (University of Miami, Coral Gables, Florida 1976) p. 3B–31–39
5.8 L.Green: "Energy Storage Via Calcium Hydride Production" in 11th Intersociety Energy Conversion Engineering Conference Proceedings, Vol. 1 (A.I.Ch.E., New York 1976) pp. 949–953
5.9 T.R.P.Gibb: "Primary Solid Hydrides", in *Progress in Inorganic Chemistry*, Vol. 3, ed. by F.A.Cotton (Interscience, New York, London 1962) pp. 315–509
5.10 G.G.Libowitz: J. Phys. Chem. **62**, 296 (1958)
5.11 J.J.Reilly: Unpublished research
5.12 W.M.Mueller, J.P.Blackledge, G.G.Libowitz: *Metal Hydrides* (Academic Press, New York, London 1968)
5.13 E.A.Owen: Phil. Mag. **35**, 50 (1944)
5.14 C.Wagner: Z. Phys. Chem. (Leipzig) **193**, 386 (1944)
5.15 D.H.Everett, P.Nordon: Proc. Roy. Soc. (London) A**259**, 341 (1960)
5.16 N.A.Scholtus, W.Keith Hall: J. Chem. Phys. **39**, 868 (1963)
5.17 E.Wicke, K.Otto: Z. Physik Chem. (N.F.) **31**, 222 (1962)
5.18 E.Wicke, G.H.Nernst: Ber. Bunsen-Gesellschaft Physik. Chem. **68**, 224 (1964)
5.19 M.v.Stackelberg, P.Ludwig: Z. Naturforsch. **19**a, 93 (1964)
5.20 J.B.Condon: J. Phys. Chem. **79**, 42 (1975)
5.21 O.M.Katz, E.A.Gulbransen: J. Nucl. Mater. **5**, 269 (1962)
5.22 C.A.Bechman, A. Goudy, T.Takeshita, W.E.Wallace, R.S.Craig: Inorg. Chem. **15**, 2184 (1976)
5.23 F.A.Kuijpers, H.H. van Mal: J. Less-Common Metals **23**, 395 (1971)
5.24 J.J.Reilly, R.H.Wiswall: Inorg. Chem. **7**, 2254 (1968)

5.25 G.C.Sinke, L.C.Walker, F.L.Oetting, D.R.Stull: J. Chem. Phys. **47**, 2759 (1967)
5.26 H.H.van Mal, K.H.J.Buschow, A.R.Miedema: J. Less-Common Metals **35**, 65 (1974)
5.27 J.J.Reilly, R.H.Wiswall: Inorg. Chem. **13**, 218 (1974)
5.28 E.Veleckis, R.K.Edwards: J. Phys. Chem. **73**, 683 (1969)
5.29 J.J.Reilly, R.H.Wiswall: Inorg. Chem. **9**, 1678 (1970)
5.30 T.B.Flanagan, J.F.Lynch: J. Phys. Chem. **79**, 444 (1975)
5.31 J.C.Warf, W.L.Korst, K.I.Hardcastle: Inorg. Chem. **5**, 1726 (1966)
5.32 S.E.Speed: "Process for Controlled Production of Hydrogen Gas, etc.", United States Patent 3,676,071 (July 11, 1972)
5.33 G.G.Libowitz, H.F.Hayes, T.R.P.Gibb: J. Phys. Chem. **62**, 76 (1958)
5.34 (a) K.H.J.Buschow, H.H.van Mal, A.R.Miedema: J. Less-Common Metals **42**, 162 (1975)
(b) A.R.Miedema, K.H.J.Buschow, H.H.van Mal: J. Less-Common Metals **49**, 463 (1976)
5.35 J.H.N.van Vucht, F.A.Kuijpers, H.C.A.M.Bruning: Philips Res. Rep. **25**, 133 (1970)
5.36 F.A.Kuijpers: J. Less-Common Metals **27**, 27 (1972)
5.37 F.A.Kuijpers: Ber. Bunsen-Gesellschaft Physik. Chem. **76**, 1220 (1972)
5.38 J.D.Fast: *Interaction of Metals and Gases*, Vol. 1 (Academic Press, New York, London 1965) p. 289
5.39 A.L.Reimann: Phil. Mag. [7] **16**, 673 (1933)
5.40 G.A.Atkinson, S.Coldrick, J.P.Murphy, N.Taylor: J. Less-Common Metals **49**, 439 (1976)
5.41 J.A.Pryde, I.S.T.Tsong: Trans. Faraday Soc. **65**, 2766 (1969); **67**, 297 (1971)
5.42 C.W.Schoenfelder, J.H.Swisher: J. Vac. Sci. Techn. **10**, 862 (1973)
5.43 D.M.Gualtieri, K.S.V.L.Narasimhan, T.Takeshita: J. Appl. Phys. **47**, 3432 (1976)
5.44 O.Boser: J. Less-Common Metals **46**, 91 (1976)
5.45 J.B.Condon, E.A.Larson: J. Chem. Phys. **59**, 855 (1973)
5.46 D.T.Peterson, D.G.Westlake: J. Phys. Chem. **63**, 1514 (1959)
5.47 E.A.Gulbransen, K.F.Andrew: Metals Trans. **185**, 741 (1949)
5.48 E.A.Gulbransen, K.F.Andrew: Metals Trans. **185**, 515 (1949)
5.49 J.J.Reilly, K.C.Hoffman, G.Strickland, R.H.Wiswall: "Iron Titanium Hydride as a Source of Hydrogen Fuel for Stationary and Automotive Applications", in 26th Power Sources Symposium (PSC Publications Committee, P.O.Box 891, Red Bank, New Jersey 1974) pp. 11–17
5.50 E.Wiberg, H.Göltzer, R.Bauer: Z. Naturforsch. **6**b, 394 (1951)
5.51 T.N.Dymova, Z.K.Sterlyadkina, V.G.Safronov: Zh. Neorg. Khim. **6**, 763 (1971)
5.52 Farbenfabriken Bayer A.G.: British Patent 777,097 (June 19, 1957)
5.53 J.P.Faust, E.D.Whitney, H.D.Batha, T.L.Heying, C.E.Fogle: J.Appl. Chem. (London) **10**, 187 (1960)
5.54 A.Küssner, E.Wicke: Z. Physik. Chem. (N.F.) **24**, 152 (1960)
5.55 H.Brodowsky, E.Wicke: Engelhard Ind. Techn. Bull. **7**, 41 (1966)
5.56 D.H.W.Carstens, J.D.Farr: J. Inorg. Nucl. Chem. **36**, 461 (1974)
5.57 G.Strickland, J.Reilly, R.Wiswall: "An Engineering-Scale Energy Storage Reservoir of Iron Titanium Hydride", in Proceedings of the Hydrogen Economy Miami Energy Conference (THEME). (Univ. of Miami, Coral Gables, Florida 1974) p. S4–9
5.58 C.E.Lundin, F.E.Lynch, C.B.Magee: "Solid-State Hydrogen Storage Materials for Application to Energy Needs", Rpt. AFOSR-TR-76-1124, Denver Research Institute, Denver, Colorado (1976)
5.59 S.Aronson, F.J.Salzano: Inorg. Chem. **8**, 1541 (1969)
5.60 C.E.Messer, J.C.Eastman, R.G.Mers, A.J.Maeland: Inorg. Chem. **3**, 776 (1964)
5.61 C.E.Messer, K.Hardcastle: Inorg. Chem. **3**, 1327 (1964)
5.62 A.P.Graefe, R.K.Robeson: J. Inorg. Nucl. Chem. **29**, 2917 (1967)
5.63 J.D.Farr: J. Inorg. Nucl. Chem. **14**, 202 (1960)
5.64 R.O.Moyer, C.Stanitski, J.Tanaka, M.I.Kay, R.Kleinberg: J. Solid State Chem. **3**, 541 (1971)
5.65 A.P.Ginsberg: Inorg. Chem. **3**, 567 (1964)
5.66 S.C.Abrahams, A.P.Ginsberg, K.Knox: Inorg. Chem. **3**, 558 (1964)
5.67 N.A.Bell, G.E.Coates: J. Chem. Soc. A 628 (1968)

5.68 C.E.Messer: "A Survey Report on Lithium Hydride", Rpt. NYO-9470 (October 27, 1960); also
K.Moers: Z. Inorg. Chem. **113**, 191 (1920)
5.69 F.C.Chang: "A Study of Ternary Hydride Phases Formed by Reaction of LiH and Selected Materials", Dissertation 68–17, 837, University Microfilms, Ann Arbor, Michigan (1969)
5.70 C.E.Lundin, F.E.Lynch: "The Safety Characteristics of FeTi Hydride", in Tenth Intersociety Energy Conversion Engineering Conference (I.E.E.E., New York 1975) pp. 1386–1390
5.71 J.J.Reilly, R.H.Wiswall: Inorg. Chem. **6**, 2220 (1967)
5.72 V.I.Mikheeva, Z.K.Sterlyadkina, A.J.Konstantinova, O.N.Kryukova: Russian J. Inorg. Chem. (English transl.) **8**, 682 (1963)
5.73 R.L.Beck: "Investigation of Hydriding Characteristics of Intermetallic Compounds", Rpt. DRI-2059, Denver Research Institute, Denver, Colorado (1962)
5.74 G.Sandrock: Private communication on research at International Nickel Co., Sterling Forest, New York. This was presented at the 1977 Intersociety Energy Conversion Engineering Conference
5.75 H.H.van Mal, K.H.J.Buschow, A.R.Miedema: J. Less-Common Metals **49**, 473 (1976)
5.76 C.E.Messer: J. Less-Common Metals **19**, 284 (1969)
5.77 F.M.Brower, N.E.Matzek, P.F.Reigler, H.W.Rinn, C.B.Roberts, D.L.Schmidt, J.A.Snover, K.Terada: J. Am. Chem. Soc. **98**, 2450 (1976)
5.78 H.Clasen: Angew. Chem. **73**, 322 (1961)
5.79 E.C.Ashby, G.J.Brendel, H.E.Redman: Inorg. Chem. **2**, 499 (1963)
5.80 E.C.Ashby, P.Kobetz: Inorg. Chem. **5**, 1615 (1966)
5.81 J.Mayet, S.Kovacevic, J.Tranchant: Bull. Soc. Chim. France 1973 (2), 503
5.82 T.N.Dymova, Yu. M.Dergachëv, V.A.Sokolov, N.A.Grechanaya: Dokl. Akad. Nauk SSSR **224**, 591 (1975)
5.83 K.I.Hardcastle, J.C.Warf: Inorg. Chem. **5**, 1728, 1736 (1966)
5.84 H.H.van Mal, K.H.J.Buschow, F.A.Kuijpers: J. Less-Common Metals **32**, 289 (1973)
5.85 H.H.van Mal: Stability of Ternary Hydrides and Some Applications. Philips Res. Rep. Suppl. (1976) No. 1
5.86 S.A.Steward, J.F.Lakner, F.Uribe: "Storage of Hydrogen Isotopes in Intermetallic Compounds", Rpt. UCRL 77455, Lawrence Livermore Laboratory (1976)
5.87 R.H.Wiswall, J.J.Reilly: "Metal Hydrides for Energy Storage", in 7th Intersociety Energy Conversion Engineering Conference (American Chemical Society, Washington, D.C. 1972) pp. 1342–1348
5.88 J.L.Anderson, T.C.Wallace, A.L.Bowman, C.L.Radosevich, M.L.Courtney: "Hydrogen Absorption by AB_5 Compounds", Rpt. LA-5320-MS, Los Alamos Scientific Laboratory (1973)
5.89 C.E.Lundin, F.E.Lynch: "A Detailed Analysis of the Hydriding Characteristics of $LaNi_5$", in 10th Intersociety Energy Conversion Engineering Conference (I.E.E.E., New York 1975) pp. 1380–1385
5.90 K.H.J.Buschow, H.H.van Mal: J. Less-Common Metals **29**, 203 (1974)
5.91 T.Takeshita, W.E.Wallace, R.S.Craig: Inorg. Chem. **13**, 2283 (1974)
5.92 A.J.Maeland, A.F.Andresen, K.Videm: J. Less-Common Metals **45**, 347 (1976)
5.93 J.Clinton, H.Bittner, H.Oesterreicher: J. Less-Common Metals **41**, 187 (1975)
5.94 J.J.Reilly, J.R.Johnson: "Titanium Alloy Hydrides; their Properties and Applications", in Proceedings, First World Hydrogen Energy Conference (University of Miami 1976) p. 8B–3–26
5.95 K.Yamanaka, H.Saito, M.Someno: Nippon Kagaku Kaishi (1975) p. 1267
5.96 D.W.Jones, N.Pessall, A.D.McQuillan: Phil. Mag. [8] **6**, 455 (1961)
5.97 M.J.Trzeciak, D.F.Dilthey, M.W.Mallett: "Study of Hydrides", Rpt. BMI-1112, Battelle Memorial Institute (1956)
5.98 V.V.Grushina, A.M.Rodin: Sov. J. Phys. Chem. (English transl.) **37**, 288 (1963); Zh. Fiz. Khim. **42**, 466 (1968)
5.99 T.D.McKinley: J. Electrochem. Soc. **102**, 117 (1955)
5.100 B.Stalinski, B.Nowak, O.J.Zogal: J. Less-Common Metals **19**, 289 (1969)

5.101 A. Pebler, E. A. Gulbransen: Electrochem. Tech. **4**, 211 (1966); Trans. Met. Soc. AIME **239**, 1593 (1967)
5.102 J. J. Reilly, R. H. Wiswall: "The Effect of Minor Constituents on the Properties of Vanadium and Niobium Hydrides", in International Meeting on Hydrogen in Metals (Jülich 1972): Preprints, Vol. 1, pp. 38–64; also Inf. Rpt. BNL-16546 Brookhaven National Laboratory
5.103 L. Kirschfeld, A. Sieverts: Z. Elektrochem. **36**, 123 (1930)
5.104 F. A. Lewis: *The Palladium-Hydrogen System* (Academic Press, New York, London 1967)
5.105 T. Takeshita, W. E. Wallace, R. S. Craig: Inorg. Chem. **13**, 2282 (1974)
5.106 W. L. Korst: "Observations on the Thorium-Hydrogen and Cobalt-Thorium-Hydrogen Systems", Rpt. NAA-SR-6881, Atomics International (1962)
5.107 R. van Houten, S. Bartram: Met. Trans. **2**, 527 (1971)
5.108 B. S. Baker, M. Onischak, R. Tripp: "Sixty-Watt Hydride-Air Fuel Cell System", in 9th Intersociety Energy Conversion Engineering Conference Proceedings (ASME, New York 1974) pp. 830–835. R. N. Camp, B. S. Baker, E. H. Reiss: Milliwatt Fuel Cell for Sensors, in 9th Intersociety Energy Conversion Engineering Conference Proceedings (ASME, New York 1974) pp. 841–845
5.109 J. M. Burger, P. A. Lewis, R. J. Isler, F. J. Salzano: "Energy Storage for Utilities Via Hydrogen Systems", in 9th Intersociety Energy Conversion Engineering Conference Proceedings (ASME, New York 1974) pp. 428–434
5.110 A. H. Beaufrere, F. J. Salzano, R. J. Isler, W.-S. Yu: International J. Hydrogen Energy **1**, 307 (1976)
5.111 G. Strickland, J. Milau, W.-S. Yu: "The Behavior of Iron Titanium Hydride Test Beds: Long-Term Effect, Heat Transfer and Modelling", in Proceedings, First World Hydrogen Energy Conference, Vol. 2 (University of Miami, Coral Gables, Florida 1976) p. 8B–41–71
5.112 G. D. Sandrock, J. J. Reilly, J. R. Johnson: "Metallurgical Considerations in the Production and Use of FeTi Alloys for Hydrogen Storage", in 11th Intersociety Energy Conversion Engineering Conference Proceedings Vol. 1 (A.I.Ch.E., New York 1976) pp. 965–971
5.113 M. A. Pick, H. Wenzl: "Physical Metallurgy of FeTi Hydride and its Behavior in a Hydrogen Storage Container", in 1st First World Hydrogen Energy Conference Proceedings, Vol. 2 (University of Miami, Coral Gables, Florida 1976) p. 8B–27–39
5.114 R. L. Meijer: Philips Techn. Rev. **31**, 169 (1970)
5.115 R. L. Woolley: "Performance of a Hydrogen Powered Transit Vehicle", in 11th Intersociety Energy Conversion Engineering Conference (A.I.Ch.E., New York 1976) pp. 992–996
5.116 H. Buchner, H. Säufferer: Automobilt. Z. **78**, 161 (1976)
5.117 H. Buchner, H. Säufferer: Automobilt. Z. **79** (2), 45 (1977)
5.118 W.-S. Yu, E. Suuberg, C. Waide: "Modelling Studies of Fixed-Bed Metal Hydride Storage Systems", in Proceedings The Hydrogen Economy Miami Energy Conference (THEME) (University of Miami 1974) p. S4–22–35
5.119 D. L. Cummings, A. J. Powers: Ind. Eng. Chem., Process Des. Develop **13**, 182 (1974)
5.120 M. Onischak, D. Dharia, D. Gidaspow: "Heat Transfer Analysis of Metal Hydrides in Metal-Hydrogen Secondary Batteries", in Proceedings, First World Hydrogen Energy Conference, Vol. 2 (University of Miami, Coral Gables, Florida 1976) pp. 7B–37–50
5.121 W. Ron: "Metal Hydrides of Improved Heat Transfer Characteristics", in 11th Intersociety Energy Conversion Engineering Conference Proceedings, Vol. 1 (A.I.Ch.E., New York 1976) pp. 954–960
5.122 E. Ecklund, F. Kester: "Hydrogen Storage on Highway Vehicles: Update '76", in 1st World Hydrogen Energy Conference Proceedings, Vol. 2 (University of Miami, Coral Gables, Florida 1976) p. 3B–3–29
5.123 (a) E. W. Justi, H. H. Ewe, A. W. Kalberlah, N. M. Saridakis, M. H. Schaefer: Energy Conversion **10**, 183 (1970)
(b) H. Ewe, E. W. Justi, K. Stephan: Energy Conversion **13**, 109 (1973)
5.124 G. Bronoël, J. Sarradin, M. Bonnemay, A. Percheron, J. C. Achard, L. Schlapbach: International J. Hydrogen Energy **1**, 251 (1976)
5.125 M. M. Earl, J. D. Dunlop: "Chemical Storage of Hydrogen in Ni/H_2 Cells", in 26th Power Sources Symposium (PSC Publications Committee, Red Bank, N.J. 1974) pp. 24–27

5.126 J.R.Powell, F.J.Salzano, W-S.Yu, J.S.Milau: Science **193**, 314 (1976)
5.127 H.H.van Mal: Chemie Ingenieur Technik **45**, 80 (1973)
5.128 O.Boser, D.Lehrfeld: "A Novel Hydrogen Compressor", in 26th Power Sources Symposium (PSC Publications Committee, Red Bank, N.J. 1974) pp. 3–6
5.129 J.J.Reilly, A.Holtz, R.H.Wiswall: Rev. Sci. Instr. **42**, 1485 (1971)
5.130 G.Alefeld: Energie **27**, 180 (1975)
5.131 G.G.Libowitz, Z.Blank: "An Evaluation of the Use of Metal Hydrides for Solar Thermal Energy Storage", in 11th Intersociety Energy Conversion Engineering Conference, Vol. 1 (A.I.Ch.E., New York 1976) pp. 673–680
5.132 D.M.Gruen, R.L.McBeth, M.Mendelsohn, J.M.Nixon, F.Schreiner, I.Sheft: "Hycsos: A Solar Heating, Cooling and Energy Conversion System Based on Metal Hydrides", in 11th Intersciety Energy Conversion Engineering Conference, Vol. 1 (A.I.Ch.E., New York 1976) pp. 681–687
5.133 J.G.Cottingham: A Hydride Heat Pump to Enhance Solar Energy Collection and Storage and for Waste Heat Scavenging. Inf. Rpt. BNL-19914, Brookhaven National Laboratory (1975)
5.134 S.Wolf: Hydrogen Sponge Heat Pump, in 10th Intersociety Energy Conversion Engineering Conference Record (I.E.E.E., New York 1975) pp. 1348–1355
5.135 J.Chen, J.J.Reilly, R.H.Wiswall, G.Schoener: Temperature Sensor. Rpt. BNL-50205, Brookhaven National Laboratory (1969) pp. 44–46

6. Superconductivity in Metal-Hydrogen Systems

By B. Stritzker and H. Wühl

With 13 Figures

The exciting properties of metal-hydrogen systems, which led to their intense study, awoke interest in a new field of physics when superconductivity was found in Th_4H_{15} [6.1] and PdH [6.2] in 1970 and 1972, respectively. Because of the high density of hydrogen in these systems, it was argued that metallic hydrogen might be responsible for the superconductivity. The reason for this speculation was simply the fact that people thought of metallic H as a superconductor with a very high transition temperature of possibly more than 100 K [6.3–6]. This speculation was one of the stimuli for a large variety of experiments and theories. But the findings contradicted the assumption of metallic H being responsible for superconductivity in metal-hydrogen systems. Nevertheless it was found that another feature of hydrogen is important, i.e., its optical vibrations inside the metal lattice. It will be shown that the strong coupling of these optic phonons to the electron system is the essential cause for the occurrence of high transition temperatures.

6.1 Superconductivity and Normal-State Parameters

Before discussing the superconducting properties of metal-hydrogen systems, we give a short review of the theory of superconductivity [6.7] as it concerns the special aspects of metal-hydrogen systems. The most obvious superconducting phenomenon is the disappearance of the electrical resistance below a transition temperature T_c characteristic for each material. The microscopic mechanism leading to the superconducting state has been proved to be the attractive interaction between conduction electrons via the vibrations of the lattice. This electron-phonon interaction is fundamental to the theory of *Bardeen*, *Cooper*, and *Schrieffer* (BCS) [6.8], who obtained the expression

$$T_c = 1.14\,\theta_D \exp[-1/N(0)V]\,, \tag{6.1}$$

where θ_D is the Debye temperature, $N(0)$ the electron density of states at the Fermi energy, and V an effective pairing potential considered to be constant.

The subsequent theories (Eliashberg equations) take into account the details of the electron-phonon interaction [6.9]. The central quantities in these theories are the function $\alpha^2(\omega)F(\omega)$ and the Coulomb pseudopotential μ^*, where $\alpha^2(\omega)$ is an averaged electron-phonon-coupling function, $F(\omega)$ the phonon density of states, and μ^* describes the repulsive electron-electron interaction. $\alpha^2(\omega)F(\omega)$ and μ^* can, in principle, be calculated provided that the normal-state parameters are known in enough detail, which at present is not the case.

For some few superconductors, however, $\alpha^2(\omega)F(\omega)$ and μ^* have been obtained by numerically analyzing the results of superconducting tunneling experiments [6.10]. With the knowledge of these quantities, the superconducting properties can be calculated.

From the exact, but complex, Eliashberg integral equations, *McMillan* [6.11] has derived the useful T_c equation

$$T_c = \frac{\langle\omega\rangle}{1.2} \exp\left[\frac{-1.04(1+\lambda)}{\lambda - \mu^* - 0.62\lambda\mu^*}\right]. \tag{6.2}$$

This equation describes T_c quite accurately if the following definitions are used for the average phonon frequency $\langle\omega\rangle$ and the coupling constant λ:

$$\langle\omega\rangle = 2\int \alpha^2(\omega)F(\omega)\,d\omega/\lambda\,, \tag{6.3}$$

$$\lambda = 2\int [\alpha^2(\omega)F(\omega)/\omega]\,d\omega\,. \tag{6.4}$$

As for the majority of the superconductors $\alpha^2(\omega)F(\omega)$ is not known, $\langle\omega\rangle$ is approximated by θ_D and (6.4) is replaced by

$$\lambda = \frac{N(0)\langle J^2\rangle}{M\langle\omega^2\rangle} = \frac{\eta}{M\langle\omega^2\rangle}, \tag{6.5}$$

where $\langle J^2\rangle$ is an average of the electron-phonon interaction over the Fermi surface, M is the ion mass, and $\langle\omega^2\rangle$ is an averaged squared phonon frequency. It has been found that for some classes of materials η varies much less than its factors $N(0)$ and $\langle J^2\rangle$. Thus λ and therefore T_c are governed by the phonon factor $M\langle\omega^2\rangle$. The dependence of T_c on $N(0)$ given by the simple BCS equation (6.1) is in these cases implicitly contained in $M\langle\omega^2\rangle$ because the high-density-of-states materials are elastically softer [6.11]; electron and phonon properties are interconnected.

Depending on the material, the influence of $N(0)$ or that of $F(\omega)$ is more obvious or dominant. Generally speaking, high transition temperatures may result from a large $N(0)$ and contributions of soft phonon modes in $F(\omega)$ [i.e., more precisely, low-frequency contributions in $\alpha^2(\omega)F(\omega)$].

A further possibility to enhance T_c consists of adding new vibrational modes [6.12], as is possible by dissolving light elements interstitially in a host lattice. Provided that the ratio of the heavy mass to the light mass is large, the electron-phonon-coupling constant λ can be decomposed into an acoustical and an optical part

$$\lambda = \lambda_{ac} + \lambda_{opt} . \tag{6.6}$$

It is, however, not possible to restrict the contribution of the heavy mass to λ_{ac} and that of the light mass to λ_{opt}, for an interference term has to be taken into account [6.13]. The electron-phonon coupling through optic phonons should always lead to an enhancement of the transition temperature unless an intrinsically favorable λ_{ac} is reduced by the addition of the interstitials.

The replacement of atoms by their isotopes generally influences T_c. In a harmonic approximation, $M\langle\omega^2\rangle$ is independent of the mass and therefore λ will not be changed, (6.5). According to (6.2), T_c is then influenced only by the mass dependence of the prefactor $\langle\omega\rangle$. Hence $T_c \propto M^{-0.5}$. Deviations from this dependence can be ascribed to a mass dependence of μ^* or to changes of λ due to anharmonic effects.

Besides the disappearance of the electrical resistivity, a superconductor has further properties. An energy gap Δ appears in the electronic excitation spectrum. This energy gap is temperature dependent. The ratio of the gap (for $T \to 0$) to the transition temperature, $2\Delta_0/k_B T_c$, is a measure for the strength of the electron-phonon coupling. Weak-coupling BCS superconductors have a value of 3.52 which is exceeded by strong-coupling superconductors. A further important property of a superconductor is that a magnetic field smaller than a critical field H_c is expelled of a superconductor of type I. An applied field stronger than H_c destroys the superconductivity in superconductors of type I, or penetrates in a quantized way into a superconductor of type II. Such a superconductor is driven normal at an upper critical field H_{c2}. High values of H_{c2} are obtained with short electron mean-free paths. Alloys and metals with a high degree of lattice disorder mainly belong to the class of type II superconductors. Not only magnetic fields but also magnetic impurities affect superconductivity. Hence, very pure materials are needed for the investigation of the optimal superconducting properties.

6.2 Influence of Hydrogen on the Superconductivity of Various Metals and Metal Alloys

Most of the transition metals dissolve hydrogen. Only the first three groups of the periodic system: Sc, Y, Lu; Ti, Zr, Hf; V, Nb, Ta; as well as the rare earths and Pd form exothermic solutions with hydrogen [6.14] occupying interstitial sites in large amounts at ambient conditions. In this section we give a survey of those metal-hydrogen systems, including the nontransition metal Al, in which an influence of hydrogen on superconductivity has been reported.

The superconducting properties of many transition metals are very sensitive to dissolved gases. Therefore, contradictory results on the transition temperatures of elements like V, Nb, and Ta ($T_c = 5.3$, 9.2, and 4.4 K, respectively) and their alloys can possibly be explained by the strong influence of gaseous impurities. Therefore the determination of the specific H influence is rather difficult. In recent experiments, *Welter* [6.15] confirmed that for extremely pure Nb samples (<40 ppm C, N, O), the β phase of Nb–H (H/Nb > 0.7) is not superconducting above 1.2 K [6.16], whereas T_c of the α phase is not influenced [6.17], presumably due to the extremely low H solubility below 200 K. In the $\alpha - \beta$ mixed region the transition temperature is determined by the α phase which short circuits the whole sample. In contrast, an initial decrease of T_c of about 0.17 K/at. % H had been reported earlier [6.18, 19]. Besides the negative influence of O- or N-impurities, the decrease of T_c can be explained by an additional destructive effect of hydrogen [6.15] trapped at these gaseous impurities [6.20, 21]. A similar T_c depression of about 0.11 K/at. % H was observed [6.19] for the $\alpha - \beta$ mixed region of Ta. In V the pure β phase was investigated by *Westlake* and *Ockers* [6.22] who also detected no superconductivity above 1.7 K. In Mo ($T_c = 0.92$ K) dissolving H endothermically, a T_c decrease of 8 mK at 14 ppm H was reported by *Bhardwaj* and *Rorschach* [6.23].

Satterthwaite and *Peterson* [6.24] searched in vain for superconductivity above 1.2 K in the compounds VH_2, NbH_2, and TaH. Superconductivity has also been investigated in some higher hydrides of the early transition metals. In TiH_2, ZrH_2, ThH_2 [6.25], and $ThZr_2H_7$ [6.24] no superconductivity was found above 1.02 K. *Duffer* et al. [6.26] observed a T_c decrease in HfV_2 from 9.7 K down to 4.9 K in HfV_2H and to 2.8 K in HfV_2D. No superconductivity above 0.33 K was obtained by *Merriam* and *Schreiber* [6.27] in nearly stoichiometric LaH_2 with a ratio H/La = 1.96. A further increase of the H concentration exceeding H/La = 2.4 leads to a semiconductor. Such a transition from the metallic to the semiconducting state also occurs in Ce and Gd for H/Ce $\geqq 2.7$ and H/Gd $\geqq 2.3$ [6.28]. A further example of a nonsuperconducting H compound is UH_3 which is ferromagnetic [6.29].

Many experiments have been performed, taking advantage of the fact that hydrides usually exist over a broad range of composition. Thus the average number of valence electrons per atom (e/a), which is related in some way to the electronic density of states $N(0)$, can be easily varied by changing the H content. In this way the influence of $N(0)$ was investigated in superconductors with A15 structure being most favorable for high T_c values. *Sahm* [6.30] reported a shallow maximum of the transition temperature in Nb_3SnH_x for small x. With increasing x, a decrease of T_c was found, in accordance with measurements of *Vieland* et al. [6.31]. The transition temperature decreased from 18 K for Nb_3Sn to 12 K for $x = 0.5$. Also in $Nb_{0.78}Ge_{0.22}$ (probably the tetragonal-distorted A15 phase), a T_c depression from 5.3 K to below 4.2 K was measured by *Reed* et al. [6.32] after addition of 16 at. % H. The variation of the ratio e/a from 5.7 to 3.8 in the A15 compound Ti_3AuH_x with $0 \leqq x \leqq 2.8$ by *Vetrano* et al. [6.33] did not lead to superconductivity above 1.6 K.

In the intercalated system H_xTaS_2, *Murphy* et al. [6.34] reported a positive influence of H on T_c. While TaS_2 becomes superconducting at 0.8 K, T_c increases up to 4 K after intercalation with hydrogen between 5 and 15 at. %. The authors claim that the electron donation of H hinders the phase transition at 80 K of the H-free system. Weak phonon modes related to this delayed phase transformation could be responsible for the enhancement of the transition temperature.

The following section deals with those superconducting metal hydrides to which a specific H influence has been ascribed. In 1970 *Satterthwaite* and *Toepke* [6.1] observed superconductivity in the higher hydrides and deuterides of Th with ratios H(D)/Th$\cong$3.75. They found 0.5 K broad superconducting transition curves with onsets up to 9 K for the stoichiometric compounds Th_4H_{15} and Th_4D_{15}. These T_c values are much larger than $T_c = 1.37$ K of pure Th metal.

In 1972 *Skoskiewicz* [6.2] discovered superconductivity in the intensively studied Pd–H system at ratios H/Pd > 0.8, whereas pure Pd is a normal metal down to 0.1 K. By means of the implantation technique, *Stritzker* and *Buckel* [6.35] could raise the H concentration above H/Pd = 1.0 and obtained maximum transition temperatures (0.5 K broad) of 8.8 K in PdH and 10.7 K in PdD. They achieved a further increase of T_c after substitution of Pd atoms by noble metals [6.36]. Thus a maximum of T_c of 16.6 K was found in the Pd–Cu–H system [6.37]—an exciting high T_c in an alloy consisting of three nonsuperconducting components.

The occurrence of superconductivity in the Th–H and Pd–H systems stimulated some related experiments. *Robbins* et al. [6.38, 39] studied the effect of H on T_c of Nb–Pd, Nb–Pd–Mo, Nb–Pd–W, and Nb–Ru alloys. Nb based alloys were chosen because Nb is the superconducting element with the highest T_c (9.2 K) and dissolves large amounts of H. The transition temperatures of the alloys investigated varied between 0.4 and 3 K. A remarkable T_c increase of about 2–4 K is observed in these alloys after addition of H. Whereas the unhydrogenated alloys crystallize in the bcc structure, the hydrogenated samples with the higher T_c ($\approx$5 K) exhibit an fcc structure. The authors argued that the appearance of an fcc host lattice, where the H atoms occupy octahedral sites, is necessary for the occurrence of the elevated transition temperatures. This conclusion was not confirmed by *Oesterreicher* and *Clinton* [6.40] who examined the H influence on NbRh and NbPd. They found also an increase of T_c with increasing H content, but no change of the lattice structure. The authors explained their T_c increase on the basis of the changed lattice parameter or the altered electronic properties.

Oesterreicher et al. [6.41] searched in vain for superconductivity above 1.8 K in hydrides of ThPd compounds. The reason for this might be an insufficient H content or a high amount of magnetic impurities in their starting material (purity 99.9). For example, *Mackliet* et al. [6.42] had indications that T_c is reduced by 50–100 K/at. % Fe in the Pd–H system.

During the last few years another system, AlH, has been regarded as a metal-hydride superconductor. *Deutscher* and *Pasternak* [6.43] investigated the effect of Ar or H coating on T_c of Al films quench-condensed onto 4 K substrates. Whereas Ar lowers T_c, H leads to an increase of about 1 K up to $T_c = 4.5$ K. *Lamoise* et al. [6.44] implanted Al, O, He as well as H and D into Al at temperatures below 6 K. Besides the transition temperatures, the authors investigated the annealing behavior of the resistivity [6.45] of these Al films. The T_c values ($\leqq 4$ K) of the Al-, O-, and He-implanted Al films were explained, in connection with the resistivity annealing behavior, by phonon softening due to lattice disorder. In contrast, T_c of H(D)-implanted Al is 6.75 K, and there are indications that an ordered compound has been produced in which the enhanced transition temperature could be caused by a coupling of the electrons to optical phonons as in the Pd–H system (see Sect. 6.6). Recent tunneling experiments of *Dumoulin* et al. [6.46] on these H(D)-implanted Al films, however, indicated that changes of the phonon spectrum are not responsible for the T_c enhancement. The authors proposed that H influences mainly the electron-phonon-coupling parameter $\alpha^2(\omega)$ by an electronic or disorder effect.

On the other hand, high transition temperatures were also obtained by *Minnigerode* and *Rothenberg* [6.47] in quench-condensed Al films with impurities of O and Cu ($T_c = 5.8$ K) and by *Lamoise* et al. [6.48] after implantation of C, Ge, and Si into Al films ($T_c = 4.2$, 7.35, and 8.35 K, respectively). According to these results, a specific influence of the hydrogen as in the Pd–H system must not necessarily be taken into account, especially since no optic phonons could be detected in the tunneling experiment.

At present only three metal-hydrides (Th–H, Pd–H, and Al–H) with elevated transition temperatures exist. Because conclusive data are not available for the Al–H system, we restrict ourselves in the following section to the discussion of the Th–H and Pd–H systems.

6.3 Special Preparation Methods for Superconducting Thorium and Palladium Hydrides

6.3.1 Preparation of Bulk Th_4H_{15}

Although powdered samples of Th_4H_{15} can be easily achieved by the conventional H_2 gas pressure technique (see Chap. 2), it is not easy to prepare bulk samples. The reason for the powdering is the large lattice expansion of 11% and the presumably large inner stresses due to the take-up of H at 1 atm H_2 and temperatures between 600 and 200 °C. By charging at 850 °C, *Satterthwaite* and *Peterson* [6.24] avoided the powdering. At this elevated temperature the volume change from Th to the intermediate phase ThH_2 is smaller and the lattice distortions introduced by the H may anneal immediately. To overcome the dissociation pressure of about 750 bar at 850 °C, the authors used a very elegant method. A quartz tube, which contains at one end a large amount of powdered Th_4H_{15} and a bulk Th piece at the other end, was sealed after pumping to 10^{-9} bar. After heating the bulk Th to 850 °C, the required H_2-pressure was achieved by also heating the Th_4H_{15} powder to 850 °C. Some hours later the bulk Th was slowly cooled to room temperature. Thus bulk Th_4H_{15} specimens of high purity were obtained. The H content was slightly above the stoichiometric composition. The Th_4H_{15} samples are stable at room temperature, but they must be handled in an inert atmosphere to avoid the formation of ThO_2.

6.3.2 Preparation of Superconducting Pd Hydrides (H/Pd > 0.8)

High Pressure or Electrolytic Charging at Low Temperatures

High H/Pd ratios >0.8, which are necessary for the occurrence of superconductivity, can be achieved by increasing the gas pressure or the applied voltage during electrolysis. The H concentration can be further raised by lowering the temperature [6.49]. The advantage of the higher solubility, however, is coupled with the disadvantage of the exponential decrease of the diffusion constant. So one has to compromise between these two opposite effects. *Harper* [6.50] achieved ratios H/Pd ≅ 1.0 by electrolysis at 193 K. *Shirber* [6.51] and *Shirber* and *Northrup* [6.52] obtained ratios H/Pd = 0.97 and D/Pd = 0.95 by cooling their pressure cell slowly to 77 K while a H_2 pressure of 4–5 kbar was applied. *Skoskiewicz* et al. [6.53] prepared Pd–H alloys with even higher H content by the same procedure using H_2 pressures of about 12.5 kbar. The experiments show that the temperature range between 220 and 170 K is optimal. The Pd–H samples prepared in this way can be stored at 77 K without any loss of H.

H Implantation at He Temperatures

While the decreasing diffusion constant limits the take-up of H with decreasing temperature for the above mentioned techniques, the absence of diffusion at 4 K is used to prepare even higher H contents by implantation [6.35]. H diffusion is not necessary because the H ions accelerated by a high voltage are shot directly into the metal. There they come to rest after having passed their mean projected range. It is the lack of diffusion at 4 K which makes it possible for the implanted H atoms to be locally accumulated in the metal. Figure 6.1 shows the H distribution in a crosscut through a Pd foil. The Pd foil was precharged to a ratio H/Pd = 0.7, which can be easily done by 4 bar H_2 gas at 300 °C. Then H_2^+ ions are implanted with about 100 kV into this foil at 4 K. Hitting the surface, every H_2^+ ion is split into two parts, each having an energy of 50 keV. They come to rest in a region 6000 Å below the metal surface and about 1500 Å wide [6.54]. The attainable H concentration is only determined by the number of free interstitial sites. The advantage of the implantation method is its total independence of the equilibrium solubility of the special host metal [6.55]. The disadvantage consists of the fact that at constant implantation energy only narrow regions can be charged with H. In thin samples a

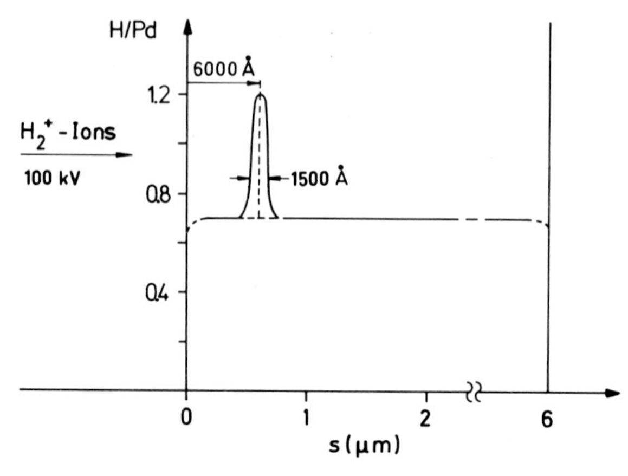

Fig. 6.1. H concentration versus depth *s* in a crosscut through a Pd foil, after homogeneous pre-charging and low-temperature implantation (calculated after [6.54])

homogeneous distribution can only be achieved by superimposing different concentration profiles using different implantation energies.

But a homogeneous distribution is not necessary for the detection of superconductivity. If the implanted region becomes superconducting, it short-circuits the whole foil resistance. If the thin superconducting region is thicker than the superconducting coherence length, it represents the properties of the bulk material. This condition is fulfilled for the Pd–H(D) system [6.37].

By means of the implantation method[1], *Stritzker* and *Buckel* achieved H/metal ratios exceeding 1.0 not only in Pd [6.35] but also in Pd–noble metal alloys [6.36], which have a low H solubility.

Codeposition of Pd and Hydrogen

A very unusual method for the fabrication of a metal-gas alloy was used by *Sansores* and *Glover* [6.56]. These authors condensed Pd and H_2 simultaneously onto a substrate held at 4 K. They obtained transition temperatures of about 5 K which correspond to ratios H/Pd $\approx$ 0.9. The H_2 molecules which are buried among the Pd atoms apparently dissociate under these conditions and form highly disordered Pd–H alloys. Even if the palladium is condensed on top of a precondensed solid H_2 film, superconducting Pd–H is formed. This hydride formation and some peculiar effects during annealing are still unexplained.

6.4 Superconducting Properties of the Pd–H System

This section deals with the properties of superconducting samples of highly concentrated Pd–H alloys which were prepared by different methods as described above.

[1] *Lamoise* et al. [6.44] used the implantation technique for the study of the Al–H system.

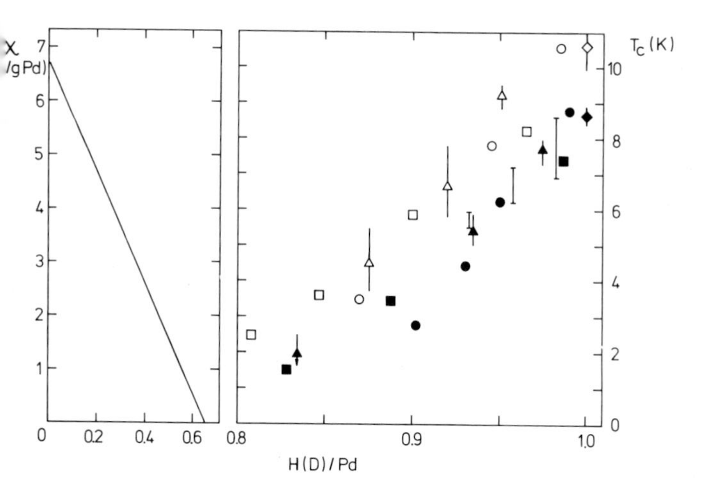

Fig. 6.2. Susceptibility χ [6.60] and transition temperature T_c versus ratio H(D)/Pd. T_c after different authors (filled symbols Pd–H, open symbols Pd–D): ◇ *Stritzker* and *Buckel* [6.35] plotted at H(D)/Pd = 1.0, ○ *Skoskiewicz* et al. [6.53], □ *Shirber* and *Northrup* [6.52], △ *Miller* and *Satterthwaite* [6.57], and I *McLachland* et al. [6.58]. Recently *McLachlan* et al. [6.68] published a T_c value of 10.4 K for PdH

6.4.1 T_c Dependence on H(D) Concentration

Maxima of the transition temperature $T_{c,max}$ with respect to the H and D concentration were first observed with the H(D)-implantation method at low temperatures by *Stritzker* and *Buckel* [6.35]. These maxima of 8.8 K in Pd–H and 10.7 K in Pd–D occurred at an estimated concentration H(D)/Pd between 1.0 and 1.2. A precise determination of the H(D) content can be performed by degassing of samples which have been homogeneously charged using the high pressure or the electrolytic technique at lowered temperatures. Measurements on such samples demonstrate that $T_{c,max}$ occurs at ratios H/Pd = D/Pd = 1.0. The maximum T_c of 8.5 K for stoichiometric PdH was observed by *Harper* [6.50]. *McLachlan* et al. [6.58] measured a T_c of 9.6 K for PdH. Other authors obtained T_c values only for H(D)/Pd ratios slightly below 1.0. Extrapolated values of $T_{c,max}$ in PdH and PdD of 9.5 and 11.7 K were reported by *Skoskiewicz* et al. [6.53], of 8 and 10 K by *Shirber* and *Northrup* [6.52], and of 9.4 and 11.8 K by *Miller* and *Satterthwaite* [6.57]. The 2 K higher transition temperature of PdD compared to PdH represents a remarkable inverse isotope effect.

Figure 6.2 shows T_c versus H(D) concentration resulting from different experiments. One can see the rather good agreement between the various Pd–H(D) samples. Apparently, the lattice defects introduced by the implantation process do not influence T_c. The width of the superconducting transition of the different samples can be explained by an inhomogeneous H(D) distribution. The sharpest transitions ($\Delta T_c < 0.06$ K) were observed by *Shirber* and *Northrup*

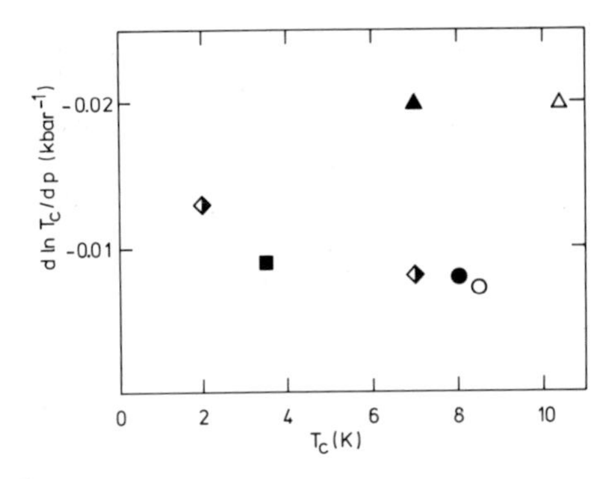

Fig. 6.3. Pressure coefficient $d\ln T_c/dp$ versus transition temperature in Pd–H (filled symbols) and Pd–D (open symbols): □ *Buckel* et al. [6.62], ◇ *Shirber* [6.63], △ *Skoskiewicz* et al. [6.53], and ○ *Wühl* [6.64]

[6.52], who charged Pd powder ($\phi < 35\,\mu$m) under 5 kbar $H_2(D_2)$ by lowering the temperature very slowly. Thus, they achieved a very homogeneous distribution throughout the powder grains due to the favorable surface/volume ratio. Similar sharp transitions were obtained by *Wühl* [6.59] in freshly condensed thin Pd films loaded with H by the same procedure. On the other hand, the superconducting transitions of foils ($>10\,\mu$m) charged at high pressure, electrolytically, or by implantation are between 0.3 and 1.0 K broad.

Besides T_c, the susceptibility χ is plotted for H(D) concentrations below H(D)/Pd$\cong$0.7 [6.60]. As can be seen, the strong paramagnetic behavior of the pure Pd decreases in the $\alpha-\beta$ mixed region with addition of H(D). At H(D)/Pd$>$0.6, the pure β phase is diamagnetic ($\chi \lesssim 0$) independent of the H(D) concentration. That means that superconductivity occurs in the region where the strong spin fluctuations of the pure Pd are totally depressed. As magnetism and superconductivity normally exclude each other, the following assumption is reasonable and well accepted: The vanishing paramagnetism is the necessary precondition for the occurrence of superconductivity in Pd [6.61]. This statement is supported by experiments on H-charged substitutional Pd alloys, as will be seen later.

6.4.2 Pressure Effect on T_c

Figure 6.3 shows the effect of hydrostatic pressure on T_c of the Pd–H system. These experiments were done by different methods. *Buckel* et al. [6.62] charged their Pd samples electrolytically before they applied pressure up to 21.6 kbar in a pressure cell filled with a pressure-transmitting liquid. The pressure was measured at He temperatures by means of a superconducting Pb manometer. *Shirber* [6.63] as well as *Wühl* [6.64] used their gas pressure equipment to charge their samples with H(D), and also to apply He pressure up to 4 and 7 kbar, respectively. The pressure was measured with a manometer at room temperature during careful isobaric cooling to 4 K, or by a superconducting Pb manometer inside the pressure cell, respectively. *Skoskiewicz* et al. [6.53] used a H_2 pressure of 12.5 kbar for the H charging as well as for the pressure experiment. As they could not maintain a constant pressure during cooling to He temperature, they had to estimate their final pressure. This might be the reason for the deviation of their results from those of the above mentioned authors as shown in Fig. 6.3. These latter measurements give a negative pressure coefficient of $d \ln T_c/dp \simeq -0.01\ \mathrm{kbar}^{-1}$. This value is comparable with the pressure effect of soft nontransition metals [6.7]. The decrease of T_c with increasing pressure can be easily understood by a stiffening of the phonon spectrum causing a decrease of the electron-phonon-coupling parameter λ. The negative pressure effect clearly contradicts the assumption [6.65] of a superconducting metallic hydrogen sublattice inside the Pd lattice since a smaller H–H distance should be more favorable for superconductivity.

6.4.3 Critical Field

Measurements of the critical field were carried out on Pd–H samples prepared by electrolysis at lowered temperatures. The experiments showed that Pd–H is a superconductor of type II. Thus the upper field $H_{c2}(0)$ at $T=0$ K depends strongly on the mean free path of the electrons, i.e., on lattice distortions and impurities. This seems to be the reason why different H_{c2} values have been reported. *Skoskiewicz* [6.66] measured $H_{c2}(0)=0.18$ Tesla, and *Alekseevskii* et al. [6.67] 0.25 Tesla, independent of the H concentration. *McLachlan* et al. [6.58, 68] observed H_{c2} values between 0.08 and 0.16 Tesla, dependent either on the H content or the number of lattice distortions. *Meservey* and *Tedrow* [6.69] achieved much higher critical fields in Pd–H prepared by the method of co-deposition at He temperature. According to the extremely high distortion in these thin films, they found $H_{c,\perp} \approx 8$ Tesla and $H_{c,\parallel} \approx 13.5$ Tesla for fields applied perpendicular and parallel to the films.

6.4.4 T_c Variation in the H(D) Charged Systems Ni–Pd–Pt and Rh–Pd–Ag

In order to check the influence of the Pd host matrix, Pd atoms were substituted by the neighboring elements Ni, Pt and Rh, Ag. Generally, these substitutions reduce the H solubility considerably [6.49]. Only $Pd_{1-x}Rh_x$ alloys with $x \leqq 8$ at.% have a higher solubility than pure Pd. The diminished H solubility in the Pd alloys is a severe problem with respect to the study of their superconducting properties. In comparison to pure Pd, it is much more difficult to increase the H concentration sufficiently in the Pd alloys by means of solubility-dependent methods like high pressure or electrolysis. Here the method of H implantation at low temperatures has a decisive advantage. Again the H concentration can be increased above the optimum content where T_c exhibits a maximum value $T_{c,\max}$. Figures 6.4 and 6.5 show $T_{c,\max}$ with respect to the H or D concentration versus the composition of the host matrix for the systems Rh–Pd–Ag and Ni–Pd–Pt. Besides the results of H or D implantation [6.70], the highest T_c values achieved by *Skoskiewicz* et al. [6.71, 72] and *Shirber* [6.51]

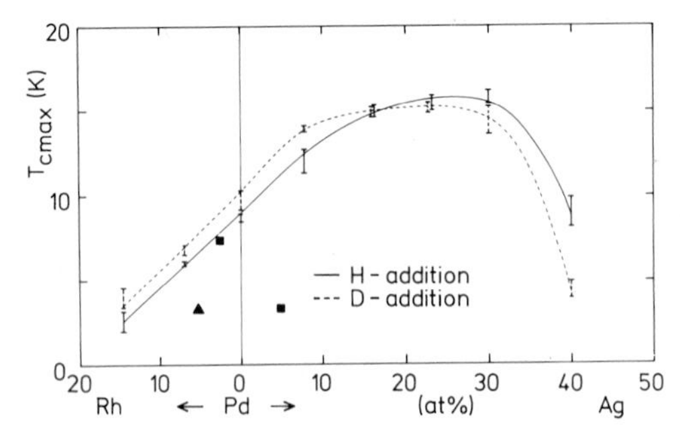

Fig. 6.4. $T_{c,max}$ with respect to the implanted H(D) content (curves) versus composition of the host matrix Rh–Pd–Ag [6.70]. The "error bars" indicate the width of the superconducting transition. The highest T_c values achieved by ■ *Skoskiewicz* et al. [6.71] and ▲ *Shirber* [6.51] with H_2 gas pressure charging are added

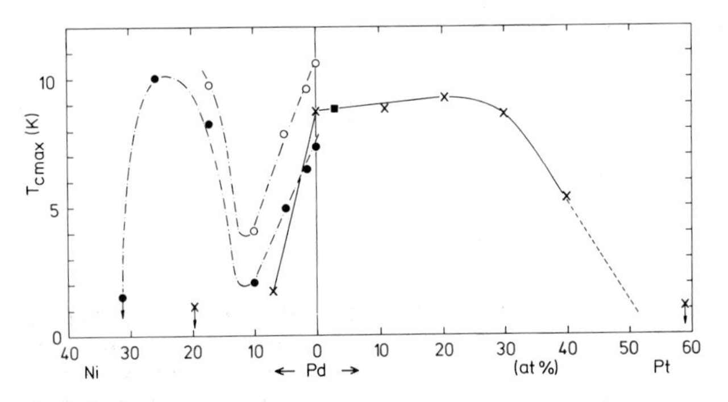

Fig. 6.5. $T_{c,max}$ with respect to the implanted H content versus composition of Ni–Pd–Pt: × *Stritzker* [6.70]; and the highest T_c values obtained by charging with high pressure of H_2 (filled symbols) and D_2 (open symbols): ■ *Skoskiewicz* et al. [6.71] and ● *Skoskiewicz* [6.72]. A symbol marked with an arrow indicates that no superconductivity was found above this temperature

with other methods are plotted. At low concentrations of the substitute elements, the results of the different experiments deviate most of all in the case of $Pd_{0.95}Ag_{0.05}$ (Fig. 6.4). The reason is that the H solubility of Pd is more reduced by substitution with Ag than with Rh, Ni, and Pt. At higher Ni concentrations (Fig. 6.5) the results of *Stritzker* [6.70] were quite different from those of *Skoskiewicz* [6.72], who detected a re-increase of T_c. At present an explanation for this contradiction is not available.

Whereas in the system Rh–Pd–Ag, $T_{c,max}$ varies smoothly from the Rh to the Ag side, this is not the case in the system Ni–Pd–Pt. The substitution of Pd with Pt does not influence $T_{c,max}$ very much up to about 25 at.% Pt, but $T_{c,max}$ is strongly depressed in Pd–Ni alloys for concentrations of less than 10 at-% Ni. The variation of $T_{c,max}$ in both systems is not correlated with the variation of the susceptibility χ and the electronic specific heat coefficient γ in the unhydrogenated Ni–Pd–Pt and Rh–Pd–Ag systems. χ and γ vary smoothly in Ni–Pd–Pt [6.73] but show a pronounced maximum at 3 at.% Rh in Rh–Pd–Ag [6.74]. The electronic properties of the host lattice seem to play no important role for the superconductivity of the hydrogenated alloy.

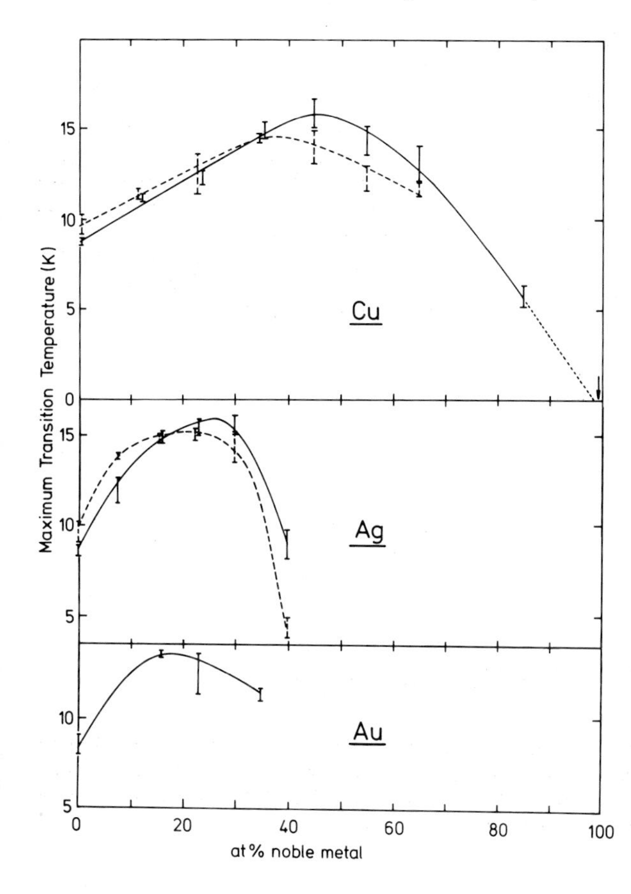

Fig. 6.6. $T_{c,max}$ for H (solid curves) and D (broken curves) implanted Pd-noble metal alloys as a function of the noble metal content [6.37]. The "error bars" indicate the width of the superconducting transition

Figure 6.4 shows another important property of the host matrix: The inverse isotope effect in pure Pd, i.e., the higher T_c of Pd–D with respect to Pd–H, is reversed into a normal isotope effect by the substitution of Pd by.more than 20 at.% Ag.

6.4.5 T_c Variation in H(D) Charged Pd–Noble Metal Alloys

Systematic variations of $T_{c,max}$ in H(D) charged Pd–noble metal alloys with similar electronic properties were observed by *Stritzker* [6.37]. Figure 6.6 shows $T_{c,max}$ achieved by H(D) implantation into Pd–(Cu, Ag, Au) alloys versus the noble metal content. In the Pd–Cu–H system, an astonishing high $T_{c,max} \simeq 17$ K was achieved in an alloy with the composition $H/Pd_{55}Cu_{45} \approx 0.7$. The corresponding values are $T_{c,max} = 15.6$ K in $H/Pd_{70}Ag_{30} \approx 0.8$ and $T_{c,max} = 13.6$ K in $H/Pd_{84}Au_{16} \approx 0.9$. A systematic increase of $T_{c,max}$ with decreasing mass of the noble metal can be seen. This is a strong hint that the acoustic phonon spectrum of these hydrides has a strong influence on T_c. In spite of the

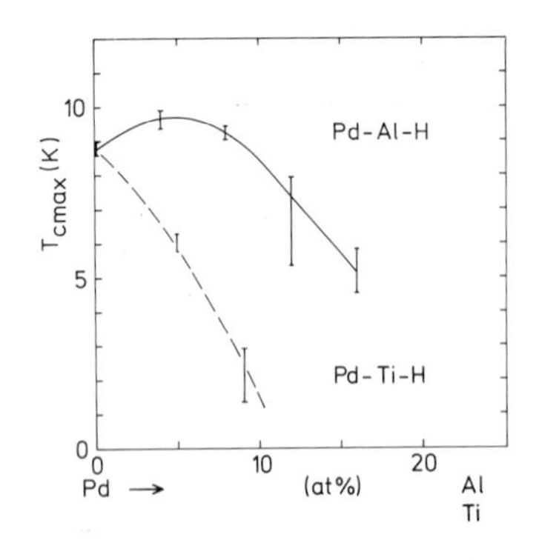

Fig. 6.7. $T_{c,max}$ for H implanted Pd–Al and Pd–Ti alloys as a function of the Al or Ti content [6.77]

uncertainty in the determination of the H concentration with the implantation method, it is obvious that $T_{c,max}$ occurs at smaller H(D) content when the noble metal concentration increases. It is a well-known fact that the substitution with a noble metal has an effect on the susceptibility χ of Pd similar to that caused by the addition of H. In both cases χ is reduced and the system becomes diamagnetic at 60 at.% noble metal [6.75] or at H/Pd$\simeq$0.6 [6.60]. So the following conclusion is reasonable: The noble metal content helps the H addition to diminish the harmful spin fluctuations of the Pd. This is confirmed by the fact that $T_{c,max}$ occurs at lower H concentrations than in PdH. The disappearance of the enhanced paramagnetism of pure Pd seems to be a necessary precondition for the occurrence of superconductivity in the Pd–H system. The further increase of T_c after partial substitution of Pd by noble metal atoms is somewhat astonishing, as the noble metal itself as well as unhydrogenated Pd-noble metal alloys do not become superconducting.

In Pd–noble metal–H(D) alloys with $T_c \geqq 15$ K, superconductivity occasionally vanishes during additional H implantation. By means of x-ray diffraction this pecularity was explained by *Becker* [6.76] as a sudden loss of all H(D) even at 4 K.

6.4.6 T_c in Other Hydrogenated Pd Alloys

The increase of $T_{c,max}$ with decreasing mass of the substituted noble metal in Pd suggested a similar experiment on $Pd_{1-x}Al_x$ alloys [6.77]. These alloys (miscible for $x \leqq 0.15$) were charged with H by implantation at low temperatures. Figure 6.7 shows $T_{c,max}$ versus the Al concentration. Only at low Al concentrations can a slight increase of T_c with increasing Al content be seen. The hope of even higher T_c values as in Pd–Cu–H was not fulfilled. In this experiment not only changes of the mass but also of the electronic structure have apparently to be taken into account.

Another implantation experiment was done on an alloy, namely PdTi [6.77], where χ decreases very rapidly after substitution of Pd by Ti [6.78]. But in this experiment only a decrease of $T_{c,max}$ was observed with increasing Ti content (Fig. 6.7).

Table 6.1. Maximum transition temperature $T_{c,max}$ of Pd after implantation of light elements

Implanted element	H	D	Li	B	C	N
$T_{c,max}$ [K]	8.8	10.7	<0.1	3.8	1.3	<0.2
Concentration at $T_{c,max}$ (ion/Pd)	1	1	>1.3	1.5	0.6	>2

6.4.7 T_c in Pd Alloys with Interstitial Elements

The inverse isotope effect in the Pd–H(D) system, i.e., the increase of T_c with increasing mass of the interstitial H(D), immediately leads to the question whether T_c increases even more with increasing mass. Unfortunately the next heavier isotope, tritium, is radioactive, which makes such an experiment very difficult. This is the reason why even heavier elements have been used. Whereas the inert He does not influence the spin fluctuations and Li is not interstitially built into the palladium, B, C, N are good candidates for this experiment. *Stritzker* and *Becker* [6.79] implanted these elements into Pd. They indeed found superconductivity, but not at higher temperatures. Table 6.1 shows the results including those of H, D, and Li implantation. A general increase of T_c with increasing mass of the interstitial is not observed. The reason for this is the change of the electronic properties of the heavier elements. The authors claim a special influence of H(D) on the electron-phonon coupling which is enhanced due to the prevailing screening of the protons (deuterons) by the conduction electrons. On the other hand, the electron-phonon coupling is reduced for the heavier elements as the nuclei are more strongly screened by their core electrons.

6.5 Superconducting Properties of the Th–H System

6.5.1 T_c of the Th–H(D) System

In contrast to the Pd–H system where the pure metal is a normal conductor, Th itself becomes superconducting at 1.37 K. T_c is lowered below 1.1 K in the dihydride ThH_2 as shown by *Matthias* et al. [6.25] and *Satterthwaite* and *Peterson* [6.24]. Further hydrogenation leads to the high hydride Th_4H_{15} with a ratio H/Th = 3.75. This compound is a good superconductor. *Dietrich* et al. [6.80], as well as *Miller* et al. [6.81], reported T_c values in the ranges 7.5–8.0 K and 8.5–9.0 K, depending upon subtle differences in sample preparation. But different modifications could not be distinguished either by the analysis of the H content or by x-ray diffraction. The crystalline structure of Th_4H_{15} consists of a simple cubic lattice with the symmetry I $\bar{4}3d$ containing 16 Th and 60 H atoms in the unit cell. *Caton* and *Satterthwaite* [6.82] found that there exists, in

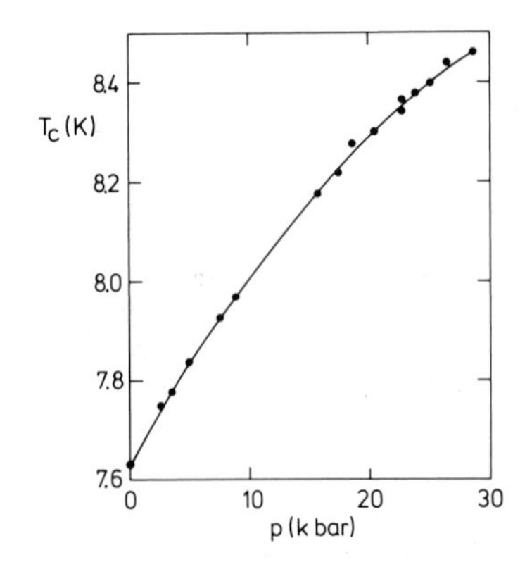

Fig. 6.8. Transition temperature of Th_4H_{15} versus applied pressure [6.80]

addition, a tetragonal phase of Th_4H_{15} which is not superconducting. T_c is observed to be very sensitive to deviations from the stoichiometric ratio H/Th = 3.75. Small deviations result in a decrease of T_c. The authors presume that this sensitivity is also the reason for rather broad transition curves ($\gtrsim 0.5$ K) and probably for the difference in T_c of different samples. Within the experimental uncertainties, no isotope effect between Th_4H_{15} and Th_4D_{15} was detected. The absence of a usual isotope effect contradicts an assumption that the superconductivity of Th_4H_{15} may be due to metallic H(D) inside the Th lattice. Metallic H should show a pronounced normal isotope effect, i.e., a lower T_c for Th_4D_{15}.

6.5.2 Pressure Effect on T_c

Dietrich et al. [6.80] succeeded in measuring the pressure effect in spite of the difficulties in handling the Th_4H_{15} which reacts with the air as well as with the pressure medium. Figure 6.8 shows the result. In contrast to the Pd–H system, T_c increases steeply with increasing pressure. The initial slope is $d\ln T_c/dp = 5.5\ 10^{-3}\ \mathrm{kbar}^{-1}$. No saturation of T_c occurs up to a pressure of 28 kbar.

6.5.3 Critical Field

Satterthwaite and *Toepke* [6.1] measured also the behavior of powdered Th_4H_{15} in a magnetic field up to 1 Tesla. From these experiments, they concluded that Th_4H_{15} is a type II superconductor with a critical field H_{c2} between 2.5 and 3.0 Tesla.

6.5.4 T_c of Th Alloys with Other Interstitials

Besides Th_4H_{15} and Th_4D_{15}, which do not differ significantly in their superconducting properties, some other Th alloys with interstitial elements

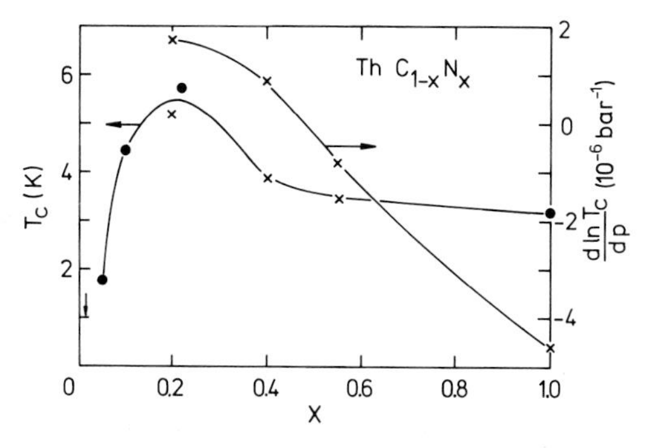

Fig. 6.9. Transition temperature T_c (● *Giorgi* et al. [6.84], × *Dietrich* [6.85]) and pressure coefficient $d \ln T_c/dp$ (× *Dietrich* [6.85]) versus composition of $ThC_{1-x}N_x$ alloys

were examined for superconductivity. ThC remains normal above 1.2 K according to *Hardy* and *Hulm* [6.83]. This result was confirmed by *Giorgi* et al. [6.84], who extended these investigations to ThN and $ThC_{1-x}N_x$ alloys. ThN was found to become superconducting at 3.2 K. Replacing N by C enhances T_c up to 5.8 K in $ThC_{0.78}N_{0.22}$. With further increase of the C content, T_c drops sharply. Figure 6.9 shows T_c of $ThC_{1-x}N_x$ and the corresponding pressure effect measured by *Dietrich* [6.85] in a liquid pressure cell. As can be seen, the pressure effect changes from negative values for large x to positive values for small x. But the positive pressure effect remains clearly below $5.5\ 10^{-3}\ \text{kbar}^{-1}$, the value observed for Th_4H_{15}.

6.6 Electron and Phonon Properties of the Pd–H System

In this section a summary is given on the electron [Ref. 6.86, Chap. 5] and phonon properties of Pd–H(D) alloys which were obtained by measurements of the specific heat, inelastic neutron scattering, the superconducting tunneling effect, and the temperature dependence of the resistivity. Following this, the theoretical models on the occurrence of superconductivity in this system are discussed.

6.6.1 Low Temperature Specific Heat

The electronic specific heat coefficient γ for pure Pd is 9.5 mJ/mol K^2 and decreases when small amounts of hydrogen are added [6.87]. For H/Pd ratios of 0.83 to 0.88, *Mackliet* et al. [6.42] reported that γ is about six times smaller than in pure Pd. Figure 6.10 shows the electronic specific heat for different H concentrations. The broad transitions of the superconducting state are typical for bulk Pd–H samples and are explained by macroscopic compositional inhomogeneity (see also Sect. 6.4.1). *Zimmermann* et al. [6.88] extended the specific heat measurements to H/Pd = 0.96. Their results fit to the above γ values. For concentrations between $c = 0.88$ and 0.96, they found a further

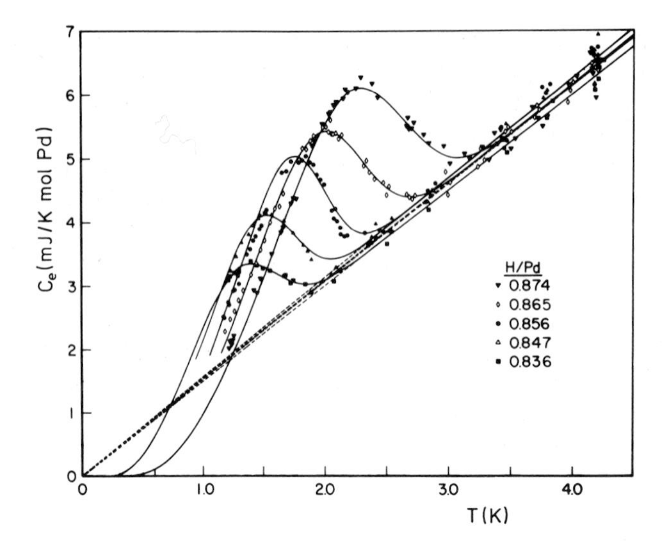

Fig. 6.10. Electron specific heat versus temperature for different Pd–H samples. The solid lines were obtained by assuming an inhomogeneous H distribution [6.42]

decrease of γ by a factor of about 2. This result agrees qualitatively with the concentration dependence of the electronic density of states $N(0)$ calculated by *Faulkner* [6.89], but it contradicts calculations of *Zbasnik* and *Mahnig* [6.90], who found a weakly increasing density of states above H/Pd = 0.8. The specific heat experiment [6.88], however, is not conclusive as there might be a large uncertainty in the determination of γ due to the extrapolation of the normal-state specific heat from $T > 5$ K to 0 K.

Values between 266 K and 278 K have been reported [6.87] for the Debye temperature θ_D of pure Pd. At hydrogen concentrations H/Pd = 0.86, $\theta_D = 270$ K has been measured [6.42, 88]. Although the lattice constant is increased by 4% due to the dissolved hydrogen [6.91], the analysis of the experiments did not show any significant change of θ_D up to this concentration. Only for H/Pd > 0.89 have *Zimmermann* et al. [6.88] reported a decrease to $\theta_D = 220$ K at H/Pd = 0.96.

6.6.2 Neutron-Scattering Experiments

Since a detailed discussion on neutron-scattering experiments is given in another Reference ([6.86], Chap. 4), only the essential results are summarized here.

Rowe et al. [6.92] studied the lattice dynamics of a single crystal with D/Pd = 0.63. They obtained a 20–30% decrease of the frequencies of the acoustic modes compared to the corresponding modes of pure Pd. The authors claimed that this frequency shift is consistent with the observed lattice expansion. Recent calculations by *Magerl* [6.93], however, show that the observed frequency shift is about four times larger than the value expected for the expanded lattice. This means a remarkable softening of the acoustic phonons due to H charging. The analysis of the scattering data for the optic modes due to D is complicated since the conventional lattice dynamics treatment cannot be applied to a nonstoichiometric sample [6.94]. From a comparison with incoherent-neutron-scattering data on polycrystalline samples with H/Pd = 0.63, *Rahman* et al. [6.95] concluded that the Pd–H force constants are about 20% stronger than the corresponding Pd–D force constants for D/Pd = 0.63. This anharmonicity is believed to be the main reason for the inverse isotope effect in the superconductivity of Pd–H(D) alloys [6.96] (see Sect. 6.6.5).

Investigations of hydrogenated Pd–Ag alloys by *Chowdhury* and *Ross* [6.97] showed that alloying with silver does not change the optic mode frequency, but does increase the bandwidth. A lowering of the acoustic frequencies is observed for Ag concentrations of below 50 at. %. This may be responsible for the high T_c in this system.

6.6.3 Tunneling Experiments

Tunneling experiments have been performed on the Pd–H(D) system by four groups. The basic configuration of the tunneling junction consists of the type Al-oxide-(Pd–H). First an Al film was condensed onto a substrate. Then its surface was oxidized to form the tunneling barrier. Different procedures have been used to complete the junction with Pd–H. *Dynes* and *Garno* [6.98] condensed the Pd film at 4 K and then exposed it to a plasma discharge in hydrogen gas at 77 K. *Eichler* et al. [6.99] and *Wühl* [6.59] charged the Pd films which were condensed at room temperature with high gas pressure. A destruction of the tunneling barrier by the increase of the lattice constant in Pd–H(D) could be avoided by a very slow charging procedure at about 150 K. After completion of the H(D) take-up, the junction was cooled to 77 K and the gas pressure was released. *Silverman* and *Briscoe* [6.100] prepared the Pd–H film after the method of *Sansores* and *Glover* [6.56], condensing Pd at 8 K on top of a previously condensed layer of H_2. *Igalson* et al. [6.101] tried to overcome the destruction of the barrier by condensing an underlayer of Nb. The junctions Nb–Al–oxide–Pd were charged at room temperature at a pressure of 3.5 kbar.

The current-voltage $(I-V)$ characteristics and their first and second derivatives were measured as a function of the voltage applied to the junctions. The superconducting energy gap Δ can be taken directly. The ratio $2\Delta_0/k_B T_c$ is a measure of the electron-phonon coupling strength. From the first derivative the quantity $\alpha^2(\omega)F(\omega)$ can be derived with the help of the Eliashberg equations, provided that the electron-phonon coupling is strong enough [6.10]. A direct experimental observation of the structure of $\alpha^2(\omega)F(\omega)$ is possible by the second derivative, since the minima in d^2I/dV^2 correspond to maxima in $\alpha^2(\omega)F(\omega)$. Only such phonons are observed which couple to the conduction electrons and therefore contribute to superconductivity.

Table 6.2. Transition temperature T_c, ratio of the energy gap Δ_0 to T_c, and optic phonon energies obtained from Pd–H(D) tunneling experiments

	T_c [K]	$2\Delta_0/k_B T_c$	ω_D [meV]	ω_H [meV]
Dynes and *Garno* [6.98]	6.3	3.83		48
Eichler et al. [6.99]	5.5	3.6	33.5	
Silverman and *Briscoe* [6.100]	4.6	3.8		48
Igalson et al. [6.101]	4.6	2.6		
Wühl [6.59]	7.5	3.7	31.0, 33.5	45.5, 50.0

The numerical results of the tunneling experiments are listed in Table 6.2. The ratio $2\Delta_0/k_B T_c$ of 3.6 to 3.8 shows that Pd–H and Pd–D are medium-coupling superconductors in the concentration range investigated. The unusual small value obtained by *Igalson* et al. [6.101] may be related to difficulties with junction preparation. Structures have been observed in d^2I/dV^2 at ω_H and ω_D in the energy region of the optic H(D) vibrations. These structures clearly demonstrate that the optic phonons substantially contribute to the attractive electron-electron interaction leading to superconductivity. The Pd–H(D) system is the first superconducting system in which the contribution of optic phonons could be experimentally established.

The charging method used by *Eichler* et al. [6.99] for their junctions yielded well-defined energy gaps and well-pronounced structures in $\alpha^2(\omega)F(\omega)$. The result is shown in Fig. 6.11. Assuming $\alpha^2(\omega)$ to be a smoothly varying function, the positions of peaks in $\alpha^2(\omega)F(\omega)$ can be compared with those in $F(\omega)$ obtained from neutron-scattering experiments [6.92]. Whereas the low-energy part of $\alpha^2(\omega)F(\omega)$ and $F(\omega)$ for Pd–D coincide rather well, the high-energy part of $\alpha^2(\omega)F(\omega)$ is shifted to lower energies compared to $F(\omega)$. The reason is the higher D content of Pd in the tunneling experiment as will be shown later (Fig. 6.12). Despite some uncertainty in the amplitude of the calculated $\alpha^2(\omega)F(\omega)$, the results shown in Fig. 6.11 reveal the important contributions of optic phonons to superconductivity in Pd–D. The recent application of gas pressure up to 7 kbar by *Wühl* [6.59] enhanced T_c and improved the phonon-induced structures in the tunneling characteristic, but it did not diminish erratic changes in the tunneling conductance due to instabilities in the junction barrier. Therefore, no further attempt has been made to calculate $\alpha^2(\omega)F(\omega)$ from the first-derivative data. Only the directly measured second derivatives d^2I/dV^2 for Pd–H and Pd–D are shown in the lower part of Fig. 6.11. Irreproducible structures owing to barrier instabilities have been omitted. Keeping in mind that minima of d^2I/dV^2 correspond to maxima of $\alpha^2(\omega)F(\omega)$, d^2I/dV^2 reveals a well-resolved acoustic (Pd–) phonon structure for the deuterided junctions (D/Pd = 0.94). This structure coincides excellently with $F(\omega)$ obtained from neutron experiments [6.92] for D/Pd = 0.63 (Fig. 6.11, upper part). The decrease of the acoustic Pd-phonon energies observed for D/Pd between 0 and

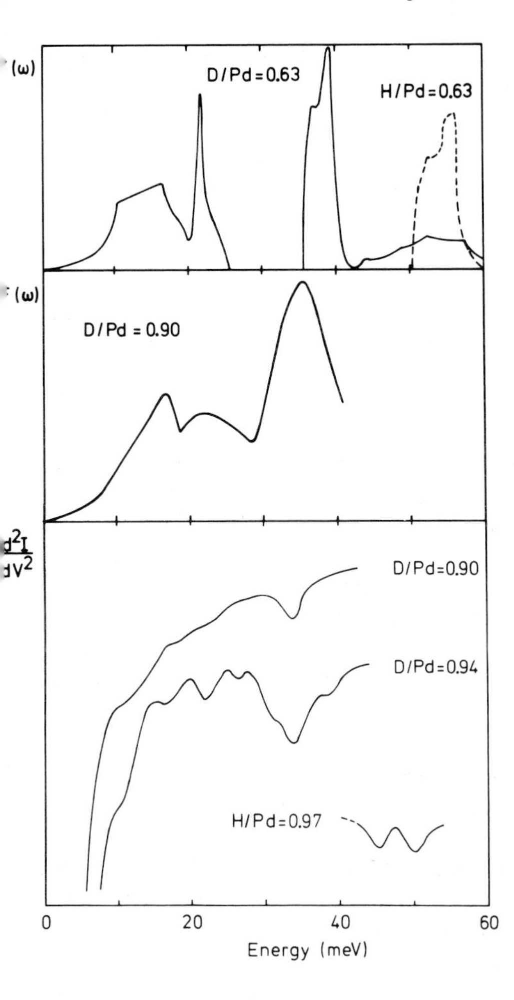

Fig. 6.11. Comparison of the phonon density of states $F(\omega)$ from inelastic neutron scattering with $\alpha^2(\omega)F(\omega)$ and d^2I/dV^2 obtained from tunneling experiments.

$F(\omega)$ is derived for PdD and PdH from a Born-von Karman fit to the dispersion relations of a D/Pd=0.63 single crystal assuming stoichiometric composition; $F(\omega)$ for PdH was calculated by replacing the deuterium mass by the hydrogen mass [6.92]. $\alpha^2(\omega)F(\omega)$ calculated for D/Pd =0.90 is a qualitative result to demonstrate the contributions of the optic modes to superconductivity [6.99]. The second derivatives of the tunneling characteristics d^2I/dV^2 are directly measured: minima correspond to maxima in $\alpha^2(\omega)F(\omega)$. Results for the higher D and H content after *Wühl* [6.59]. The structure in d^2I/dV^2 for D/Pd=0.94 at 38 meV can be attributed to phonons in the Al counter film of the tunneling junction

0.63 in the neutron scattering experiment does apparently not continue when the D concentration is increased to 0.94 in the tunneling experiment.

In the optic part of the spectrum (energies >27 meV), d^2I/dV^2 shows a double-minima structure for D/Pd=0.94 as well as for H/Pd=0.97. This structure is well pronounced in the latter sample. The structure can be attributed to the transverse optic (TO)-phonon modes. The longitudinal optic modes which are smaller in amplitude and reach to about 80 meV could not be measured in the tunneling experiment since the junctions became too noisy at high applied voltages. For Pd–H, only the optic part of the spectrum is presented, since in the acoustic part no reproducible structure of d^2I/dV^2 could be obtained. This result gives the impression that the electron-phonon coupling of the acoustic phonons is stronger for Pd–D than for Pd–H. Possibly, $\alpha^2(\omega)$ of the acoustic phonons benefits from an enhanced contribution of the vibrations

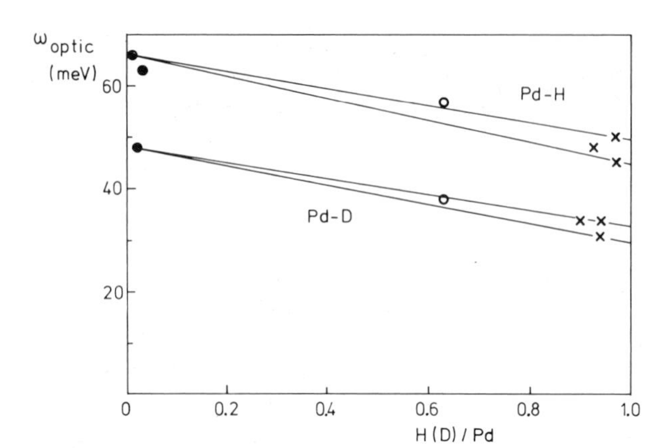

Fig. 6.12. Position of the H(D) local modes for dilute Pd–H(D) (● *Drexel* et al. [6.102]); and the position of the transverse optic modes for H(D)/Pd = 0.63 samples from neutron scattering experiments (○ *Rahman* et al. [6.95], *Rowe* et al. [6.92]) as well as for samples with H(D) concentrations of 0.9 to 0.97 (×) from tunneling experiments. The tunneling data are taken from Table 6.2. The H(D) concentrations are derived from T_c values using the experimental results of [6.57] (see Fig. 6.2)

of the D atoms to the acoustic spectrum owing to the heavier mass of deuterium compared to hydrogen.

In Fig. 6.12 the central frequencies of the TO modes are plotted as a function of H(D) concentration. Data from neutron experiments on alloys with H(D)/Pd = 0.63 [6.92, 95] and $\lesssim 0.03$ [6.102] are also added. Straight lines are tentatively drawn to indicate the decrease of the phonon frequencies and the splitting of the TO modes. This splitting can be explained by an increasing H–H(D–D) interaction with growing concentration. The frequency ratio of the H to D modes is about 1.5 instead of $\sqrt{m_D/m_H} = \sqrt{2}$ at high H(D) concentrations, pointing to anharmonic effects which were first proposed by *Ganguly* [6.96].

6.6.4 Temperature Dependence of the Electrical Resistivity

McLachlan et al. [6.103] measured the temperature-dependent resistivity of nearly stoichiometric PdH(c = 0.995). A change in the slope $d\varrho/dT$ at about 140 K was interpreted as the onset of an additional scattering of the conduction electrons with optic phonons. For $T < 80$ K, the experimental data were fitted to a Grüneisen function with the Debye temperature $\theta_D = 210$ K. The resistivity could be determined only up to 280 K because of the desorption of hydrogen. In the temperature range between 80 and 280 K, the excess resistivity could be described by a Debye spectrum with $\theta_D = 1000$ K as well as by an Einstein spectrum with the characteristic temperature $\theta_E = 550$ K. The latter is in close agreement with the results of the tunneling experiments. Assuming $d\varrho/dT$ to be proportional to the electron-phonon coupling constant λ, *McLachlan* et al.

[6.103] obtained $\lambda_{opt} \approx 3\lambda_{ac}$. A similar analysis of $\varrho(T)$ was made by *Chiu* and *Devine* [6.104] for different H/Pd ratios. They deduced a reduction of the electron density of states by a factor of 6 in varying H/Pd from 0 to 0.9.

6.6.5 Theories

It is widely believed that Pd is not superconducting due to the presence of spin fluctuations. The quenching of the spin fluctuations by adding hydrogen was assumed by *Bennemann* and *Garland* [6.61] to be already sufficient for the occurrence of superconductivity in Pd–H. The high T_c in Pd–H was ascribed to the coupling between the acoustic phonons and the *d* electrons. The authors predicted a positive pressure effect for T_c which, however, was disproved by experiment. *Hertel* [6.105] and *Brown* [6.106] came to the conclusion that the coupling of the acoustic phonons to the *sp* electrons rather than to the *d* electrons is responsible for the superconductivity. The optic phonons should not play any important role in disagreement with the tunneling data (see Sect. 6.6.3). *Ganguly* [6.96, 107] pointed out that the quenching of spin fluctuations alone is not sufficient to account for the high transition temperatures in Pd–H(D), and he considered the possibility that the optic modes of H(D) contribute to the electron-phonon interaction. He explained the inverse isotope effect by the stiffening of the force constants due to a larger zero point motion of the hydrogen compared to that of the deuterium. This prediction is supported by the results of the tunneling experiments.

An alternative approach is the electronic model proposed by *Miller* and *Satterthwaite* [6.57]. This model is based on energy-band calculations by *Switendick* [6.108] and photoemission studies by *Eastman* et al. [6.109] and *Antonangeli* et al. [6.110] for the Pd–H system [Ref. 6.86, Chap. 5]. These calculations show that the electrons of the added hydrogen atoms fill three different classes of states: 1) 0.36 holes in the *d* band, 2) *s* states which are lowered and hybridized with hydrogen *s* states forming Pd–H bonds, and 3) *sp* band states which lead to a noticeable increase of the Fermi energy. *Miller* and *Satterthwaite* [6.57] argued that the difference in the amplitude of the zero point motion of H and D causes a larger overlap of wave functions between hydrogen and palladium than between deuterium and palladium. Thus, more electrons are in Pd–H bonding states. The number of *sp* electrons is therefore smaller in Pd–H alloys than in Pd–D alloys, though the H(D)/Pd ratio is the same. Since superconductivity is assumed to depend critically on the occupation of the *sp* band, Pd–H alloys have smaller transition temperatures than do the corresponding Pd–D alloys. The experimentally determined transition temperatures versus composition for Pd–H and Pd–D coincide when the curve for Pd–H is shifted 0.05 units to higher H/Pd ratios. In this model the inverse isotope effect is explained by an electronic effect.

This electronic model does not seem to be at variance with the phonon model proposed by *Ganguly* [6.107]. From the larger zero point motion of H, one would expect a larger lattice constant of PdH compared to PdD. Since the lattice constants of PdH and PdD are almost identical ($a_{PdH}=4.090$ Å, $a_{PdD}=4.084$ Å) [6.91], the conclusion is confirmed that there are stronger bonds between Pd–H than between Pd–D. The stiffening of the force constants when D is replaced by H is the cause for the anharmonic effects discussed by *Ganguly*.

Papaconstantopoulos and *Klein* [6.111] calculated T_c for stoichiometric PdD. They assumed that the Pd atoms mainly vibrate in the acoustic modes and the D atoms in the optic modes. The electronic factors of λ are determined using the APW method for a self-consistent energy-band calculation, and the phonon factor is obtained from neutron-scattering experiments on D/Pd = 0.63 samples [6.92]. The authors showed that the major contributions to λ come from the optic branches of the D-atom vibrations which are located at rather low energies. They obtained $\lambda_{opt} \approx 3\lambda_{ac}$, in excellent agreement with the $\varrho(T)$ measurements [6.103]. In order to reproduce the experimental $T_c = 11$ K, they had to assume the Coulomb pseudopotential $\mu^* = 0.22$. This value is unusually high for a superconductor and is thought to be a remainder of the high value of μ^* for pure Pd which is due to spin fluctuations.

Pindor [6.112], however, pointed out that a large value of μ^* for PdH may also be caused by a small bandwidth and the high phonon frequencies which enter the equation for μ^*. An improbable high μ^* was obtained by *Nakajima* [6.113], who analyzed the inverse isotope effect. He considered the influence of the volume change on T_c due to the different zero point motions of the interstitials, neglecting, however, anharmonicity in the force constants.

Burger and *McLachlan* [6.114] calculated the electron-phonon parameter λ using an isotropic phonon spectrum with nearest neighbor and next-nearest neighbor interactions. They showed that λ can be decomposed into acoustic and optic contributions to a good approximation because of the large mass ratio M_{Pd}/M_H. It turned out, however, that there is an enhancement of λ_{ac} due to an interference term which takes into account the simultaneous motion of the Pd and H ions in the acoustic mode. It was confirmed that λ_{opt} is larger than λ_{ac} because of the very low optic phonon frequencies.

At first sight, alloys of Pd with noble metals do not differ from Pd–H alloys. The susceptibility and the specific heat are influenced in the same way by adding noble metals [6.75, 115] as by charging with hydrogen [6.60, 116]. The spin fluctuations in Pd, for example, are also suppressed for concentrations of 60 at. % noble metal. But these Pd alloys do not become superconducting. The occurrence of superconductivity when charged with H was explained by *Ganguly* [6.117] to be based on the coupling of the electrons to the optic phonons as in PdH(D). The small differences in T_c are ascribed to the individual properties of Cu, Ag, and Au, which enter through the coupling to the acoustic phonons. The content of noble metals at which the maximum T_c occurs is speculated to be associated with the disappearance of the $\alpha-\beta$ phase boundary. *Gomersall* and *Gyorffy* [6.118] argued that the large oscillations of a light component with a large electron-phonon interaction cause a high T_c and signal the beginning of a lattice instability. The lattice expansion with the addition of Ag should give rise to larger oscillations of the H atoms and an enhancement of T_c.

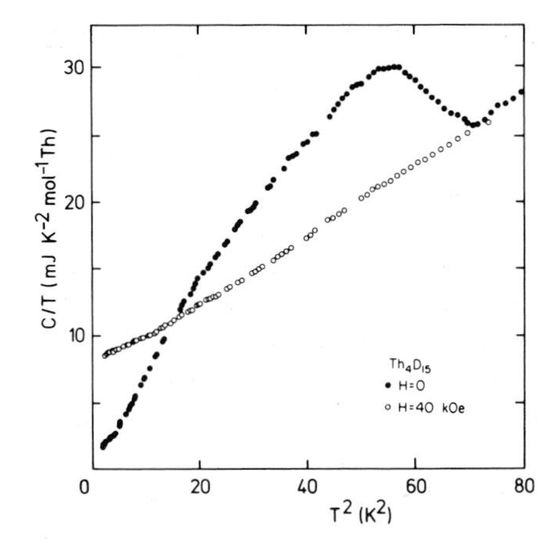

Fig. 6.13. Specific heat c of Th_4D_{15} plotted as c/T versus T^2 [6.81]

6.7 Electron and Phonon Properties of the Th–H System

There is much less information on the Th–H system than on the Pd–H system. The reason may be that the fabrication is more difficult and that the complicated lattice structure is not very attractive to theorists.

6.7.1 Low Temperature Specific Heat

The heat capacity of Th_4H_{15} and Th_4D_{15} was measured by *Miller* et al. [6.81] from 1.4 to 20 K in both the normal and superconducting states (Fig. 6.13). The Debye temperatures are 163 K for pure Th [6.119] and 211 K for Th_4H_{15}, although the Th–Th interatomic distance is about 11% larger in the hydride. The electron specific-heat coefficient γ increases from 4.31 to 8.07 mJ/mol K^2. No definite isotope effect could be established in γ or θ_D within experimental uncertainty. From the electron specific heat, the thermodynamic critical field of 0.099 Tesla and the energy gap of 1.16 meV were derived for $T=0$ K. The ratio $2\Delta_0/k_B T_c$ was calculated to be 3.42 for Th_4H_{15}. In a previous paper, *Schmidt* and *Wolf* [6.120] had reported $\theta_D = 200$ K and $\gamma = 3.4$ mJ/mol K^2 for Th_4H_{15}. The much smaller value of γ is probably due to an uncertain extrapolation from above T_c.

6.7.2 Neutron-Scattering Experiments

The optic phonon densities of states in ThH_2 and Th_4H_{15} were measured by *Dietrich* et al. [6.121] using incoherent neutron scattering. They observed a pronounced difference between ThH_2 and Th_4H_{15}. While the optic spectrum of nonsuperconducting ThH_2 is relatively narrow and centered around 125 meV, the optic spectrum of superconducting Th_4H_{15} exhibits two broad peaks centered around 150 and 80 meV, overlapping with the acoustic spectrum. Qualitatively similar results for Th_4H_{15} were observed by *Miller* et al. [6.122]. The softening of at least one optic branch is considered [6.121] as a possible precurser of a lattice instability. Since Th_4H_{15} has an acoustic phonon spectrum less favorable for superconductivity than the nonsuperconducting ThH_2, it is assumed that the high transition temperature of Th_4H_{15} is essentially caused by the optic phonons.

6.7.3 Theory

The contributions of the acoustic and optic phonon modes to $\alpha^2(\omega)F(\omega)$ were calculated by *Winter* and *Ries* [6.123]. In view of the complicated lattice structure of Th_4H_{15}, an attempt was made to treat the electronic quantities entering $\alpha^2(\omega)F(\omega)$ in a cluster approximation. The presence of the hydrogen vibrations is found to be decisive for the occurrence of superconductivity in Th_4H_{15}. The calculations show a significant contribution of the f band to the density of states at the Fermi energy, which may account for the high γ value observed in the specific-heat experiment. Owing to the steep rise of the f band near the Fermi energy, its position critically determines T_c. The large positive pressure dependence of T_c is explained by a small shift of the f band to higher energies relative to the Fermi energy.

6.8 Conclusion

At present only two metal-hydrogen systems appear to exist in which hydrogen contributes constructively to superconductivity, Th–H and Pd–H. Since gaseous impurities usually lower T_c of the transition metals, the superconductivity in these two metal hydrides is surprising. The reason for this behavior must be an additional electron-phonon coupling in PdH and Th_4H_{15}.

In the Pd–H(D) system there is strong evidence that if the harmful spin fluctuations are suppressed, the coupling of the conduction electrons to the optic phonons is the main reason for the high T_c values. The inverse isotope effect can be explained by an anharmonic potential for the H atoms having a larger vibrational amplitude than D, owing to the difference in zero point energy. The rather large negative pressure effect on T_c is caused by a stiffening of the force constants as the lattice spacing is reduced. The further increase of T_c of the Pd–H system on alloying with noble metals is not yet fully understood. As the optic H modes are almost unchanged in the Pd–Ag–H alloys, it is speculated that a softening of the acoustic phonons is responsible for the further enhancement of T_c. However, a change of the electronic properties may just as well account for the higher T_c.

The Th–H system differs from Pd–H in the way that superconductivity is restricted to the stoichiometric compound Th_4H_{15}. As in the Pd–H system, there also exist some low-lying optic H-phonon modes, which presumably are responsible for the high T_c of this compound. The absence of the regular negative isotope effect can be regarded as a positive (inverse) one and thus be explained similar to the Pd–H system. The positive pressure effect could be a manifestation of the contributions of f electrons to $N(0)$ dominating the negative influence of the stiffening of the force constants.

Besides these two superconducting metal-hydrides (PdH, Th_4H_{15}), there are a few other systems (Al–H [6.43, 44], TaS_2–H [6.34], and some hydrogenated Nb-based alloys [6.38–40]) in which a positive influence of the H could be detected. But in these cases the essential role of the H for the superconductivity is not yet settled. In the predominant number of metal-H systems, hydrogen exerts a negative influence on T_c. An intensively studied system is Nb–H, in which superconductivity is suppressed below 1.3 K in the H-dissolving β phase [6.15]. The normal conductor Nb–H and the superconductor Pd–H have in common that H is dissolved nearly as a proton [6.124, 125]. That means that an essential share of the H core electron is transferred to the *sp* conduction band. So the proton is a strong scatterer for the screening conduction electrons. Nb–H and Pd–H differ from each other in the positions of H in the host lattice. In Nb (bcc) the hydrogen occupies tetrahedral sites and in Pd (fcc) octahedral sites (see Chap. 2). The energies of the corresponding vibrational H modes are much higher in Nb–H than in Pd–H, hence being less favorable for possible superconductivity of Nb–H. Another pecularity is that after hydrogenation, the shape of the acoustic phonon spectrum is more drastically changed in Nb than in Pd. Anomalies in the acoustic spectrum of pure Nb [6.126, 127] are thought to be responsible for the high T_c of 9.2 K [6.128]. The reduction of these anomalies by the hydrogen addition may compensate the T_c-enhancing coupling of the optic H modes resulting in a suppression of T_c.

References

6.1 C.B.Satterthwaite, I.L.Toepke: Phys. Rev. Lett. **25**, 741 (1970)
6.2 T.Skoskiewicz: Phys. Status Solidi (a) **11**, K 123 (1972)
6.3 A.A.Abrikosov: Sov. Phys. JETP **14**, 408 (1962)
6.4 N.W.Ashcroft: Phys. Rev. Lett. **21**, 1748 (1968)
6.5 T.Schneider, E.Stoll: Physica **55**, 702 (1971)
6.6 A.C.Switendick: Proceedings of the 2nd Rochester Conf. on Superconductivity in *d*- and *f*-band metals, ed. D.H.Douglass (Plenum Press, New York, London 1976) p. 593
6.7 R.D.Parks, ed.: *Superconductivity* (Marcel Dekker, New York 1969)
6.8 J.Bardeen, L.N.Cooper, J.R.Schrieffer: Phys. Rev. **108**, 1175 (1957)
6.9 D.J.Scalapino, J.R.Schrieffer, J.W.Wilkins: Phys. Rev. **148**, 263 (1966)
6.10 W.L.McMillan, J.M.Rowell: In *Superconductivity*, ed. by R.D.Parks (Marcel Dekker, New York 1969) p. 561
6.11 W.L.McMillan: Phys. Rev. **167**, 331 (1968)
6.12 G.Bergmann, D.Rainer: Z. Physik **263**, 59 (1973)
6.13 H.Rietschel: Z. Physik B **22**, 133 (1975)
6.14 R.B.McLellan, C.G.Harkins: Mater. Sci. Eng. **18**, 5 (1975)
6.15 J.-M.Welter, F.J.Johnen: Z. Physik B**27**, 227 (1977)

6.16 E.Schröder: Z. Naturforsch. **12**a, 247 (1957)
6.17 G.C.Rauch, R.M.Rose, J.Wulff: J. Less-Common Metals **8**, 99 (1965)
6.18 C.D.Wiseman: J. Appl. Phys. **37**, 3599 (1966)
6.19 F.H.Horn, W.F.Bruksch, W.T.Ziegler, D.H.Andrews: Phys. Rev. **61**, 738 (1949)
6.20 G.Pfeiffer, H.Wipf: J. Phys. F **6**, 167 (1976)
6.21 D.Richter, J.Töpler, T.Springer: J. Phys. F **6**, L 93 (1976)
6.22 D.G.Westlake, S.T.Ockers: Phys. Rev. Lett. **25**, 1618 (1970)
6.23 B.D.Bhardwaj, H.E.Rorschach: Proc. Low Temp. Phys. LT13, Vol. 3 (Plenum Press, New York, London 1974) p. 517
6.24 C.B.Satterthwaite, D.T.Peterson: J. Less-Common Metals **26**, 361 (1972)
6.25 B.T.Matthias, T.H.Geballe, V.B.Compton: Rev. Mod. Phys. **35**, 1 (1963)
6.26 P.Duffer, D.M.Gualtieri, V.U.S.Rao: Phys. Rev. Lett. **37**, 1410 (1976)
6.27 M.F.Merriam, D.S.Schreiber: J. Phys. Chem. Sol. **24**, 1375 (1963)
6.28 R.C.Heckman: J. Chem. Phys. **40**, 2958 (1964)
6.29 B.T.Matthias: J. Phys. (Paris) C 1, **32**, 607 (1971)
6.30 P.R.Sahm: Phys. Lett. **26** A, 459 (1968)
6.31 L.J.Vieland, A.Wicklund, J.G.White: Phys. Rev. B **11**, 3311 (1975)
6.32 T.B.Reed, H.C.Gatos, W.J.LaFleur, J.T.Roddy: In *Meth. of Adv. Electr. Materials*, ed. by G.E.Brock (Interscience, New York 1963) p. 71
6.33 J.B.Vetrano, G.L.Guthrie, H.E.Kissinger: Phys. Lett. **26** A, 45 (1967)
6.34 D.W.Murphy, F.J.DiSalvo, G.W.Hull, J.V.Waszczak, S.F.Mayer, G.R.Stewart, S.Early, J.V.Acrivos, T.H.Geballe: J. Chem. Phys. **62**, 967 (1975)
6.35 B.Stritzker, W.Buckel: Z. Physik **257**, 1 (1972)
6.36 W.Buckel, B.Stritzker: Phys. Lett. **43** A, 403 (1973)
6.37 B.Stritzker: Z. Physik **268**, 261 (1974)
6.38 C.G.Robbins, J.Muller: J. Less-Common Metals **42**, 19 (1975)
6.39 C.G.Robbins, M.Ishikawa, A.Treyvard, J.Muller: Solid State Commun. **17**, 903 (1975)
6.40 H.Oesterreicher, J.Clinton: J. Solid State Chem. **17**, 443 (1976)
6.41 H.Oesterreicher, J.Clinton, H.Bittner: J. Solid State Chem. **16**, 209 (1976)
6.42 C.A.Mackliet, D.J.Gillespie, A.I.Schindler: J. Phys. Chem. Sol. **37**, 379 (1976)
6.43 G.Deutscher, M.Pasternak: Phys. Rev. B **10**, 4042 (1974)
6.44 A.M.Lamoise, J.Chaumont, F.Meunier, H.Bernas: J. Physique Lett. **36**, L-271 (1975)
6.45 A.M.Lamoise, J.Chaumont, F.Meunier, H.Bernas: J. Physique Lett. **36**, L-305 (1975)
6.46 L.Dumoulin, P.Nédellec, J.Chaumont, D.Gilbon, A.M.Lamoise, H.Bernas: C.R. Acad. Sc. Paris B **283**, 285 (1976)
6.47 G.v. Minnigerode, J.Rothenberg: Z. Physik **213**, 397 (1968)
6.48 A.M.Lamoise, J.Chaumont, F.Lalu, F.Meunier, H.Bernas: J. Physique Lett. **37**, L-287 (1976)
6.49 F.A.Lewis: *The Palladium Hydrogen System* (Academic Press, London 1967)
6.50 J.M.E.Harper: Phys. Lett. **47** A, 69 (1974)
6.51 J.E.Shirber: Phys. Lett. **45** A, 141 (1973)
6.52 J.E.Shirber, C.J.M.Northrup,Jr.: Phys. Rev. B **10**, 3818 (1974)
6.53 T.Skoskiewicz, A.W.Szafranski, W.Bujnowski, B.Baranowski: J. Phys. C: Solid State Phys. **7**, 2670 (1974)
6.54 B.J.Smith: *Ion Implantation* (Appendices II, III), ed. by G.Dearnaley (North-Holland, Amsterdam 1973)
6.55 G.Heim, B.Stritzker: Appl. Phys. **7**, 239 (1975)
6.56 L.E.Sansores, R.E.Glover: Proc. 14th Internat. Conf. on Low Temperature Physics, Vol. 2, ed. by M.Krusius, M.Vuorio (North-Holland, Amsterdam-Oxford; American Elsevier, New York 1975) p. 36
6.57 R.J.Miller, C.B.Satterthwaite: Phys. Rev. Lett. **34**, 144 (1975)
6.58 D.S.McLachlan, T.B.Doyle, J.P.Burger: see Ref. 6.56, p. 44
6.59 H.Wühl: To be published
6.60 H.C.Jamieson, F.D.Manchester: J. Phys. F **2**, 323 (1972)
6.61 K.H.Bennemann, J.W.Garland: Z. Physik **260**, 367 (1973)

6.62 W. Buckel, A. Eichler, B. Stritzker: Z. Physik **263**, 1 (1973)
6.63 J. E. Shirber: Phys. Lett. **46** A, 285 (1973)
6.64 H. Wühl: Unpublished
6.65 S. Auluck: Nuovo Cimento Lett. **7**, 545 (1973)
6.66 T. Skoskiewicz: Phys. Status Solidi (b) **59**, 329 (1973)
6.67 N. E. Alekseevskii, Yu. A. Samarskii, H. Wolf, V. I. Tsebro, V. M. Zakosarenko: JETP Lett. **19**, 350 (1974)
6.68 D. S. McLachlan, T. B. Doyle, J. P. Burger: J. Low Temp. Phys. **26**, 589 (1977)
6.69 R. Meservey, P. M. Tedrow: Bull. Am. Phys. Soc. **21**, 340 (1976)
6.70 B. Stritzker: See Ref. 6.56, p. 32
6.71 T. Skoskiewicz, A. W. Szafranski, B. Baranowski: Phys. Status Solidi (b) **59**, K 135 (1973)
6.72 T. Skoskiewicz: Proc. 5th Internat. Conf. on High Pressure and Technology, Moscow 1975; High Temp.—High Pressures **7**, 684 (1977)
6.73 M. Gibson, D. E. Moody, R. Stevens: J. Phys. (Paris) Suppl. **32**, C 1–990 (1971)
6.74 J. E. van Dam: Thesis University of Leiden (1973) and references therein
6.75 T. Tsuchida: J. Phys. Soc. Japan **18**, 1016 (1963)
6.76 J. Becker: Techn. Rpt. Jül-1358, KFA Jülich (1976)
6.77 B. Stritzker: Unpublished
6.78 D. Gerstenberg: Ann. Physik **2**, 237 (1958)
6.79 B. Stritzker, J. Becker: Phys. Lett. **51** A, 147 (1975)
6.80 M. Dietrich, W. Gey, H. Rietschel, C. B. Satterthwaite: Solid State Commun. **15**, 941 (1974)
6.81 J. F. Miller, R. H. Caton, C. B. Satterthwaite: Phys. Rev. B **14**, 2795 (1976)
6.82 R. H. Caton, C. B. Satterthwaite: Bull. Am. Phys. Soc. **19**, 348 (1974)
6.83 G. F. Hardy, J. K. Hulm: Phys. Rev. **93**, 1004 (1954)
6.84 A. L. Giorgi, E. G. Szklarz, M. C. Krupka: Proc. of the 1st Rochester Conf. on Superconductivity in *d*- and *f*-band metals, ed. by D. H. Douglass (American Institute of Physics, New York 1972) p. 147
6.85 M. Dietrich: Ges. f. Kernforschung Karlsruhe, KFK-2098 (1974)
6.86 G. Alefeld, J. Völkl (eds.): *Hydrogen in Metals I. Basic Properties*, Topics in Applied Physics, Vol. 28 (Springer, Berlin, Heidelberg, New York 1978)
6.87 U. Mizutani, T. B. Massalski, J. Bevk: J. Phys. F: Metal Phys. **6**, 1 (1975) and references therein
6.88 M. Zimmermann, G. Wolf, K. Bohmhammel: Phys. Status Solidi (a) **31**, 511 (1975)
6.89 J. S. Faulkner: Phys. Rev. B **13**, 2391 (1976)
6.90 J. Zbasnik, M. Mahnig: Z. Physik B **23**, 15 (1976)
6.91 J. E. Shirber, B. Morosin: Phys. Rev. B **12**, 117 (1975)
6.92 J. M. Rowe, J. J. Rush, H. G. Smith, M. Mostoller, H. E. Flotow: Phys. Rev. Lett. **33**, 1297 (1974)
6.93 A. Magerl: Private communication
6.94 C. J. Glinka, J. M. Rowe, J. J. Rush, A. Rahman, S. K. Sinha, H. E. Flotow: Proc. Conf. Neutron Scattering, Gatlinburg (1976)
6.95 A. Rahman, K. Sköld, C. Pelizzari, S. K. Sinha, H. Flotow: Phys. Rev. B **14**, 3630 (1976)
6.96 B. N. Ganguly: Z. Physik **265**, 433 (1973)
6.97 M. R. Chowdhury, D. K. Ross: Solid State Commun. **13**, 229 (1973)
M. R. Chowdhury: J. Phys. F: Metal Phys. **4**, 1657 (1974)
6.98 R. C. Dynes, J. P. Garno: Bull. Am. Phys. Soc. **20**, 422 (1975)
6.99 A. Eichler, H. Wühl, B. Stritzker: Solid State Commun. **17**, 213 (1975)
6.100 P. J. Silverman, C. V. Briscoe: Phys. Lett. **53** A, 221 (1975)
6.101 J. Igalson, L. Sniadower, A. J. Pindor, T. Skoskiewicz, K. Blüthner, F. Dettmann: Solid State Commun. **17**, 309 (1975)
6.102 W. Drexel, A. Murani, D. Tocchetti, W. Kley, I. Sosnowska, D. K. Ross: J. Phys. Chem. Sol. **37**, 1135 (1976)
6.103 D. S. McLachlan, R. Mailfert, J. P. Burger, B. Souffaché: Solid State Commun. **17**, 281 (1975)
6.104 J. C. H. Chiu, R. A. B. Devine: J. Phys. F **6**, L 33 (1976)
6.105 P. Hertel: Z. Physik **268**, 111 (1974)

6.106 J.S.Brown: Phys. Lett. **51** A, 99 (1975)
6.107 B.N.Ganguly: Phys. Rev. B **14**, 3848 (1976)
6.108 A.C.Switendick: Ber. Bunsenges. Physik. Chem. **76**, 535 (1972)
6.109 D.E.Eastman, J.K.Cashion, A.C.Switendick: Phys. Rev. Lett. **27**, 35 (1971)
6.110 F.Antonangeli, A.Balzarotti, A.Bianconi, P.Perfetti, P.Ascarelli, N.Nistico: Solid State Commun. **21**, 201 (1977)
6.111 D.A.Papaconstantopoulos, B.M.Klein: Phys. Rev. Lett. **35**, 110 (1975)
6.112 A.J.Pindor: Phys. Status Solidi (b) **74**, K 19 (1976)
6.113 T.Nakajima: Phys. Status Solidi (b) **77**, K 147 (1976)
6.114 J.P.Burger, D.S.McLachlan: J. Phys. (Paris) **37**, 1227 (1976)
6.115 H.Montgomery, G.P.Pells, E.M.Wray: Proc. Roy. Soc. London A **301**, 261 (1967)
6.116 C.A.Mackliet, A.I.Schindler: Phys. Rev. **146**, 463 (1966)
6.117 B.N.Ganguly: Z. Physik B **22**, 127 (1975)
6.118 I.R.Gomersall, B.L.Gyorffy: J. Phys. F: Metal Phys. **3**, L 138 (1973)
6.119 J.E.Gordon, H.Montgomery, R.J.Noer, G.R.Pickett, R.Tobin: Phys. Rev. **152**, 432 (1966)
6.120 H.G.Schmidt, G.Wolf: Solid State Commun. **16**, 1085 (1975)
6.121 M.Dietrich, W.Reichardt, H.Rietschel: Solid State Commun. **21**, 603 (1977)
6.122 J.F.Miller, C.B.Satterthwaite, T.O.Brun, J.D.Jorgensen, K.Sköld: To be published
6.123 H.Winter, G.Ries: Z. Physik B **24**, 279 (1976)
6.124 P.Pattison, M.Cooper, J.R.Schneider: Z. Physik B **25**, 155 (1976)
6.125 R.Lässer, B.Lengeler: Phys. Rev. Lett. (to be published)
6.126 J.M.Rowe, N.Vagelatos, J.J.Rush, H.E.Flotow: Phys. Rev. B **12**, 2959 (1975)
6.127 V.Lottner, A.Kollmar, T.Springer, H.Bilz, W.Kress, W.Teuchert: Proc. Int. Conf. Lattice Dynamics, Paris (1977)
6.128 S.K.Sinha, B.N.Harmon: Phys. Rev. Lett. **35**, 1515 (1975)

7. Electro- and Thermotransport of Hydrogen in Metals

H. Wipf

With 7 Figures

7.1 Background

To apply forces on mobile lattice defects—as hydrogen interstitials in a metal—and to study resulting transport processes is an experimental challenge to any investigator. Both electrotransport and thermotransport are leading examples for atomic transport caused by such forces; they arise in the presence of an electric current in a metal and in the presence of a temperature gradient, respectively. Figure 7.1 demonstrates the effects to be observed in thermotransport. The figure shows the hydrogen concentration profiles that were measured in two titanium samples containing hydrogen after a temperature gradient has been applied (*Sawatzky* and *Duclos* [7.1]). The concentrations are plotted versus the (reciprocal) temperature of the sample sections in which they were found. The figure indicates that the effects can be considerable. The concentration variations set up by the temperature gradient can be seen to amount to almost an order of magnitude.

Electrotransport and thermotransport provide a method to create atomic fluxes and concentration shifts within metals. Thus, they allow the study of atomic mobilities and diffusivities. Concentration profiles can be adjusted

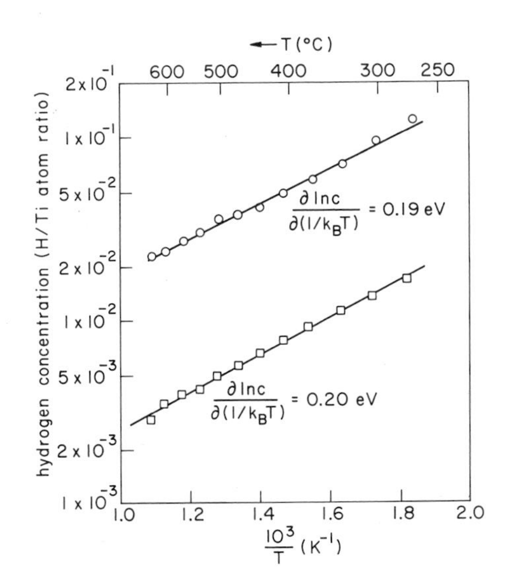

Fig. 7.1. Thermotransport of hydrogen in titanium (*Sawatzky* and *Duclos* [7.1]). For two samples with different hydrogen content (○, □), the hydrogen concentration profiles are logarithmically plotted versus the reciprocal temperature

within a sample as desired for experimental investigations. The forces causing the transport are interesting also in themselves. They arise from complex nonequilibrium interactions within a metal. Their size and direction reflect details of both electronic band-structure and phonon properties. From a theoretical point of view, the forces are by no means completely understood yet, not quantitatively and often not even qualitatively.

According to the intention of the present book, the electro- and thermotransport only of hydrogen in metals will be considered; for a discussion of these effects in other systems, the reader is referred to the review articles or monographs in [7.2–13]. Two properties should be emphasized which are significant for transport processes of hydrogen interstitials.

a) The hydrogen atoms are interstitials with an extremely high mobility even at low temperatures. For that reason, transport experiments can be performed over a wide temperature range extending far below room temperature.

b) Hydrogen is the lightest of all elements; it is the element whose isotopes are available with the highest possible mass ratios. Quantum effects in diffusion and transport are, therefore, most likely to occur, and isotope effects as a tool to check theoretical predictions will most easily be observable in experiments.

Both properties make metal-hydrogen systems outstandingly suitable and interesting for diffusion and transport studies.

7.2 Phenomenological Description

7.2.1 Forces and Hydrogen Flux

Electrotransport and thermotransport represent irreversible transport processes of atoms which phenomenologically are described within the framework of nonequilibrium thermodynamics [7.3, 14–16]. In the present case, the transport of interstitial hydrogen, with its extremely high mobility, is considered. Compared with the hydrogen, the host-lattice atoms can be taken as immobile. However, they provide a suitable reference lattice in which to describe the hydrogen transport. Counter to the phenomenological description usually necessary for transport processes, the transport only of one atomic species within a fixed lattice needs to be considered, and properties of the host-lattice atoms do not enter explicitly into the resulting equations. (The hydrogen can be considered to be a lattice gas [7.17].) The phenomenological equations will, therefore, become particularly elementary and transparent. For reasons of simplicity, an isotropic behavior will additionally be presupposed.

The transport processes are caused by forces on the hydrogen interstitials.

a) *Electrotransport:* The force $\boldsymbol{F}$ arises in the presence of an electric current; it is described phenomenologically by an effective charge number Z^*

$$\boldsymbol{F} = -eZ^* \operatorname{grad} \Phi . \tag{7.1}$$

In this equation, e is the (positive) elementary charge and $-\mathrm{grad}\,\Phi$ is the applied electric field that causes the electric current. The quantity eZ^* characterizes that electric charge that, if exposed solely to the applied electric field, would feel the same force as the hydrogen interstitials in the metal. According to this definition, a positive or negative effective charge number indicates a force acting towards the cathode or anode, respectively. The effective charge number is a quantity introduced purely phenomenologically. Thus, its value can be very different from that characterizing the actual ionic state of a hydrogen interstitial.

b) *Thermotransport:* The force $\boldsymbol{F}$ arises in the presence of a temperature gradient grad T; it is described by the heat of transport Q^*

$$\boldsymbol{F} = -Q^* \frac{\mathrm{grad}\,T}{T}. \tag{7.2}$$

Positive and negative values for the heat of transport indicate forces towards colder and warmer sample sections, respectively. The physical significance of the name "heat of transport" will become clearer later as a consequence of the Onsager relations (Sect. 7.2.5).

If both an electric field and a temperature gradient act on the hydrogen, a linear superposition occurs of the forces given in (7.1, 2).

In the presence of a force $\boldsymbol{F}$, a hydrogen drift flux $\boldsymbol{J}_{\mathrm{drift}}$ arises in addition to the diffusive flux $\boldsymbol{J}_{\mathrm{diff}}$ that is caused by gradients in the hydrogen concentration. Hence, the total hydrogen flux $\boldsymbol{J}$ can be written

$$\boldsymbol{J} = \boldsymbol{J}_{\mathrm{diff}} + \boldsymbol{J}_{\mathrm{drift}} = -nD\,\mathrm{grad}\,c + nc\,M\boldsymbol{F}. \tag{7.3}$$

The fluxes are defined as particle currents of the hydrogen interstitials measured relative to the host-atom lattice. c is the hydrogen concentration, defined as the hydrogen/host-metal atom ratio. According to this definition, n is the number of lattice atoms per unit volume (so that nc is the hydrogen-atom density). D is Fick's diffusion coefficient and M is the mobility of the hydrogen interstitials.

Without the force $\boldsymbol{F}$, (7.3) is reduced to Fick's equation describing the flux $\boldsymbol{J}_{\mathrm{diff}}$ in a concentration gradient. The drift flux $\boldsymbol{J}_{\mathrm{drift}}$ results from a drift velocity $\boldsymbol{v}_{\mathrm{drift}}$ of the hydrogen interstitials that is caused by the force $\boldsymbol{F}$

$$\boldsymbol{v}_{\mathrm{drift}} = \frac{\boldsymbol{J}_{\mathrm{drift}}}{nc} = M\boldsymbol{F}. \tag{7.4}$$

In electrotransport literature, an electric mobility u is sometimes used describing the drift velocity in an unit electric field. Hence, u is defined by

$$u = MeZ^*. \tag{7.5}$$

The diffusion coefficient D and the mobility M are related to each other by the chemical potential μ of the hydrogen interstitials

$$D = c\frac{\partial \mu}{\partial c} M. \tag{7.6}$$

This equation is the generalized Einstein relation (correlation factors need not be considered in a transport process of interstitials). In the case of thermodynamically ideal behavior of the hydrogen interstitials, the following equation applies (k_B is Boltzmann's constant):

$$c\frac{\partial \mu}{\partial c} = k_B T. \tag{7.7}$$

This equation can be expected to be valid as long as the hydrogen concentrations are low [up to which concentrations (7.7) actually can be applied will be discussed in *The Chemical Potential* μ on p. 285].

7.2.2 The Influence of Stresses

Stress gradients within a sample cause an additional force on the hydrogen interstitials [7.3, 14–16, 18–20]. The force can be written [7.18–20]

$$\boldsymbol{F} = P \,\mathrm{grad}\,(s_{iikl}\sigma_{kl}). \tag{7.8}$$

P characterizes the dipole-moment tensor of the hydrogen interstitials; it is the mean value of the trace components of this tensor. A typical value for P is about 3 eV [7.19–23]. s_{ijkl} and σ_{kl} are tensor components of the elastic coefficients and of the stresses, respectively.

Stress gradients can be caused by forces applied from outside on a sample. The resulting transport processes were found to be a powerful method to study hydrogen diffusion (Gorsky effect [7.18–20, 24, 25]). However, stress gradients can also be caused by inhomogeneous hydrogen or temperature distributions within a sample; they arise because fluctuations in both hydrogen concentration and temperature modulate the lattice parameter thus leading to coherency stresses [7.20, 25–30]. In this case, the local value of the stress gradients depends on the hydrogen and temperature distribution within the entire sample; it depends also on the shape and dimension of the sample and, in addition, on the boundary conditions imposed on the sample surface. In general, therefore, the force $\boldsymbol{F}$ arising according to (7.8) cannot be described within a local theory solely by the local concentration or temperature gradients; it has to be calculated according to the individual experimental conditions.

From the size of the dipole-moment tensor, the stress-induced forces can be seen to remain usually small, particularly when compared with those in electro- and thermotransport [7.20]. However, they can become important [7.20, 25, 30] for some sample geometries (for instance, foils with concentration gradients in normal direction) and in the neighborhood of a critical point in the metal-hydrogen phase diagram (where small forces can cause large concentration shifts). In such cases, the possible influence of stresses should be taken into consideration also for electrotransport and thermotransport processes.

7.2.3 Steady States

In a steady state, the total hydrogen flux $\boldsymbol{J}$ in (7.3) is zero. The flux $\boldsymbol{J}_{\text{drift}}$ caused by the force is, thus, exactly compensated by an opposite flux $\boldsymbol{J}_{\text{diff}}$ caused by a concentration gradient. Under these conditions, the hydrogen distribution within a sample is given by

a) *Electrotransport*

$$\frac{\partial \ln c}{\partial \Phi} = -\frac{eZ^*}{c \cdot \dfrac{\partial \mu}{\partial c}}. \tag{7.9}$$

b) *Thermotransport*

$$\frac{\partial \ln c}{\partial (1/T)} = T\frac{Q^*}{c \cdot \dfrac{\partial \mu}{\partial c}}. \tag{7.10}$$

The two equations follow from (7.1–3, 6). The hydrogen concentration in a given sample section is a function of the local value of the electric potential Φ or the temperature T. The hydrogen distribution does not depend on transport quantities such as the mobility or the diffusion coefficient.

For low hydrogen concentrations, $c(\partial\mu/\partial c)$ can be replaced by $k_{\text{B}}T$, (7.7). If, in addition, Z^* and Q^* are assumed to be concentration independent, and Q^* also temperature independent, the two foregoing equations can be integrated (c_0 being an integration constant).

a) *Electrotransport*

$$c = c_0\, \mathrm{e}^{-\frac{eZ^*\Phi}{k_{\text{B}}T}}. \tag{7.11}$$

b) *Thermotransport*

$$c = c_0\, \mathrm{e}^{\frac{Q^*}{k_{\text{B}}T}}. \tag{7.12}$$

The concentrations are found proportional to a Boltzmann factor.

7.2.4 Time-Dependent Transport Processes

The time dependence of the local concentration c is described by the equation

$$n\mathring{c} = -\operatorname{div} \boldsymbol{J}, \tag{7.13}$$

where the hydrogen flux $\boldsymbol{J}$ is given by (7.3). For simplicity, the quantities n, D, M, and $\boldsymbol{F}$ in (7.3) will be assumed not to depend on the spatial coordinates. In this case, (7.3) and (7.13) together yield

$$\mathring{c} = D\Delta c - M\boldsymbol{F} \operatorname{grad} c. \tag{7.14}$$

The above assumptions will be certainly not correct in thermotransport, for example, where the temperature varies with the coordinates. However, the qualitative aspects of the subsequent discussion will not be impaired by these assumptions.

The quantity to be considered is the time required to establish steady-state conditions in a transport process. This time is characterized differently according to which of the two terms on the right side of (7.14) has the decisive influence on the transport process.

First, the diffusion term $D\Delta c$ may be considered. It dominates the transport process if the force $\boldsymbol{F}$ is small. In the extreme case of a zero force, (7.14) is reduced to the ordinary diffusion equation. The time τ_{diff} required for a diffusion process to decay is approximately [7.3]

$$\tau_{\text{diff}} = \frac{L^2}{D}, \tag{7.15}$$

where L represents the distance over which concentration shifts occur. L is, therefore, a sample dimension characteristic of the transport process.

The second term on the right-hand side of (7.14) is the drift term $-M\boldsymbol{F}$ grad c. It dominates the transport process for high values of the force $\boldsymbol{F}$. In the extreme case of a negligible diffusion term, the hydrogen interstitials are all drifting in the force direction with a drift velocity $\boldsymbol{v}_{\text{drift}} = M \cdot \boldsymbol{F}$. The time τ_{drift} required for the drift process to finish is determined by the drift velocity and by the length L over which the drift process occurs

$$\tau_{\text{drift}} = \frac{L}{v_{\text{drift}}} = \frac{L}{MF}. \tag{7.16}$$

Here again, L is the sample dimension characteristic of the drift process.

In general, the transport process will be governed by the faster of the two processes discussed. The shorter of the two times τ_{diff} and τ_{drift} will, therefore, represent the time period of the transport process. From the above equations, the time τ_{diff} can be seen to be the shorter one for small forces $\boldsymbol{F}$ (diffusion

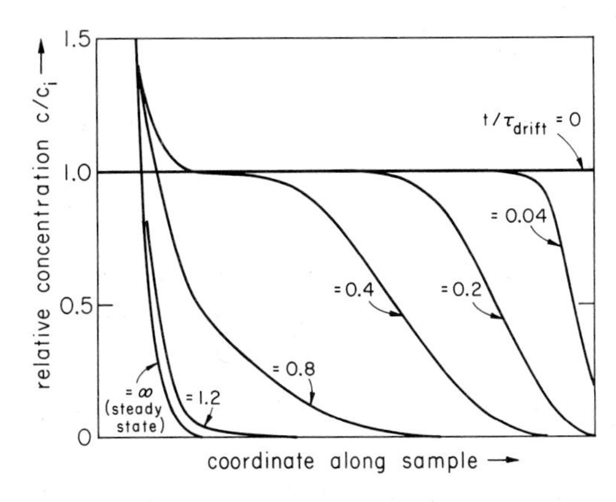

Fig. 7.2. One-dimensional drift process according to (7.14) (*Peterson* [7.33]). Concentration profiles are shown as calculated at different times t (indicated by the ratio $t/\tau_{\rm drift}$). L is the length of the sample and $c_{\rm i}$ the homogeneous initial concentration at time $t=0$. The concentration profiles apply to a ratio $\tau_{\rm diff}/\tau_{\rm drift}=40$ [see (7.17)]

behavior), whereas $\tau_{\rm drift}$ becomes shorter in the case of very large forces (drift behavior). The passage from a diffusion to a drift behavior occurs when both times $\tau_{\rm diff}$ and $\tau_{\rm drift}$ become approximately equally large. This condition can be written

$$\frac{\tau_{\rm diff}}{\tau_{\rm drift}}=1=\frac{LMF}{D}=\frac{LF}{c\cdot\dfrac{\partial\mu}{\partial c}}\left(=\frac{LF}{k_{\rm B}T}\right), \tag{7.17}$$

where the last term (in parentheses) is only valid in the case of thermodynamically ideal behavior. LF is the potential difference built up over the sample by the force $\boldsymbol{F}$. The transport process exhibits, therefore, a diffusion behavior as long as the potential difference LF is smaller than $c(\partial\mu/\partial c)$ (or $k_{\rm B}T$); it shows a drift behavior if LF is greater than $c(\partial\mu/\partial c)$ (or $k_{\rm B}T$).

The foregoing considerations are confirmed by exact solutions of (7.14). A solution can be given, for instance, for a one-dimensional region with impermeable surfaces and with a homogeneous initial value $c_{\rm i}$ for the concentration c [7.3, 31–33]. Figure 7.2 shows, as an example, concentration profiles calculated accordingly in the case of a distinct drift behavior with $\tau_{\rm diff}/\tau_{\rm drift}=40$ (*Peterson* [7.33]). The concentration profiles shown are found after time periods equal to the indicated fractions of the drift time $\tau_{\rm drift}$. From this figure, the drift behavior of the transport process can clearly be recognized.

7.2.5 The Onsager Relations

In (7.1, 2) the effective charge number Z^* and the heat of transport Q^* were introduced as parameters describing the force in the presence of an electric field or a temperature gradient. However, an additional and important physical meaning can be attributed to both quantities using the Onsager relations.

a) *Electrotransport:* The effective charge number describes the total flux $\boldsymbol{J}_{\mathrm{E}}$ of electric charge that accompanies a hydrogen flux $\boldsymbol{J}$. In the absence of an electric field (grad $\Phi=0$), the electric charge flux $\boldsymbol{J}_{\mathrm{E}}$ is given by

$$\boldsymbol{J}_{\mathrm{E}}=Z^{*}\,\mathrm{e}\,\boldsymbol{J}\,. \tag{7.18}$$

To what extent the flux $\boldsymbol{J}_{\mathrm{E}}$ is caused by fluxes of electronic or ionic charge cannot be decided within the present phenomenological description. If both an electric field and a concentration gradient are present, the charge flux can be written (7.1, 3)

$$\boldsymbol{J}_{\mathrm{E}}=-Z^{*}\,\mathrm{e}\,nD\,\mathrm{grad}\,c-\frac{1}{\varrho}\,\mathrm{grad}\,\Phi\,, \tag{7.19}$$

where ϱ stands for the electric resistivity in the absence of a concentration gradient.

An important consequence of (7.19) is the occurrence of a diffusion potential. It is observed in the case of a vanishing electric charge flux $\boldsymbol{J}_{\mathrm{E}}$ (for instance, in the case of an electrically isolated sample). Under these conditions, the relation

$$\frac{\partial\Phi}{\partial c}=-\varrho Z^{*}\,\mathrm{e}\,nD \tag{7.20}$$

is valid, yielding the diffusion potential Φ as a function of the hydrogen concentration.

b) *Thermotransport:* The heat of transport Q^{*} describes the flux $\boldsymbol{J}_{Q}$ of heat accompanying a hydrogen flux $\boldsymbol{J}$. For a uniform temperature (grad $T=0$), both fluxes are related by the equation

$$\boldsymbol{J}_{Q}=Q^{*}\boldsymbol{J}\,. \tag{7.21}$$

The heat flux in the presence of concentration and temperature gradients is given by (7.2, 3)

$$\boldsymbol{J}_{Q}=-Q^{*}nD\,\mathrm{grad}\,c-\varkappa\,\mathrm{grad}\,T\,. \tag{7.22}$$

In this equation, $\varkappa$ is the thermal conductivity measured in the absence of concentration gradients.

The heat flux $\boldsymbol{J}_{Q}$ is often called the reduced heat flux in order to discriminate it from other heat fluxes that can be defined [7.3, 14–16]. It is, essentially, the total energy flux reduced by the enthalpy flux $\boldsymbol{J}\cdot h$ that accompanies the hydrogen flux $\boldsymbol{J}$ (h is the partial enthalpy of the hydrogen interstitials).

Similar to electrotransport, the steady state arising for a zero heat flux $\boldsymbol{J}_Q$ may separately be considered. In such a case, temperature gradients are caused by concentration gradients (Dufour effect). The relation between temperature T and concentration c can be written

$$\frac{\partial T}{\partial c} = -\frac{l}{\varkappa} Q^* n D . \tag{7.23}$$

7.3 Atomistic Origin of the Forces

7.3.1 Electrotransport

The force acting on an atom is commonly considered to consist of two parts. The first of these is the direct field force. It characterizes the force on the ionic charge of the considered atom that is directly caused by the applied electric field (ionic charge, in this context, stands for the electric charge of the nucleus and the core electrons); this force also includes any static screening or polarization effects. The second force (electron wind or electron drag force) arises from interaction effects with the conduction electrons drifting in a preferential direction as a consequence of the applied electric field. According to the separation into two forces, the effective charge number Z^* consists of two terms

$$Z^* = Z^*_{\text{dir}} + Z^*_{\text{ele}} , \tag{7.24}$$

where Z^*_{dir} and Z^*_{ele} describe the direct field force and the electron wind force, respectively.

In first calculations of the electron wind force (*Fiks* [7.34], *Huntington* and *Grone* [7.35]), the force was derived within a free-electron model from the momentum transfer that takes place in the scattering processes between conduction electrons and an atom. The result for Z^*_{ele} is given by

$$Z^*_{\text{ele}} = -n_{\text{e}} \cdot S \cdot l , \tag{7.25}$$

where n_{e} denotes the electron density, S the scattering cross section of the considered atom, and l the mean free path of the electrons. This equation can, according to *Fiks*, be understood in the following way. An electron with the Fermi speed v_{F} moves for a time interval $t = l/v_{\text{F}}$ between two collisions, and acquires, in the electric field $-\text{grad}\,\Phi$, an additional momentum $l/v_{\text{F}} \cdot \text{e}\,\text{grad}\,\Phi$. This momentum is transferred to a scattering atom thus yielding, for an electron density n_{e}, a force $\boldsymbol{F}_{\text{ele}} = n_{\text{e}} S l\,\text{e}\,\text{grad}\,\Phi$, which is the result of (7.25). With $\varrho = m v_{\text{F}}/(\text{e}^2 n_{\text{e}} l)$ for the electric resistivity (m is the electron mass and v_{F} the Fermi speed), and with the residual resistivity $\Delta\varrho = m v_{\text{F}} N S/(\text{e}^2 n_{\text{e}})$ for that part of the resistivity which is caused by electron scattering at N of the considered atoms per unit volume, (7.25) can also be written

$$Z^*_{\mathrm{ele}} = -\frac{m v_{\mathrm{F}} S}{e^2 \varrho} = -\frac{n_{\mathrm{e}}}{\varrho} \frac{\Delta\varrho}{N}. \tag{7.26}$$

Bosvieux and *Friedel* [7.36] obtained essentially the same result using an alternative calculatory method. They calculated the electric field created at the position of the considered atom due to the scattering of conduction electrons. The field arises since the spatial conduction electron density is redistributed around the scattering atom. The action of the field on the ionic charge of the atom causes then the electron wind force.

Electronic band-structure effects on the electron wind force were considered in subsequent calculations [7.37–42]. For a simple two-band metal with an electron and a hole band, the extension of (7.25) is found to be

$$Z^*_{\mathrm{ele}} = -n_{\mathrm{e}} S_{\mathrm{e}} l_{\mathrm{e}} + n_{\mathrm{h}} S_{\mathrm{h}} l_{\mathrm{h}}, \tag{7.27}$$

where the subscripts e and h refer to electrons and holes, respectively. In agreement with the picture of momentum transfer from carriers of negative and positive charge, electrons and holes yield negative and positive contributions to the electron wind force.

The direct field force turned out to be a subject of theoretical controversy. *Fiks* [7.34] and *Huntington* and *Grone* [7.35] assumed the applied electric field to act unchanged upon the ionic charge of the considered atom. Thus, a temperature-independent contribution Z^*_{dir} should arise for the effective charge number equivalent to the ionic charge. If, in the case of hydrogen interstitials, the electron of an hydrogen atom is assumed to enter the conduction bands, Z^*_{dir} should be $+1$. *Bosvieux* and *Friedel* [7.36], on the other hand, held a contrary view. They assumed (for an interstitial atom) the applied field to be totally screened due to a polarization of the conduction-electron screening cloud around the impurity charge of an interstitial. Therefore, Z^*_{dir} would be zero for an interstitial atom. In spite of a great amount of theoretical work [7.38, 42–52], this controversy is not yet completely settled. However, most of the authors assume that a screening as proposed by *Bosvieux* and *Friedel* does not exist.

An exact result for the electrotransport force on an impurity charge was found within a free-electron (jellium) model (*Das* and *Peierls* [7.45, 46]). The result was obtained by momentum-conservation arguments. It was confirmed by several authors ([7.42, 51, 53], see also [7.54]) under even more general conditions. The result for Z^* is

$$Z^* = -\frac{n_{0\mathrm{e}}}{\varrho} \frac{\varrho - \varrho_0}{N}. \tag{7.28}$$

In this equation, n_{0e} stands for the electron density in the impurity-free jellium and N for the density of the impurities. ϱ_0 and ϱ denote the resistivities of the impurity-free jellium and the jellium with impurities, respectively. The above result describes the total force, thus including the effects of both the direct force and the electron wind force. It was extended to a two-band model by *Das* [7.55].

The expression in (7.28) seems very similar to the result for the electron wind force in (7.26). The actual difference can, however, be realized from the different meaning of both the electron densities (n_{0e} and n_e) and the resistivity differences ($\varrho - \varrho_0$ and $\Delta\varrho$).

7.3.2 Thermotransport

Three terms of different origin are discussed to contribute to the heat of transport Q^* [7.12, 39, 40, 56–63],

$$Q^* = Q^*_{\text{ele}} + Q^*_{\text{pho}} + Q^*_{\text{int}} . \tag{7.29}$$

Q^*_{ele} and Q^*_{pho} are contributions to Q^* arising from interaction effects with electrons and phonons, respectively. Q^*_{int} is the so-called intrinsic heat of transport. It must be noted that the existence and significance of the above three terms are not unanimously recognized. Additional contributions to Q^* are also discussed (e.g., in [7.63]).

The electronic contribution Q^*_{ele} to the heat of transport consists itself of two parts: The first part stems from the thermoelectric field acting on the ionic charge of an atom [7.58]. As in the case of electrotransport, screening effects are also discussed for this field [7.59]. The second part of Q^*_{ele} arises from forces caused by electron scattering at the considered atom. Similar to electrotransport, these forces can be considered to result from a momentum transfer during scattering or, alternatively, from an electric field caused in the scattering processes [7.58–60]. However, the condition of a zero net electronic current must be maintained. This allows the following conclusions to be made [7.60, 61]: For a metal with a single electron band, any flux of energy-rich electrons from hot sample sections must be compensated by a low-energy electron flux of equal size coming from cold sample sections. The net force results only from differences between "hot" and "cold" electrons in momentum and scattering rate. The force can, therefore, be assumed to be small and of either sign. However, for a metal with electron and hole bands, codirectional fluxes of both charge carriers can arise towards the low-temperature sample sections. The zero net-current condition remains fulfilled, but the scattering forces of electrons and holes both act in the same direction. Large positive contributions should, therefore, be observed to the heat of transport.

The term Q^*_{pho} in (7.29) takes account of the force that is caused by phonon scattering [7.40, 57, 62]. To what extent scattering processes of phonons can cause momentum transfer and thus a force was a subject of controversy, particularly when the force was calculated from the phonon quasi-momentum change during scattering [7.40, 57]. However, it was shown by *Sorbello* [7.62] that, in the presence of anharmonic effects, the scattering of phonons can be considered to transfer real momentum.

The intrinsic heat of transport Q^*_{int} results from the fact that, even in the absence of the forces above, the probability for an atom to jump in a given direction depends on the temperature value and on the temperature distribution around the atom. That this causes atomic fluxes to arise in a temperature gradient can already be seen in the most simplified case of isotropic jumps. Namely, for a homogeneous concentration, the flux of atoms crossing a lattice plane from the direction of higher jump rates (and thus higher temperatures) is larger than that coming from the direction of lower jump rates. As a result, a net atomic flux will be observed.

Q^*_{int} is usually discussed within the model of *Wirtz* [6.56] in which this quantity is closely related to the activation enthalpy for diffusion. For an interstitial atom, Q^*_{int} should have a smaller absolute value than the activation enthalpy; its sign can be either positive or negative.

7.4 Experimental Considerations and Techniques

7.4.1 The Physical Quantities of Interest

Z^* and Q^*

The value of an experimental technique depends, to a certain extent, on the specific physical quantity one would like to measure. With respect to the following discussions, such a consideration seems necessary because both the effective charge number Z^* and the heat of transport Q^* are not the only quantities that can be used to describe electro- and thermotransport. In electrotransport, for example, the electric mobility u [$u = \text{Me}\,Z^*$, see (7.5)] is an alternative quantity. It is the drift velocity in a unit applied electric field, and it can be used instead of Z^* to describe electrotransport processes. The two quantities u and Z^* are, however, measured with different experimental precision depending on which experimental technique is applied. Therefore, according to which of the quantities is desired, different experimental techniques might be preferred.

From a theoretical point of view, Z^* and Q^* are the quantities of primary interest. They are the parameters that define the forces in electro- and thermotransport, (7.1, 2), and they achieve also a second characteristic physical meaning as a consequence of the Onsager relations (Sect. 7.2.5). To gain an atomistic understanding, these quantities must, therefore, be considered; they were, for that reason, also the subject of the calculations and models discussed in the last section.

This does, however, not imply that Z^* and Q^* are in all cases the most convenient quantities to deal with. Two examples may be given:

a) In *electrotransport*, the electric mobility u is more suitable than Z^* when the drift velocity v_{drift} is the actual quantity of interest. For instance, in cases where electrotransport is utilized for sample purification ([7.9, 33, 64]; see also Sect. 7.6), the time period that is required for an experiment is particularly important. This time period is usually given by the drift time τ_{drift}, that is, by the time period it takes for an atom to drift over a sample dimension (Sect. 7.2.4). To calculate the drift time, solely the electric mobility u must be known (besides the length of the sample and the applied voltage), whereas the knowledge of Z^* alone would not allow a determination because the value of the mobility M would additionally be required.

b) In *thermotransport*, the concentration profile that is established in a temperature field is of technical interest (e.g., because of material failures that can be caused by hydrogen embrittlement [7.65]; see Sect. 7.6). The concentration profile can, under steady-state conditions, be calculated from (7.10). However, it can be seen that to know only Q^* is not sufficient to determine the concentration profile because the quantity $c(\partial\mu/\partial c)$ is also contained in (7.10). On the other hand, the knowledge solely of a quantity like $Q^*/[c(\partial\mu/\partial c)]$ (or $Q^*k_{\mathrm{B}}T/[c(\partial\mu/\partial c)]$) would allow a determination so that this quantity would be more suitable for the purposes above than Q^*.

The two examples show that, particularly for technical applications, the measurement of other quantities can be more important than a precise determination only of Z^* and Q^*. However, in this chapter, Z^* and Q^* are considered the subjects of primary interest because of their physical significance. Experimental techniques will, therefore, be discussed mainly in regard to a precise measurement of these quantities.

The Chemical Potential μ

The quantity $c(\partial\mu/\partial c)$, i.e., the concentration dependence of the chemical potential μ, proved to be of importance within the phenomenological description of Section 7.2. In the limit of very low hydrogen concentrations, $c(\partial\mu/\partial c)$ can be replaced by the thermal energy $k_{\mathrm{B}}T$; it is, however, important to know up to what concentrations such an assumption can be made, and also what the values of $c(\partial\mu/\partial c)$ are at high concentrations.

For concentrations not too high, the concentration-dependent part of the chemical potential μ can be fairly well described by [7.17, 66–70]

$$\mu = -\varepsilon c + k_{\mathrm{B}}T \ln \frac{c}{r-c}, \tag{7.30}$$

where ε is a mean interaction energy among the hydrogen interstitials; the quantity r takes account of a short-range repulsive interaction leading to blocking effects (e.g., an interstice can only be occupied by one hydrogen atom).

For V, Nb, Ta, and Pd, the value of ε is experimentally found to be about 0.3 eV, and that of r about 1 [7.21, 67, 69, 71]. According to (7.30), the quantity $c(\partial\mu/\partial c)$ can be written

$$c\frac{\partial\mu}{\partial c} = -\varepsilon c + k_B T\frac{r}{r-c}. \tag{7.31}$$

From this equation, the deviations of $c(\partial\mu/\partial c)$ from the ideal value $k_B T$ are found appreciable even at relatively low concentrations; for instance, at room temperature and for a concentration of only 1 at.%, the deviations amount to about 10% using values for ε and r as given above.

Essentially two experimental methods are available to determine $c(\partial\mu/\partial c)$. The first of these consists of hydrogen solubility measurements, from which $c(\partial\mu/\partial c)$ can be obtained by the equation [7.17, 68, 72]

$$c\frac{\partial\mu}{\partial c} = \frac{k_B T}{2}\frac{\partial \ln p}{\partial \ln c}, \tag{7.32}$$

where p represents the equilibrium pressure of hydrogen gas (H_2) required to maintain a hydrogen concentration c at a given temperature T. The relaxation strength, for instance in Gorsky effect measurements, offers the second experimental method to determine $c(\partial\mu/\partial c)$ [7.19–21, 25, 73–75].

Diffusion Coefficient *D* and Mobility *M*

The diffusion coefficient and the mobility are important in the case of time-dependent transport processes. They were, however, recently reviewed [7.76, 77; Ref. 7.78, Chap. 12]; a discussion in this context seems, therefore, unnecessary.

7.4.2 Drift-Velocity and Flux Measurements

This experimental method was only applied in electrotransport studies for which it represents the most widely used technique. The force $\boldsymbol{F}$, and thus the effective charge number Z^* [see (7.1)], is determined from measurements either of the drift velocity $\boldsymbol{v}_{\text{drift}} = M\boldsymbol{F}$ (7.4) or of the drift flux $\boldsymbol{J}_{\text{drift}} = ncM\boldsymbol{F}$ (7.3).

The disadvantage of the method is that the mobility M (or the quantity ncM) must be known in order to calculate the force $\boldsymbol{F}$ from the measured quantities. The value of M must, therefore, be taken from other experiments, which implies that the relatively large experimental errors of this transport quantity affect directly the accuracy with which Z^* is determined. How difficult it is to obtain reliable data for Z^* is demonstrated in Fig. 7.3. The figure shows, as a result of drift-flux measurements, the electric mobility u $(= MeZ^*)$ of hydrogen in palladium in a plot versus temperature (*Hérold* and *Rat* [7.79]).

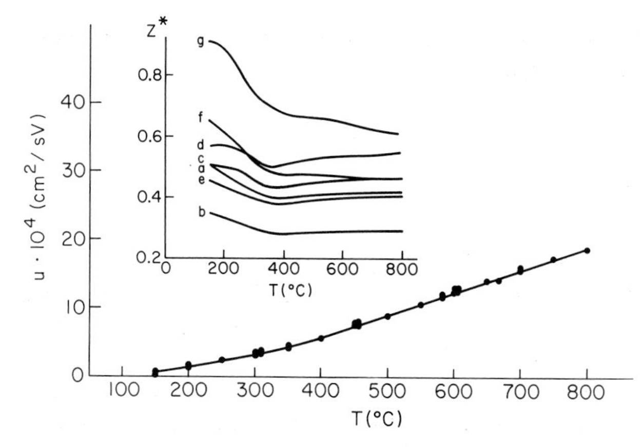

Fig. 7.3. Electric mobility u ($u = MeZ^*$) of H in Pd in a plot versus temperature (*Hérold* and *Rat* [7.79]). The inset shows values for the effective charge number Z^* which were calculated from seven different literature data of the diffusion coefficient D or the mobility M (*a Jost* and *Widmann* [7.80], *b Jost* and *Widmann* [7.81], *c Toda* [7.82], *d Katz* and *Gulbransen* [7.83], *e Toda* [7.84], *f Flanagan* and *Simons* [7.85], *g Gol'tsov* et al. [7.86])

The inset shows, in addition, values of the effective charge number which were calculated from different literature data for the diffusion coefficient or the mobility [7.80–86]. The scatter among the results for Z^* indicates very well the inaccuracy with which this quantity is obtained.

In drift-velocity measurements, the rate of the shift of a hydrogen concentration profile under the influence of the applied force is determined. One method to measure the shift rate consists of resistance measurements in which the resistance variation in a given sample section indicates the changes in the hydrogen concentration [7.87–89]. In another technique, the sample is sectioned into several pieces, and the hydrogen concentration is measured separately for each of these pieces. The hydrogen concentration profile at a given time can, thus, be determined. The hydrogen concentrations of the sample pieces were measured either by vacuum extraction [7.90] or, in a study of tritium electrotransport, by counting the tritium decay rate [7.91]. In two very early studies on palladium [7.92, 93], the shift rate of the hydrogen profile was investigated by chemical surface analysis.

The most common technique in electrotransport studies is the measurement of the drift flux $\boldsymbol{J}_{\text{drift}}$ through a sample [7.79, 94–101]. The technique was first introduced by *Wagner* and *Heller* [7.94]. Figure 7.4 shows, as an example, the apparatus used by *Knaak* and *Eichenauer* [7.97] for their measurements on palladium. The apparatus consists essentially of two capillary volumes on its right- and left-hand sides. Both volumes are filled with hydrogen gas of equal pressure. The two volumes are joined to each other by the melted-in palladium wire sample. In the experiments, an electric current provided by the two (practically hydrogen impermeable) gold wires flows along the sample so that

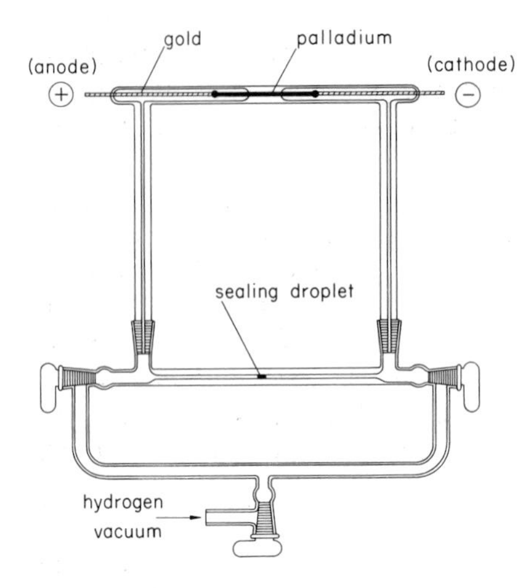

Fig. 7.4. Determination of the effective charge number by drift-flux measurements. The figure shows the apparatus used by *Knaak* and *Eichenauer* [7.97]

hydrogen is transferred from one capillary volume to the other. The transferred hydrogen causes the mobile sealing droplet at the bottom of the apparatus to shift, and the amount of hydrogen transferred is determined from the shift rate.

In two of the above cited measurements [7.96, 100], a somewhat modified version of the experimental setup shown in Fig. 7.4 was used. These measurements were the subject of a recent critique [7.102], which pointed out that the measured Z^* values were too small because the electrotransport current was shunted by electrodes at the end sections of the samples, thereby making the current only partially effective for hydrogen transport (see also [7.103]).

Finally, it is noted that a study [7.104] often considered an electrotransport investigation of hydrogen in iron (via drift-velocity measurements) actually did not investigate this phenomenon. This can be realized from the fact that the overpotential variations observed (up to 0.3 V) were not potential differences within the iron sample membrane.

7.4.3 Steady-State Measurements

Steady-state measurements were applied in both electrotransport and thermotransport investigations. The force, and thus Z^* or Q^*, is determined from the hydrogen distribution that is established in a sample under steady-state conditions. Under these conditions, the concentration depends on the electric potential or on the temperature according to (7.9) or (7.10), respectively.

The advantage of these measurements is that transport quantities like the mobility M or the diffusion coefficient D need not be known in order to determine Z^* or Q^*. The values of Z^* or Q^* are, therefore, usually determined with a higher experimental precision than in drift-velocity or flux measurements. A disadvantage is that the time periods required to establish steady-state conditions (see Sect. 7.2.4) impose limitations on the sample dimensions. However, for the extremely mobile hydrogen interstitials, these limitations are by far not so restrictive as is usual in experimental investigations on other systems.

In most steady-state measurements, the hydrogen concentration profile is determined within a sample. This can be done by sectioning the sample and measuring the hydrogen concentration of the individual parts by vacuum extraction [7.1, 105–112]. Figure 7.1 shows a leading example for such a measurement. The resistivity variations caused by the hydrogen were also used to determine the concentration profile [7.113–115]. This method has the advantage that measurements can repeatedly be made on the same sample. In one experiment [7.116], the concentration profile at high temperatures (300 °C) was estimated from the length of the precipitates that were formed after cooling the sample to room temperature.

The value of $c(\partial\mu/\partial c)$ is required in order to determine Z^* or Q^* from the measured concentration profiles [(7.9) or (7.10)]. With the exception of very low hydrogen concentrations (for which $c(\partial\mu/\partial c)$ can be replaced by $k_B T$), the values of $c(\partial\mu/\partial c)$ must, therefore, be taken from other experiments (see *The Chemical Potential* μ on p. 285). It is noted that, particularly in thermotransport measurements [7.1, 106, 112] $c(\partial\mu/\partial c)$ was practically throughout replaced by $k_B T$ even for high hydrogen concentrations. The quantity measured is, therefore, $Q^* k_B T/[c(\partial\mu/\partial c)]$ rather than Q^*.

Thermo-osmosis measurements [7.117–119] represent a second steady-state technique that has been applied for thermotransport investigations. In this technique, a temperature gradient is imposed across a sample membrane, each face of which is in contact with a volume of hydrogen gas. The hydrogen concentration at each membrane face, and thus the concentration gradient through the membrane, is monitored from the hydrogen pressure in the two gas volumes. The heat of transport Q^* is given by

$$Q^* = \frac{k_B T_1 T_2 \ln(p_2/p_1)}{2(T_1 - T_2)} - \Delta H\,, \qquad (7.33)$$

where T_1 and T_2 are the temperatures of the two membrane faces, and p_1 and p_2 the pressures in the corresponding gas volumes. ΔH is the enthalpy of solution per hydrogen atom ($\Delta H = h - \frac{1}{2} h_{H_2}$, where h and h_{H_2} are the partial enthalpies of a dissolved hydrogen atom and a hydrogen gas molecule, respectively). This quantity must, therefore, be taken from other experiments.

7.4.4 Diffusion Potential Measurements

Until now, all experimental methods discussed have been based on the action of an applied force on the hydrogen distribution. Because of the Onsager relations, a different procedure can also be applied. This was done by *Marêché* et al. [7.120] in an electrotransport study. They determined the effective charge number of hydrogen in palladium from the electric diffusion potential which is caused by concentration gradients according to (7.20). The flux of electric charge accompanying a hydrogen flux was, therefore, measured rather than the effects of an applied force.

It can readily be estimated from (7.20) that the voltages to be measured are very small. In the experiments of *Marêché* et al., these were in the μV range. The experiments are, therefore, very sensitive with respect to parasitic voltages. In order to calculate Z^* from the measured voltages, the value of the diffusion coefficient (or the mobility) is additionally required. For these reasons, this technique tends not to yield very precise values for Z^*. It is, however, an interesting example for an application of the Onsager relations.

7.5 Experimental Data

7.5.1 Effective Charge Number

Table 7.1 shows a compilation of data on experimental electrotransport studies. In addition to the values of the effective charge number, the hydrogen concentration, the temperature range, and the experimental technique of the measurements are listed.

Size of Z^*

With only one exception (H and D in Ag), the absolute values of all the effective charge numbers listed are of the order of magnitude of 1. It is interesting to compare these experimental sizes with those that can be calculated from the exact result for the jellium model of *Das* and *Peierls* (Sect. 7.3.1, [7.42, 46, 51, 53]). From (7.28), room-temperature values are found for Z^* which are, for instance in the case of V, Nb, Ta, and Pd, in the range between -20 and -40; these values derive from data for the hydrogen-induced resistivity increase [7.114, 130–134].

The difference between the calculated and experimental Z^* values amounts to an order of magnitude. It must almost entirely be attributed to differences in the wind forces (independently of the existence or nonexistence of a direct force contributing only about 1 to Z^*). These differences can be understood because the free-electron model of *Das* and *Peierls* does certainly not apply to the transition metals named above with electron and hole forces of opposite sign compensating each other at least partially. (The large and negative experimental values found for Z^* in silver are, accordingly, of the right size to indicate a wind force caused predominantly by electrons in this monovalent noble metal.) The model of Das and Peierls also does not take account of any lattice distortions around an interstitial atom. This tends to overestimate the size of the electron wind force since electron scattering processes at the lattice

Table 7.1. Effective charge numbers of hydrogen in metals

Metal and Hall coefficient[a] [in units of 10^{-11} m^3/(As)]	Hydrogen isotope	Hydrogen concentration [atom ratios]	Temperature [K]	Effective charge number Z^*	Experimental technique
Cu (−5.2, polycrystal)	H	—	1170–1270	Negative[b]	Hydrogen flux measurements [7.99, 122]
	H	—	310–350	Positive[b]	Hydrogen flux measurements [7.99]
Ag (−8.8, polycrystal)	H D	$\leqq 1.8 \cdot 10^{-6}$[c]	790–980 750–980	−6.8[d] −18[d]	Hydrogen flux measurements [7.100]
Y (−5 to −17, depending on orientation)	H	$\leqq 1.1 \cdot 10^{-2}$	1050–1220	−0.9 to −0.26	Drift velocity (sample sectioning and vacuum extraction analysis) [7.90]
Ti (+7 to −11, depending on orientation)	H D	0.32[e] (β phase, bcc)	1070	Positive ($Z^* < 1$)	Hydrogen flux measurements [7.124]
	T	$6 \cdot 10^{-2}$ (H and T)	[f]	+2.2[g]	Drift velocity (sample sectioning and T decay rate counting) [7.91]
Zr (+12, polycrystal)	H	$4.6 \cdot 10^{-3}$ (α phase, hcp)	570	Negative	Steady-state measurements (length of precipitates) [7.116]
	H D	0.89[e] (β phase, bcc)	1070	Positive ($Z^* < 1$)	Hydrogen flux measurements [7.124]
	T	$5 \cdot 10^{-2}$ (H and T)	[f]	+2.0[g]	Drift velocity (sample sectioning and T decay rate counting) [7.91]
V (+8, polycrystal)	H D	$\leqq 0.25$	720–1070	Approx. +1[h]	Hydrogen flux measurements [7.125, 126]
	H D	$9.1 \cdot 10^{-3}$ $1.2 \cdot 10^{-2}$ $5.4 \cdot 10^{-2}$	320–520	+1.6 to +1.4 +1.8 to +1.5 +0.97 to +1.1	Steady-state resistance measurements [7.115]

Table 7.1 (continued)

Metal and Hall coefficient[a] [in units of 10^{-11} m^3/(As)]	Hydrogen isotope	Hydrogen concentration [atom ratios]	Temperature [K]	Effective charge number Z^*	Experimental technique
Nb (+8.8, polycrystal)	H D	$\leqq 0.6$	720–1070	Approx. +1[h]	Hydrogen flux measurements [7.125, 126]
	H	$3.9 \cdot 10^{-3}$	320– 520	+2.6 to +2.4	Steady-state resistance measurements [7.115]
		$1.9 \cdot 10^{-2}$		+1.8 to +1.7	
	D	$4.6 \cdot 10^{-3}$	370– 520	+2.1 to +1.9	
		$9.0 \cdot 10^{-3}$	320– 520	+2.0 to +1.7	
Ta (+10, polycrystal)	H D	$\leqq 0.32$	720–1070	Approx. +1[h]	Hydrogen flux measurements [7.125, 126]
	H	$\leqq 0.1$	300	+2.5[i]	Drift velocity (resistance measurements) [7.88]
	H	$6.0 \cdot 10^{-3}$ to $2.5 \cdot 10^{-2}$	310	−5 to +5	Drift velocity (resistance measurements) [7.89]
	H	$1.3 \cdot 10^{-2}$	320– 520	+0.22 to +0.68	Steady-state resistance measurements [7.115]
		$2.9 \cdot 10^{-2}$		+0.17 to +0.59	
	D	$7.3 \cdot 10^{-3}$		+0.22 to +0.53	
Fe (+2, polycrystal)	H	$\leqq 6 \cdot 10^{-5}$[e]	710– 820	+0.24 to +0.29[j]	Hydrogen flux measurements [7.96, 103]
	D		740– 810	+0.42 to +0.43[j]	
	H	—	870–1120	Positive[b]	Hydrogen flux measurements [7.98, 99]
	H	—	310– 350	Negative[b]	Hydrogen flux measurements [7.99]
Ni (−6, polycrystal)	H	$\leqq 3 \cdot 10^{-4}$[e]	720– 900	+0.67 to +0.57[j]	Hydrogen flux measurements [7.96, 103]
	D		750– 920	+0.84 to +0.74[j]	
	H	—	310– 350	Negative[b]	Hydrogen flux measurements [7.99]
			970–1270	Positive[b]	

Table 7.1 (continued)

Metal and Hall coefficient [a] [in units of 10^{-11} $m^3/(As)$]	Hydrogen isotope	Hydrogen concentration [atom ratios]	Temperature [K]	Effective charge number Z^*	Experimental technique
Pd (−7.6, polycrystal)	H	High (α' phase)	290– 350	Positive[k]	Drift velocity (chemical surface analysis and resistance measurements) [7.87, 92, 93]
	H	$\leqq 2.8 \cdot 10^{-2}$	455– 513	+0.4 to +0.55	Hydrogen flux measurements [7.94]
	H	$\leqq 1.6 \cdot 10^{-2}$	520– 620	+0.54	Hydrogen flux measurements [7.97]
	D	$\leqq 1.0 \cdot 10^{-2}$		+0.51 to +0.59	
	H D	$\leqq 4.0 \cdot 10^{-2}$[e]	420–1070	Between +0.3 and +0.7	Hydrogen flux measurements [7.79, 95, 101, 128, 129]
	H	$\leqq 8.2 \cdot 10^{-3}$[e]	970	+0.44	Diffusion potential measurements [7.120]
	D			+0.35	
Pd–Ag alloys	H	Partially high	300	Positive	Drift velocity (resistance measurements) [7.87]
	H D	Equilibrium concentration to 1 atm H_2 pressure	470–1070	+0.4 to +1.0 positive[h]	Hydrogen flux measurements [7.101, 129]

[a] Hall coefficient values at room temperature according to [7.121]; for Fe and Ni, the ordinary Hall coefficient is listed.

[b] The values quoted for Z^* seem unreliable since the hydrogen drift flux was not found to be linear with the applied electric field.

[c] Concentration calculated from solubility data of hydrogen in silver [7.123].

[d] According to [7.102], the absolute values quoted for Z^* should be too small.

[e] Concentration calculated from solubility data given in [7.68].

[f] Temperature not quoted.

[g] During data evaluation, $c(\partial\mu/\partial c)$ was replaced by $k_B T$ (see *The Chemical Potential* μ on p. 285).

[h] In the experiments, the electric mobility u was determined.

[i] During data evaluation, $c(\partial\mu/\partial c)$ was replaced by $k_B T$; from solubility measurements [7.69, 71], the value quoted for Z^* can be estimated to be too high (see *The Chemical Potential* μ on p. 285).

[j] The original Z^* values of [7.96] are increased by 10% [7.103].

[k] In the experiments, the electric mobility u was determined; assuming a H/Pd atom ratio of about 0.6, Z^* can be estimated to be approx. +1 [7.127].

distortions contribute to the resistivity but not necessarily to the wind force [7.8, 40].

Because of the small size of the effective charge numbers, the direct field force, if existent, represents a substantial part of the entire force. Experimental efforts to separate the direct field contribution of Z^* seem, therefore, to be particularly promising in metal-hydrogen systems.

Isotope Dependence

In Ag, V, Ta, Fe, and Ni, isotopic differences in the effective charge numbers are reported which exceed the quoted experimental errors. In a theoretical analysis by *Stoneham* and *Flynn* [7.135] on isotope effects in electrotransport, a wind force was concluded to be greater the lighter the isotope. The analysis is based on the assumption that, as a consequence of lattice anharmonicities [7.136, 137], a lighter isotope causes a greater lattice expansion. However, the experimental data only partially substantiate the predicted behavior (independent of the existence or nonexistence of a direct field force).

Temperature Dependence

In (7.26), the proportionality between Z^*_{ele} and the reciprocal resistivity $1/\varrho$ introduces a characteristic temperature dependence for Z^*_{ele} through the temperature dependence of the resistivity. With temperatures becoming higher, Z^*_{ele} should continuously decrease because of the increasing resistivity; solely the direct field force should, therefore, contribute to the high-temperature final value of the total effective charge number Z^*. This temperature dependence offers an experimental possibility to separate the electron wind force from the temperature independent direct force. Precise experimental data are, however, required which cover a wide temperature range. Figure 7.5 shows results for the effective charge number Z^* of H and D in V, Nb, and Ta which seem suitable for such an analysis [7.115]. For all three metals, Z^* seems to approach high-temperature final values of about $+1$, thus indicating a direct-field force contribution of this size to the effective charge number. In Fig. 7.5, curves are shown fitting the temperature dependence of Z^* assuming a relation

$$Z^* = Z^*_{\text{dir}} + Z^*_{\text{ele}} = +1 + K/\varrho\,, \tag{7.34}$$

where K is the fitted parameter and ϱ the electrical resistivity. A value of $+1$ was, accordingly, presupposed for Z^*_{dir}, and Z^*_{ele} was assumed to be inversely proportional to the resistivity. The fits can be seen to provide a fair description of the experimental data, particularly since a proportionality between Z^*_{ele} and $1/\varrho$ cannot be expected to hold exactly for transition metals with electrons and holes. This is because deviations from a $1/\varrho$ behavior can occur if the partial resistivities characteristic of both types of charge carriers show different temperature dependences.

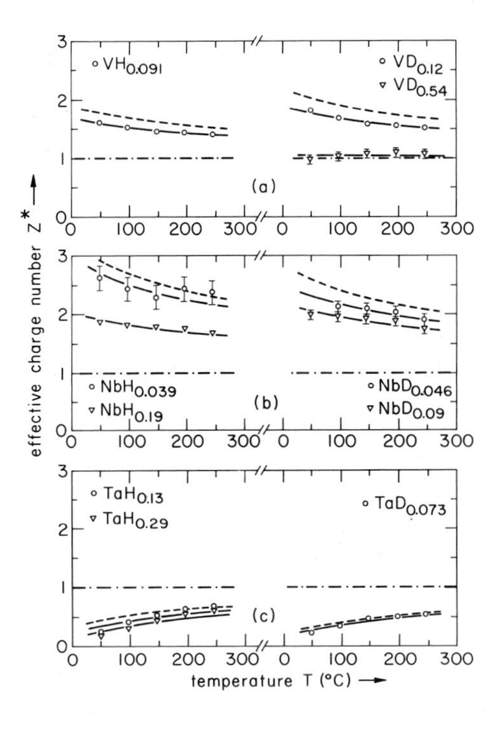

Fig. 7.5a–c. The effective charge number Z^* versus temperature for different H and D concentrations in (a) V, (b) Nb, and (c) Ta (*Erckmann* and *Wipf* [7.115]). The solid and broken lines are fits using (7.35) where the dashed lines correspond to Z^* values extrapolated to a zero H or D concentration. The dash-dot-dashed lines are for $Z^* = +1$

The result of about $+1$ for Z^*_{dir} agrees with the concept of an essentially unscreened direct field force as, for instance, assumed by *Fiks* [7.34] and *Huntington* and *Grone* [7.35]. It seems incompatible with a complete screening of this force as proposed by *Bosvieux* and *Friedel* [7.36]. In particular, the fact that for ten samples of three different metals, $+1$ appears to be approximately the high-temperature final value of Z^* makes a theory without a direct field force highly improbable (as, for example, assuming that electron and hole wind forces have very different temperature dependences).

The value for Z^* of about $+1$ also indicates that the screening effects discussed by *Landauer* [7.47, 52, 138] seem to be only a minor contribution. These effects are caused by local perturbations in the electronic band structure around an impurity atom (carrier density modulation). The perturbations can effect local variations in the electric conductivity, thus modulating the macroscopic electric field.

Concentration Dependence

The results of Fig. 7.5 demonstrate a marked concentration dependence of Z^*. For all systems shown, the Z^* values become smaller with increasing concentration. This behavior cannot be explained simply by an electron wind force decreasing proportional to $1/\varrho$ as a consequence of the hydrogen-induced increase of the resistivity ϱ (about 3% per at.% hydrogen [7.114, 130, 132–134]).

In the case of V and Nb, the effect is considerably too large for such an explanation, and it has, for Ta, even the wrong sign. Therefore, the effects of hydrogen concentration must be considered separately for the wind forces of electrons and holes. The Z^* values decreasing with concentration indicate, accordingly, that the wind force caused by holes is very rapidly reduced with increasing hydrogen concentration; a hole-band resistivity can, therefore, be concluded which increases steeply with concentration.

Correlation with the Hall Coefficient

To characterize the electronic band structure, data for the Hall coefficients of the pure host metals [7.121] are listed in Table 7.1. For a single band metal with either electrons or holes, the Hall coefficient and Z^*_{ele} have the same sign. For that reason, a trend towards a similar correlation can be suspected also for metals with a more complex electronic structure. To assume such a correlation is, however, not necessarily correct because the Hall coefficient and Z^*_{ele} are determined differently by the electronic band parameters [7.6, 13]. The data in Table 7.1 indicate also that such a correlation is only partially observed.

Transport Measurements in a Hall Field

In an investigation of *Pietrzak* and *Rozenfeld* [7.91], the Hall field electrotransport of tritium was studied in Ti and Zr. The authors report a strong nonlinearity of the drift velocity (in the Hall field direction) with the size of the applied magnetic field. After an initial increase proportional to the field, the drift velocity was found to pass through a maximum and thereupon to decrease with increasing magnetic field. The drift velocity was, however, essentially proportional to the electric current causing the Hall field. The authors did not report on the direction of the drift velocity with respect to that of the Hall field; the sign of the Hall field effective charge can, therefore, not be concluded from [7.91].

7.5.2 Heat of Transport

In Table 7.2, experimental results are compiled for the heat of transport Q^* of hydrogen in metals. In addition to the values found for Q^*, the hydrogen concentration and the temperature range of the measurements are indicated as well as the applied experimental technique.

Size of Q^*

The values of Q^* in Table 7.2 are, with the exceptions of Fe and Ni, positive, and they range up to several tenths of an eV. The mainly positive sign qualitatively can be attributed to the presence of electrons and holes in the listed metals tending to cause relatively large and positive electronic contributions Q^*_{ele} to the heat of transport (see Sect. 7.3.2, [7.60, 61]).

Table 7.2. Heat of transport of hydrogen in metals

Metal	Hydrogen isotope	Hydrogen concentration [atom ratios]	Temperature [K]	Heat of transport Q^* [eV]	Experimental technique
Ti	H	$1.7 \cdot 10^{-3}$ (α phase, hcp)	670–760	+0.23	Steady-state measurements[a] (sample sectioning and vacuum extraction analysis) [7.109]
	H	$\leqq 5 \cdot 10^{-2}$ (α phase, hcp)	550–920	+0.20[b]	Steady-state measurements (sample sectioning and vacuum extraction analysis) [7.1]
		0.48 (β phase, bcc)	590–860	+0.027[b]	
	H	$\leqq 1 \cdot 10^{-2}$ (α phase, hcp)	580–850	+0.32	Steady-state measurements (sample sectioning and vacuum extraction analysis) [7.110]
Zr	H	1.7 (δ phase, fcc)	810–970	+0.056[b]	Steady-state measurements (sample sectioning and vacuum extraction analysis) [7.106]
	H	0.44 and 0.86 (β phase, bcc)	900–1130	+0.25 to +0.5	Steady-state thermo-osmosis measurements [7.117]
	H	$5.5 \cdot 10^{-3}$ (α phase, hcp)	570–770	+0.26	Steady-state measurements (sample sectioning and vacuum extraction analysis) [7.107]
	H	1.6 (δ phase, fcc)	690–920	~0	Steady-state measurements (sample sectioning and vacuum extraction analysis) [7.108]
	H	$1 \cdot 10^{-3}$ (α phase, hcp)	470–750	+0.23	Steady-state measurements (sample sectioning and vacuum extraction analysis [7.111]

Table 7.2 (continued)

Metal	Hydrogen isotope	Hydrogen concentration [atom ratios]	Temperature [K]	Heat of transport Q^* [eV]	Experimental technique
Zircaloy-2	H	$5.5 \cdot 10^{-3}$ (α phase)	570–770	+0.23	Steady-state measurements (sample sectioning and vacuum extraction analysis) [7.105, 107]
	D			+0.28	
V	H	$1.3 \cdot 10^{-2}$	270–320	+0.017	Steady-state resistance measurements [7.114]
	H	$7.2 \cdot 10^{-2}$	540–870	+0.087[c]	Steady-state measurements (sample sectioning and vacuum extraction analysis) [7.112]
Nb	H	$1.5 \cdot 10^{-3}$	190–330	+0.15 to +0.13	Steady-state resistance measurements [7.113]
	D	$1.1 \cdot 10^{-3}$		+0.13 to +0.11	
Fe	H	$\leqq 8 \cdot 10^{-5}$[d]	680–870	−0.35 to −0.24	Steady-state thermo-osmosis measurements [7.118]
	D			−0.34 to −0.23	
Ni	H	$\leqq 3 \cdot 10^{-4}$[d]	790–900	−0.065 to −0.009	Steady-state thermo-osmosis measurements [7.118]
	D		850–890	−0.056 to −0.035	
Pd	H	$1.2 \cdot 10^{-2}$[d]	670–830	+0.065	Steady-state thermo-osmosis measurements [7.119]

[a] The experiments were discontinued before steady-state conditions were completely accomplished.
[b] During data evaluation, $c(\partial\mu/\partial c)$ was replaced by $k_B T$ (see *The Chemical Potential* μ on p. 285).
[c] During data evaluation, $c(\partial\mu/\partial c)$ was replaced by $k_B T$; from solubility measurements [7.69], the value quoted for Q^* can be estimated to be too high by about 20% (see *The Chemical Potential* μ on p. 285).
[d] Concentration calculated from solubility data given in [7.68].

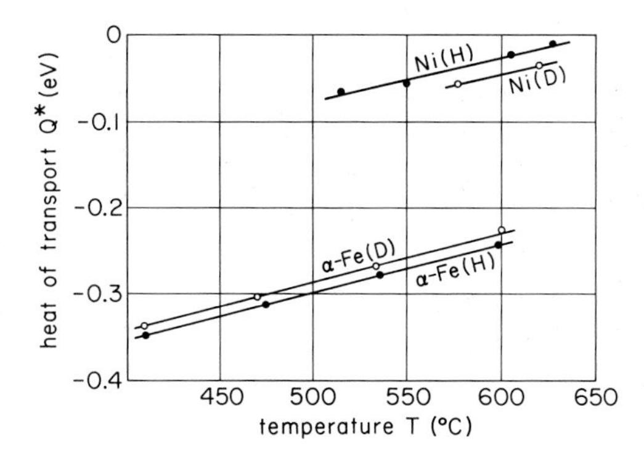

Fig. 7.6. Heat of transport of H and D in Fe and Ni in a plot versus temperature (*Gonzalez* and *Oriani* [7.118])

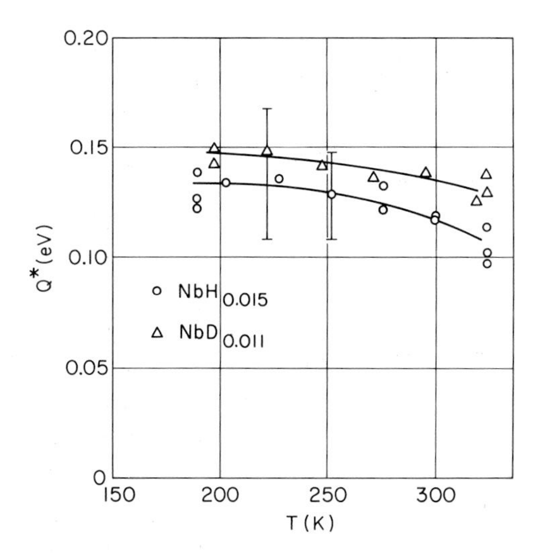

Fig. 7.7. The heat of transport Q^* of H and D in Nb in a plot versus temperature (*Wipf* and *Alefeld* [7.113])

Isotope Dependence

Isotopic differences in the Q^* values are reported for Zircaloy-2, for Fe and for Ni. Figure 7.6 shows results on Fe and Ni obtained by *Gonzalez* and *Oriani* [7.118]. The isotope effects are, however, not very significant and near the borderline of experimental accuracy. In a theoretical analysis, *Stoneham* and *Flynn* discussed the isotope effects in thermotransport [7.135].

Temperature Dependence

A temperature dependence of Q^* exceeding the experimental errors was observed in Zr [7.117] and in Fe and Ni ([7.118], Fig. 7.6). In these cases, the values of Q^* showed a trend to become positive or more positive with increasing temperature.

In the case of niobium, a comparison seems interesting between the temperature dependence of Q^* with that of the diffusion coefficient. Figure 7.7

shows values of Q^* for H and D in niobium in the temperature range between 190 and 330 K [7.113]. For both isotopes, a temperature dependence exceeding the experimental errors cannot be recognized. However, the diffusion coefficient shows a significant and unexpected temperature dependence with an activation energy for hydrogen (^{1}H) diffusion changing from about 0.106 eV above 230 K to about 0.065 eV below this temperature [7.21, 77, 78, 113, 139]; for deuterium, no change is observed for the activation enthalpy (0.125 eV) in the same temperature range.

The fact that the heat of transport of hydrogen remains unaffected by a variation of the activation enthalpy allows conclusions to be made on the intrinsic heat of transport $Q^*_{\rm int}$ (which is considered to be related to the activation enthalpy; see Sect. 7.3.2): Either $Q^*_{\rm int}$ is not markedly influenced by a 40% change of the activation enthalpy, or its contribution to the total value of Q^* is insignificant. (To assume a change in $Q^*_{\rm int}$, which is exactly compensated by variations of the other terms contributing to Q^*, seems improbable.)

Effects of Concentration and Lattice Structure

In Ti and Zr, values of Q^* were observed which varied strongly between different phases (phase diagrams can be found, e.g., in [7.68]). The phases differ in their hydrogen concentration as well as in their lattice structure. It is difficult, for that reason, to attribute the variations of Q^* appropriately to the effects of both concentration and structure. To obtain data on the concentration dependence of Q^* within a single phase seems interesting because insight can be expected from such data into the different terms contributing to the heat of transport [7.63].

7.6 Technological Aspects

In thermotransport, the concentration variations within a sample, and thus the size of the forces, are impressively demonstrated by Fig. 7.1. In electrotransport, the effects on a concentration profile can be expected to be even larger in size. Consider a voltage of 500 mV applied over a sample. With a representative value of 1 for Z^*, a ratio of e^{20} can be estimated to exist under steady-state conditions between the hydrogen concentrations at the two sample ends [assuming room temperature and $c(\partial\mu/\partial c)=k_{\rm B}T$; (7.11)]. This is an extremely large effect due to which a part of the sample is practically free of hydrogen; it illustrates also very well the size of the forces in electrotransport.

The large forces that can arise make electrotransport and thermotransport important and interesting also from a technological point of view. Electrotransport, for instance, has proven to be the reason for failures in the metal stripes of integrated circuitry. On the other hand, it has also been utilized successfully to purify metals from interstitial impurities [7.9, 33, 64]. In metal-hydrogen systems, thermotransport seems to be particularly important. Any local heating (or cooling) will effect hydrogen redistribution within short time periods because of the high mobility of the hydrogen. Such a redistribution occurs in electron microscopy investigations on metal-hydrogen systems, with the electron beam as source of local heating [7.140]. Thermotransport is also important in connection with material failures caused by hydrogen precipitation (hydrogen embrittlement). In heavy-water reactors, for instance, zirconium alloys used as fuel cladding are subjected to strong temperature gradients, and they absorb hydrogen continuously as a secondary effect of corrosion. Under these conditions, the value of the heat of transport determines the hydrogen profile in the alloy and thus, according to the phase diagram, the spatial sections of the alloy where precipitation (and embrittlement) takes place [7.65].

References

7.1 A.Sawatzky, M.Duclos: Trans. Met. Soc. AIME **245**, 831 (1969)
7.2 J.Verhoeven: Met. Rev. **8**, 311 (1963)
7.3 Y.Adda, J.Philibert: *La Diffusion dans les Solides* (Presses Universitaires de France, Paris 1966)
7.4 A.R.Allnatt, A.V.Chadwick: Chem. Rev. **67**, 681 (1967)
7.5 T.Hehenkamp: In *Vacancies and Interstitials in Metals* (North-Holland, Amsterdam 1969) p. 91
7.6 H.Wever: *Elektro- und Thermotransport in Metallen* (Joh. Ambrosius Barth, Leipzig 1973)
7.7 J.N.Pratt, R.G.R.Sellors: *Electrotransport in Metals and Alloys*, Vol. 2 of Diffusion and Defect Monograph Series, Trans. Tech. SA, Riehen 1973
7.8 F.M.d'Heurle, R.Rosenberg: *Electromigration in Thin Films, Physics of Thin Films*, Vol. 7 (Academic Press, New York, London 1973)
7.9 R.G.Jordan: Contemp. Phys. **15**, 375 (1974)
7.10 D.A.Rigney: In *Charge Transfer/Electronic Structure of Alloys*, ed. by L.H.Bennet, R.H.Willens (Metall. Soc. of AIME, New York 1974)
7.11 H.B.Huntington: In *Diffusion in Solids*, ed. by A.S.Nowick, J.J.Burton (Academic Press, New York 1975) p. 303
7.12 H.B.Huntington: Thin Solid Films **25**, 265 (1975)
7.13 R.S.Sorbello: In *Electro- and Thermotransport in Metals and Alloys*, ed. by R.E.Hummel, H.B.Huntington (Metall. Soc. of AIME, New York 1977) p.2
7.14 S.R.deGroot: *Thermodynamics of Irreversible Processes* (North-Holland, Amsterdam 1951)
7.15 R.Haase: *Thermodynamik der irreversiblen Prozesse* (Steinkopf Verlag, Darmstadt 1963)
7.16 S.R.deGroot, P.Mazur: *Non-Equilibrium Thermodynamics* (North-Holland, Amsterdam 1969)
7.17 G.Alefeld: Phys. Stat. Sol. **32**, 67 (1969)
7.18 G.Alefeld, J.Völkl, G.Schaumann: Phys. Stat. Sol. **37**, 337 (1970)
7.19 J.Völkl: Ber. Bunsenges. Physik. Chem. **76**, 767 (1972)
7.20 H.Wipf: J. Less-Common Metals **49**, 291 (1976)
7.21 G.Schaumann, J.Völkl, G.Alefeld: Phys. Stat. Sol. **42**, 401 (1970)
7.22 M.A.Pick, R.Bausch: J. Phys. F: Metal Phys. **6**, 1751 (1976)
7.23 H.Metzger, J.Peisl, J.Wanagl: J. Phys. F: Metal Phys. **6**, 2195 (1976)
7.24 G.Schaumann, J.Völkl, G.Alefeld: Phys. Rev. Lett. **21**, 891 (1968)
7.25 J.Völkl, G.Alefeld: Nuovo Cimento **33**B, 190 (1976)
7.26 J.W.Cahn: Acta Met. **9**, 795 (1961)
7.27 J.W.Cahn: Acta Met. **10**, 907 (1962)
7.28 H.Wagner, H.Horner: Advan. Phys. **83**, 587 (1974)

7.29 T.W.Burkhardt: Z. Physik **269**, 237 (1974)
7.30 R.Bausch, H.Horner, H.Wagner: J. Phys. C: Solid State Phys. **8**, 2559 (1975)
7.31 M.Mason, W.Weaver: Phys. Rev. **23**, 412 (1924)
7.32 S.R.deGroot: Physica **9**, 699 (1942)
7.33 D.T.Peterson: In *Atomic Transport in Solids and Liquids*, ed. by A.Lodding, T.Lagerwall (Verlag der Zeitschrift für Naturforschung, Tübingen 1971) p. 104
7.34 V.B.Fiks: Fiz. Tver. Tela **1**, 16 (1959) [English transl.: Sov. Phys.-Solid State **1**, 14 (1959)]
7.35 H.B.Huntington, A.R.Grone: J. Phys. Chem. Sol. **20**, 76 (1961)
7.36 C.Bosvieux, J.Friedel: J. Phys. Chem. Sol. **23**, 123 (1962)
7.37 M.D.Glinchuk: Ukrain. Fiz. Zh. **4**, 684 (1959)
7.38 R.S.Sorbello: J. Phys. Chem. Sol. **34**, 937 (1973)
7.39 V.B.Fiks: In *Atomic Transport in Solids and Liquids*, ed. by A.Lodding, T.Lagerwall (Verlag der Zeitschrift für Naturforschung, Tübingen 1971) p. 3
7.40 M.Gerl: Z. Naturforsch. **26**a, 1 (1971); or in *Atomic Transport in Solids and Liquids*, ed. by A.Lodding, T.Lagerwall (Verlag der Zeitschrift für Naturforschung, Tübingen 1971) p. 9
7.41 H.B.Huntington, W.B.Alexander, M.D.Feit, J.L.Routbort: In *Atomic Transport in Solids and Liquids*, ed. by A.Lodding, T.Lagerwall (Verlag der Zeitschrift für Naturforschung, Tübingen 1971) p. 91
7.42 L.J.Sham: Phys. Rev. B**12**, 3142 (1975)
7.43 H.B.Huntington: Trans. Met. Soc. AIME **245**, 2571 (1969)
7.44 G.Frohberg: In *Atomic Transport in Solids and Liquids*, ed. by A.Lodding, T.Lagerwall (Verlag der Zeitschrift für Naturforschung, Tübingen 1971) p. 19
7.45 A.K.Das, R.Peierls: J. Phys. C: Solid State Phys. **6**, 2811 (1973)
7.46 A.K.Das, R.Peierls: J. Phys. C: Solid State Phys. **8**, 3348 (1975)
7.47 R.Landauer, J.W.F.Woo: Phys. Rev. B**10**, 1266 (1974)
7.48 P.Kumar, R.S.Sorbello: Thin Solid Films **25**, 25 (1975)
7.49 R.S.Sorbello: Comments Solid State Phys. **6**, 117 (1975)
7.50 H.E.Rohrschach: Ann. Phys. **98**, 70 (1976)
7.51 L.Turban, P.Nozières, M.Gerl: J. Phys. Paris **37**, 159 (1976)
7.52 R.Landauer: Phys. Rev. B**14**, 1474 (1976)
7.53 R.Landauer: J. Phys. C: Solid State Phys. **8**, L389 (1975)
7.54 R.Landauer: Phys. Rev. B**16**, 4698 (1977)
7.55 A.K.Das: Solid State Commun. **18**, 601 (1976)
7.56 K.Wirtz: Phys. Zschr. **44**, 221 (1943)
7.57 V.B.Fiks: Fiz. Tver. Tela **3**, 994 (1961), [English transl.: Sov. Phys.–Solid State **3**, 724 (1961)]
7.58 V.B.Fiks: Fiz. Tver. Tela **5**, 3473 (1963), [English transl.: Sov. Phys.–Solid State **5**, 2549 (1964)]
7.59 M.Gerl: J. Phys. Chem. Sol. **28**, 725 (1967)
7.60 H.B.Huntington: J. Phys. Chem. Sol. **29**, 1641 (1968)
7.61 R.A.Oriani: J. Phys. Chem. Sol. **30**, 339 (1969)
7.62 R.S.Sorbello: Phys. Rev. B**6**, 4757 (1972)
7.63 T.Hehenkamp: Thin Solid Films **25**, 281 (1975)
7.64 J.D.Verhoeven: J. Metals **18**, 26 (1966)
7.65 R.C.Asher, F.W.Trowse: J. Nucl. Mat. **35**, 115 (1970)
7.66 J.Lacher: Proc. Roy. Soc. London A**161**, 525 (1937)
7.67 E.Wicke, G.H.Nernst: Ber. Bunsenges. Physik. Chem. **68**, 224 (1964)
7.68 W.M.Mueller, J.P.Blackledge, G.G.Libowitz (eds): *Metal Hydrides* (Academic Press, New York, London 1968)
7.69 E.Veleckis, R.K.Edwards: J. Phys. Chem. **73**, 683 (1969)
7.70 R.B.McLellan: Mat. Sci. Eng. **9**, 121 (1972)
7.71 O.J.Kleppa, P.Dantzer, M.E.Melnichak: J. Chem. Phys. **61**, 4048 (1974)
7.72 T.B.Flanagan, W.A.Oates: Ber. Bunsenges. Physik. Chem. **76**, 706 (1972)
7.73 G.Alefeld, J.Völkl, J.Tretkowski: J. Phys. Chem. Sol. **31**, 1765 (1970)
7.74 J.Tretkowski, J.Völkl, G.Alefeld: Z. Naturforsch. **26**a, 588 (1971)
7.75 D.Schnabel, G.Alefeld: Collect. Phenom. **2**, 29 (1975)

7.76 H.K.Birnbaum, C.A.Wert: Ber. Bunsenges. Physik. Chem. **76**, 806 (1972)
7.77 J.Völkl, G.Alefeld: In *Diffusion in Solids*, ed. by A.S.Nowick, J.J.Burton (Academic Press, New York 1975)
7.78 G.Alefeld, J.Völkl (eds.): *Hydrogen in Metals I. Basic Properties*, Topics in Applied Physics, Vol. 28 (Springer, Berlin, Heidelberg, New York 1978)
7.79 A.Hérold, J.-C.Rat: Bull. Soc. Chim. France **1**, 80 (1972)
7.80 W.Jost, A.Widmann: Z. Physik. Chemie B**29**, 247 (1935)
7.81 W.Jost, A.Widmann: Z. Physik. Chemie B**45**, 285 (1940)
7.82 G.Toda: J. Research Inst. Catalysis (Hokkaido Univ.) **6**, 13 (1958)
7.83 O.M.Katz, E.A.Gulbransen: Rev. Sci. Instr. **31**, 615 (1960)
7.84 G.Toda: Schokubai (Tokyo) **5**, 11 (1963)
7.85 T.B.Flanagan, J.W.Simons: J. Phys. Chem. **69**, 3581 (1965)
7.86 V.A.Gol'tsov, V.B.Demin, V.B.Vykhodets, G.Yu.Kagan, P.V.Gel'd: Fiz. Metal. Metalloved. **29**, 1305 (1970) [English transl.: Phys. Met. Met. **29**, 195 (1970)]
7.87 A.Coehn, H.Jürgens: Z. Physik **71**, 179 (1931)
7.88 B.A.Merisov, G.Ya.Khadzhay, V.I.Khotkevich: Fiz. Metal. Metalloved. **39**, 324 (1975) [English transl.: Phys. Met. Met. **39**, 88 (1975)]
7.89 Yu.K.Ivashina, V.F.Nemchenko, V.G.Charnetskiy: Fiz. Metal. Metalloved **40**, 243 (1975), [English transl.: Sov. Phys. Met. Met. **40**, 97 (1975)]
7.90 O.N.Carlson, F.A.Schmidt, D.T.Peterson: J. Less-Common Metals **10**, 1 (1966)
7.91 R.Pietrzak, B.Rozenfeld: Acta Phys. Polonica A**49**, 341 (1976)
7.92 A.Coehn, W.Specht: Z. Physik **62**, 1 (1930)
7.93 A.Coehn, K.Sperling: Z. Physik **83**, 291 (1933)
7.94 C.Wagner, G.Heller: Z. Physik. Chemie B**46**, 242 (1940)
7.95 A.Hérold: C.R.Acad. Sci. Paris **243**, 806 (1956)
7.96 R.A.Oriani, O.D.Gonzalez: Trans. Met. Soc. AIME **239**, 1041 (1967)
7.97 J.Knaak, W.Eichenauer: Z. Naturforsch. **23**a, 1783 (1968)
7.98 V.M.Sidorenko, R.I.Kripyakevich: Fiz-Khim. Mek. Mat. **4**, 335 (1968) [English transl.: Sov. Mater. Sci. **4**, 244 (1968)]
7.99 V.M.Sidorenko, R.I.Kripyakevich: Fiz-Khim. Mek. Mat. **5**, 191 (1969) [English transl.: Sov. Mater. Sci. **5**, 145 (1969)]
7.100 R.E.Einziger, H.B.Huntington: J. Phys. Chem. Sol. **35**, 1563 (1974)
7.101 J.-F.Marêché, J.-C.Rat, A.Hérold: J. Chim. Phys. **72**, 697 (1975)
7.102 H.Wipf, V.Erckmann: Scripta Met. **10**, 813 (1976)
7.103 R.A.Oriani: Scripta Met. **10**, 817 (1976)
7.104 S.Wach, A.P.Miodownik: Trans. Faraday Soc. **66**, 2334 (1970)
7.105 A.Sawatzky: J. Nucl. Mater. **2**, 321 (1960)
7.106 A.W.Sommer, W.F.Dennison: AEC Research and Development Rp. NAA-SR-5066 (1960)
7.107 A.Sawatzky: J. Nucl. Mater. **9**, 364 (1963)
7.108 U.Merten, J.C.Bokros, D.G.Guggisberg, A.P.Hatcher: J. Nucl. Mater. **10**, 201 (1963)
7.109 R.P.Marshall: Trans. Met. Soc. AIME **233**, 1449 (1965)
7.110 M.Kitada, S.Koda: Scripta Met. **3**, 583 (1969)
7.111 S.Morozumi, M.Kitada, K.Abe, S.Koda: J. Nucl. Mater. **33**, 261 (1969)
7.112 S.Morozumi, S.Goto, T.Yoshida: Scripta Met. **10**, 537 (1976)
7:113 H.Wipf, G.Alefeld: Phys. Stat. Sol. (a) **23**, 175 (1974)
7.114 R.Heller, H.Wipf: Phys. Stat. Sol. (a) **33**, 525 (1976)
7.115 V.Erckmann, H.Wipf: Phys. Rev. Lett. **37**, 341 (1976)
7.116 Y.Mishima, S.Ishino, N.Miyazaki: J. Fac. Eng. Univ. Tokyo A**11**, 46 (1973)
7.117 J.W.Droege: Battelle Rept. BMI 1502 (1961)
7.118 O.D.Gonzalez, R.A.Oriani: Trans. Met. Soc. AIME **233**, 1878 (1965)
7.119 A.Oates, J.G.Shaw: Met. Trans. **1**, 3237 (1970)
7.120 J.-F.Marêché, J.-C.Rat, A.Hérold: C. R. Acad. Sc. Paris **281**, 449 (1975)
7.121 C.M.Hurd: *The Hall Effect in Metals and Alloys* (Plenum Press, New York, London 1972)
7.122 V.M.Sidorenko, R.I.Kripyakevich, B.F.Kachmar: Fiz.-Khim. Mek. Mat. **5**, 187 (1969) [English transl.: Sov. Mater. Sci. **5**, 142 (1969)]

7.123 R.B.McLellan: J. Phys. Chem. Sol. **34**, 1137 (1973)
7.124 A.Hérold, J.-F.Marêché, J.-C.Rat: C. R. Acad. Sci. Paris **278**, 1009 (1974)
7.125 J.-F.Marêché, J.-C.Rat, A.Hérold: J. Chim. Phys. **73**, 1 (1976)
7.126 A.Hérold, J.-F.Marêché, J.-C.Rat: C. R. Acad. Sc. Paris **273**, 1736 (1971)
7.127 G.Bohmholdt, E.Wicke: Z. Physik. Chemie (N.F.) **56**, 133 (1967)
7.128 A.Hérold, J.-C.Rat: C. R. Acad. Sc. Paris **271**, 701 (1970)
7.129 J.-F.Marêché, J.-C.Rat, A.Hérold: C. R. Acad. Sc. Paris **275**, 661 (1972)
7.130 M.V.Borgucci, L.Verdini: Phys. Stat. Sol. **9**, 243 (1965)
7.131 J.W.Simons, T.B.Flanagan: J. Chem. Phys. **44**, 3486 (1966)
7.132 D.G.Westlake: Trans. Met. Soc. AIME **239**, 1341 (1967)
7.133 G.Pfeiffer, H.Wipf: J. Phys. F: Metal Phys. **6**, 167 (1976)
7.134 K.Rosan, H.Wipf: Phys. Stat. Sol. (a) **38**, 611 (1976)
7.135 A.M.Stoneham, C.P.Flynn: J. Phys. F: Metal Phys. **3**, 505 (1973)
7.136 J.A.D.Matthew: Phys. Stat. Sol. **42**, 841 (1970)
7.137 H.Wipf: Fiz. Nizkikh Temp. **1**, 645 (1975) [English transl.: Sov. J. Low. Temp. Phys. **1**, 310 (1976)]
7.138 R.Landauer: J. Phys. C: Solid State Phys. **8**, 761 (1975)
7.139 D.Richter, B.Alefeld, A.Heidemann, N.Wakabayashi: J. Phys. F.: Metal Phys. **7**, 569 (1977)
7.140 T.Schober: Phys. Stat. Sol. (a) **29**, 395 (1975)

8. Trapping of Hydrogen in Metals

Ch. A. Wert

With 11 Figures

8.1 Overview

Random solid solutions of hydrogen in metals might exist if the metal were a single crystal free of defects, and if the dissolved hydrogen were in dilute enough solution. However, all real solids contain defects which may trap hydrogen, and the mutual interaction of dissolved hydrogen atoms may permit clusters to form. Therefore, all real solid solutions are nonrandom, and hydrogen may be trapped in a variety of sites. The same is true, of course, of all atoms dissolved in a solid, but the rapid diffusion of hydrogen in metals, even at low temperatures, permits the trapping to take place at much lower temperatures than is true of any other dissolved atom. Furthermore, the presence of trapped hydrogen may be associated with many striking mechanical effects—especially embrittlement of many metals. Therefore, it is important to examine the nature of trapped hydrogen from a practical point of view, as well as studying trapping for its intrinsic intellectual value.

Historically, the embrittlement of steels by hydrogen has been of immense interest. This interest was extended to other bcc metals as they became important structural materials. Even more recently, the fcc metals and their alloys have been extensively examined, some for their intrinsic interest, others because of possible structural applications in hydrogen environments. Examination of the fantastic array of observations reported in the literature shows that a variety of ways might be used to examine trapping phenomena. One might do it by material, i.e., ferritic steel, refractory metals, stainless steels, noble metals, etc. Alternately, one might examine the effects by an observation method such as study of trapping by measurement of diffusion or internal friction, by electron microscope observations or solubility measurements, etc. Finally, one might examine the trapping by the nature of the trapping sites themselves such as voids, surfaces, boundaries, impurity atoms, self-traps, dislocations. It is the last method of examination which is employed in this chapter. Because of the enormous number of papers existing through the scientific and technological literature, it is impossible to cite them all, or even to find them all. Consequently, some of the citations are to review articles where the reader might look for additional references on individual subjects.

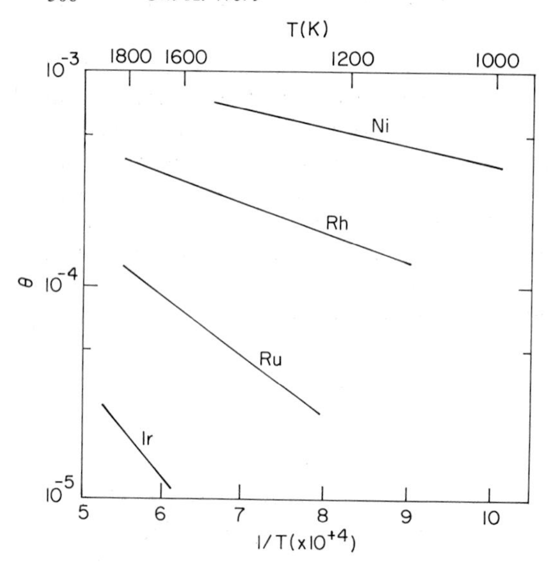

8.1. Solubility of hydrogen in four close-packed metals at high temperatures [8.4]

8.2 Thermodynamics of Random Solutions

Many studies have been made of equilibrium between hydrogen dissolved in a solid with a surrounding gas of hydrogen. An excellent review of the literature prior to 1970 was given by *Fast* [8.1, 2]. A more recent compilation of data along with details of methods of measurement has been published by *Fromm* and *Gebhardt* [8.3]. To show the general nature of the phenomenon, a few observations are cited here.

First, one can describe the solubility θ as a function of temperature by the simple expression

$$\theta = \beta \exp(-\Delta\bar{H}_u/kT)\exp(\Delta\bar{S}_u^{xs}/k)\,. \tag{8.1}$$

Here $\Delta\bar{H}_u$ is the relative partial molar enthalpy of solution and $\Delta\bar{S}_u^{xs}$ is the relative partial molar excess entropy of solution. For dilute solutions this expression fits observations very well; data for four metals which dissolve hydrogen sparingly are shown in Fig. 8.1. These data of *McLellan* and *Oates* fit (8.1) nicely and yield values of $\Delta\bar{H}_u$ and $\Delta\bar{S}_u^{xs}$ which are independent of temperature [8.4].

These metals dissolve only a relatively small amount of hydrogen. Systems for which the solubility is very large may show a temperature dependence of these partial molar quantities. *Boureau* et al. reported such measurements for alloys of palladium-hydrogen and described possible causes for variation in $\Delta\bar{H}_u$ and $\Delta\bar{S}_u^{xs}$ [8.5, 6]. They attributed a part of this variation to dual site occupancy, that is, hydrogen and its isotopes might occupy both 0 and T sites in fcc palladium. They did not, however, attribute temperature variation of the relative partial molar quantities to trapping, even though their specimens were not single crystals and contained traces of a variety of impurities. *McLellan* and *Oates* did not find any effect of grain boundaries or impurities on the solubility they reported for their metals, either. The important feature to be noted is that measurement of equilibrium solubility for both dilute absorbers or high absorbers is not an especially effective way of studying trapping, even in systems where one might suspect that it exists. Special instances in which careful measurements over a wide range of temperature have been interpreted in terms of trapping are described in succeeding sections.

8.3 General Features of Trapping

Embrittlement of steel by hydrogen has historically been associated with the view that hydrogen could be trapped in voids in highly strained regions around crack tips, at dislocations, in a variety of places where its effect might be deleterious to mechanical soundness of the material [8.7]. Later on, such measurements of solubility at high pressure and diffusion anomalies at low temperature permitted more detailed description of trapping in steels. Consequently, several groups examined the trapping from a theoretical point of view. An important step was made by *McNabb* and *Foster*, who described solutions of the diffusion equation for various conditions of trapping [8.8]. Their equation for the diffusion coefficient D_{eff} in terms of the intrinsic diffusion coefficient D in material containing no traps is the following:

$$D_{\mathrm{eff}} = D(1 + Nk/p)^{-1} . \tag{8.2}$$

Here p and k are release and trapping parameters, respectively. N is the trap density.

The general formulation of *McNabb* and *Foster* was applied to diffusion and trapping of hydrogen in steel by *Oriani* under certain conditions of local equilibrium [8.9]. He, furthermore, evaluated the relevant trapping parameters from previous experiments. He concluded that traps at interfaces were more important for trapping of hydrogen in steel which had not been cold worked than were the traps at dislocations. His expression for the effective diffusion coefficient D_{eff} is given by

$$D_{\mathrm{eff}} = Dc_{\mathrm{L}}[c_{\mathrm{L}} + c_{\mathrm{x}}(1 - \theta_{\mathrm{x}})]^{-1} . \tag{8.3}$$

In this equation c_{L} is the lattice concentration of hydrogen, c_{x} is the concentration of hydrogen in trapped sites, and θ is the fraction of trapping sites which are occupied.

The analysis of *Oriani* was based on constancy of the activation energy for diffusion right up to the trapping site. Modification of this treatment was made by *Koiwa*, who considered the expression for the effective diffusion coefficient in terms of a model which permitted attention to the effect of a change in activation energy for diffusion near the traps [8.10]. His expression for the apparent activation energy is really the same as those of *Oriani* and *McNabb* and *Foster*, and reduces to that equation when the change in saddle point energy is made equal to 0. *Koiwa* also examined differences in trapping and escape parameters when tetrahedral or octahedral site occupancy is assumed in the bcc lattice. He did not, however, apply his expressions to real cases, so one cannot decide whether a variation in saddle point energy as one approaches the traps is an important concept, practically.

8.4 Accumulation in Voids

The clustering of vacancies in metals to form voids has been observed for many materials—platinum, aluminum, gold, etc. The possibility exists, of course, that simultaneous coalescence of the vacancies and arrival of hydrogen atoms from solution might produce voids containing hydrogen gas. This was demonstrated rather conclusively by *Clarebrough* et al. for copper, silver, and gold [8.11]. They showed that quenching of these metals from temperatures near their melting point permitted both vacancies and hydrogen atoms to be retained in solution when the atmosphere from which the metal was quenched contained hydrogen or when hydrogen had been introduced into the material by some other means. Irregular voids a few thousand angstroms in diameter were produced in copper and silver after the metal was held at room temperature for a few minutes. In some instances, dislocation loops were punched out in the region surrounding the void. Gold also produced voids, but the size was a little smaller because the hydrogen solubility in gold at the melting point is somewhat less than for copper and silver at their melting points. In addition to the irregular voids, stacking-fault tetrahedra are produced as well.

In a later study *Johnston* et al. produced stacking-fault tetrahedra by quenching gold from near 950 to 0 °C following aging at 500 °C in various atmospheres [8.12]. They used atmospheres of CO, oxygen, and hydrogen. Electron microscopy showed many small tetrahedra whose densities were around 10^{14} per cm^3. Those specimens which had been held in a hydrogen atmosphere produced nearly a twentyfold increase in defect density after aging for a few minutes at room temperature. This was attributed to hydrogen causing increased combination of divacancies into larger clusters. They postulated that a single hydrogen ion could enter a divacancy, that the combination would be mobile, and that agglomeration of these divacancies could produce the tetrahedra they observed. To eliminate the possibility that oxides of other metals in the gold were reduced by the hydrogen to produce molecules of H_2O, they etched some samples in aqua regia at 55 °C after quenching. Again, an increase in number of tetrahedra was observed. If their analysis is right, then vacancies and hydrogen ions can coalesce simultaneously to form voids of microscopic dimension.

The precipitation of hydrogen in copper in voids was studied somewhat later in more detail by *Wampler* et al. [8.13]. They monitored the hydrogen concentration more carefully than had been done by the earlier investigators, and they carried out a more systematic examination of bubble formation. After aging quenched copper containing 300 atomic ppm of hydrogen for several hours at room temperature, they examined the material with transmission electron microscopy. The bubbles were again irregular (roughly spherical) and fairly uniform in diameter, around 3000 Å. They always had a dense array of dislocations punched out in the region around the bubble. The bubble density was about $10^{10}\,cm^{-3}$. They could calculate, then, the pressure within the bubbles to be about 1000–2000 atm.

Bubbles of this size should behave almost as though they possess a free, normal surface, and the hydrogen should be resorbed by the metal as the temperature is raised. At least this should be so if the inside of the bubble is not covered by an oxide film. None of the investigators reported such measurements, but one would think that the "binding energy" of hydrogen to these voids should be the same as the relative partial molar enthalpy of solution at a free surface. Re-solution of the bubbles might then occur as the temperature was again increased. It need not though—at least under all circumstances. Thermal cycling of aluminum was shown by *Baranov* and *Goncharova* to lead to continuous growth of voids in a kind of ratcheting [8.14].

Voids in bcc metals have not been so easy to observe. The difficulty of quenching vacancies into bcc metals has precluded the possibility of experiments similar to these. However, at least two groups of investigators described retardation of hydrogen diffusion in bcc metals in terms of trapping in microvoids.

The hypothesis that voids containing hydrogen could be produced by annealing or by deformation in iron and steel was examined by *von Ellerbrock* et al. [8.15]. They observed that solubility of hydrogen in iron and steel shows departure from (8.1) at lower temperatures, in particular when the surrounding gas is at high pressure [see Ref. 8.15, Fig. 1]. They also noted the departure from Arrhenius behavior of diffusion of hydrogen in iron below 150 °C (this had been a fact long known). They proposed that both effects could be explained by assuming that hydrogen could be reversibly trapped in small voids in steels. They analyzed observations on hydrogen outgassing at room temperature previously published by a number of investigators. They were able to adequately fit the data by calculating the rate of detrapping of hydrogen from microvoids. The porosity they assumed was remarkably constant through these various investigations, about 0.1%.

The trapping of hydrogen in microvoids induced by deformation in a more pure material, Armco iron, was deduced by *Kumnick* and *Johnson* [8.16]. From transient effects in permeation of hydrogen through annealed iron, they postulated that these transients were caused by trapping of hydrogen in microvoids which had been introduced during the annealing following a rolling operation by means of which they had produced their thin foils. They observed no transient for foils thinned by chemical reduction prior to annealing. The ratio of effective diffusivity to lattice diffusivity in their experiments was 0.3, a value close to that of 0.7 for *von Ellerbrock* et al.

A number of investigators examined coalescence of hydrogen in metals after bombardment with protons. The physics of slowing down of such particles is such that most particles stop in a thin layer; thereby large supersaturations can be produced. The resultant hydrogen atoms diffuse to form voids which then typically coarsen with further annealing. Such an investigation was shown for proton irradiation of aluminum by *Ells* and *Evans* [8.17].

Trapping of hydrogen in voids should not be confused with the growth of methane-containing bubbles formed in steels at higher temperatures (from 300 to 500 °C). In this later instance, termed "hydrogen attack", the simultaneous presence of carbon and hydrogen causes fissures to grow along grain boundaries, greatly reducing ductility and strength. An excellent review of the features of that attack was given by *Shewmon*, who described the kinetics of the nucleation and growth of such bubbles [8.18]. More recently, *Shih* and *Johnson* showed that nucleation of such bubbles in a 1020 steel at 600 °C occurred at grain boundaries on small particles which they tentatively identified as MnS [8.19]. The bubbles they observed were in the range 0.5 – 6 μm in diameter. Voids and blisters are observed for many alloys besides steel. A review of such observations for alloys of Al, Mg, and Cu was given by *Talbot* [8.20].

The ultimate refinement of hydrogen trapping in voids would be trapping in single vacancies or perhaps small vacancy clusters. Interaction of hydrogen with vacancies has been proposed frequently, but has been difficult to observe. However, such interactions have been reported in copper and gold, where other types of traps can be reduced in number by purification and annealing. A hydrogen-vacancy interaction was measured by recovery of electrical resistivity in copper wires quenched from high temperatures [8.21]. Retardation of mobility of deuterium in gold quenched from high temperature was measured by *Caskey* and *Derrick* [8.22]. They deduced that the trapping sites were probably vacancies. They were not, however, able to estimate any values of trapping energy. Finally, radiation may produce vacancies which will interact with hydrogen in solution (see the paper of *Budin* et al. [8.23]).

The energy of a dissolved hydrogen atom in the vicinity of vacancies in nickel and iron was calculated by *Gol'tzov* and *Podolinskaya* [8.24]. They found that the site of lowest energy is an interstitial site near the vacancy, not the center of the vacancy. They cautioned that their calculation does not take all interactions into account. Their results are in accord, though, with observations on trapping of deuterium in ion-damaged W by *Picraux* and *Vook* [8.25]. Using an ion channeling technique, they found that the dissolved D atoms remain in interstitial sites. Again, though, the interpretation is ambiguous, since the nature of the ion-induced defect is not fully known.

8.5 Accumulation of Hydrogen in High Strain Fields

Hydrogen has been assumed to segregate preferentially in dilatational strain fields. Thus, it has been assumed to segregate in the strain field near the tip of a crack and thereby assist in crack growth. Many hypotheses of this type of segregation have been proposed, but few experiments have succeeded in showing this segregation directly.

The interaction of interstitials with strain fields about dislocations has been addressed by many people. *Cottrell* and *Bilby* described the way in which atmospheres of interstitials might form about edge dislocations [8.26]. Later, *Cochardt* et al. calculated the interaction energies of interstitials with the strain fields about both edge and screw dislocations [8.27]. A recent summary and extension of such work was presented by *Li* [8.28]. The difficulty of making calculations for the large local strains near interstitials and of properly applying boundary conditions is exemplified by a series of exchanges in Scripta Metallurgica in 1969 and 1970. *Hirth* and *Cohen*, *Schoeck*, and *Bacon* exchanged a series of views concerning the interaction of interstitials with dislocations [8.29–34]. Even for an interstitial such as carbon in iron, where the local strains about the carbon atom can be inferred from the strain associated with carbon-martensite, the correctness of calculations is difficult to assure. The tetragonal nature of the strain field about C, N, and O in bcc metals has been well established. However, hydrogen may produce a displacement of lattice atoms in bcc metals of nearly cubic symmetry. Measurements by quasi-elastic scattering of neutrons by Nb containing deuterium have shown the strain field to have cubic symmetry [8.35]. This observation by *Bauer* et al. was corroborated by *Metzger* et al. for H in Nb [8.36]. Using Huang scattering, they also showed cubic symmetry of the displacement field.

The present discussion does not assess these calculations and observations. Rather, the features considered here are binding energy of hydrogen to regions of dilatational strain, rates at which segregation can occur, and the saturability of the effect. The following paragraphs note a few experiments which have quantitative conclusions.

An important demonstration of modification of solubility by an applied stress was presented by *Wriedt* and *Oriani* [8.37]. They showed, in an alloy of 75% Pd and 25% Ag, that elastic and compressive stresses much below the elastic limit of the solid could appreciably modify the hydrogen concentration in solid solution. Elastic stresses in the range 25,000 psi produced changes in the equilibrium hydrogen concentration of about 1%. Their data for a particular specimen stressed at 75 °C are shown in Fig. 8.2. First, one should note the magnitude of the effect. Secondly, one should note that the effect is linear and reversible. Thus, the phenomenon they observed is apparently not due to defects produced by the stress.

Wriedt and *Oriani* developed a simple expression to show that the change in concentration under stress may be related to a single material parameter, the partial molar volume of solution, $\bar{V}_i$. Their expression for the solubility c_i in the

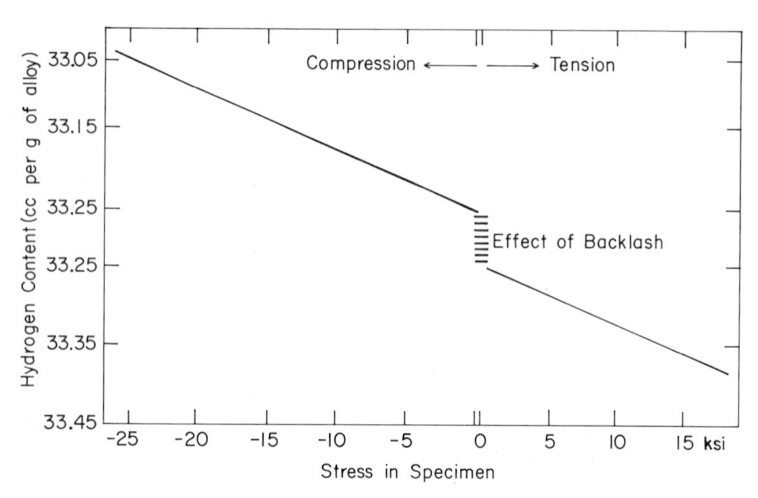

Fig. 8.2. Effect of stress on solubility of hydrogen in an alloy of Pd–Ag (measurements from [8.37])

presence of a stress σ compared to the solubility at zero stress, c_i^0, is

$$\ln(c_i/c_i^0)_f = \sigma \bar{V}_i / 3RT. \tag{8.4}$$

One of the notable features of this experiment is the relatively large change in hydrogen content which comes from a small elastic strain.

A second experiment worth noting was also carried out in palladium; it showed the effect of lattice defects (presumably dislocations) on hydrogen solubility in palladium, also near room temperature. In this work, *Flanagan* et al. showed that the equilibrium hydrogen solubility of palladium at room temperature could be increased appreciably by lattice defects caused by plastic deformation [8.38]. The effect for typical absorption isotherm is shown in Fig. 8.3. There the hydrogen in solution is plotted as a function of the square root of pressure of a surrounding hydrogen gas. The upper curve for well-annealed palladium showed excellent agreement with earlier measurements. The lower curve shows higher solubility at all pressures for a specimen which had been cold rolled 78%. The ratio of hydrogen solubility for the deformed material to that of the undeformed is of order 1.23. These authors showed that the lower curve is reversible; the hydrogen may be pumped out and reinserted. Thus, the introduction of hydrogen does not destroy the defects, nor introduce new ones, at least not of this type.

The interpretation given to this effect by *Flanagan* and *Lynch* is the accumulation of excess hydrogen in the dilatational strain field around a dislocation [8.39]. The magnitude of the effect requires that it be a large accumulation; it is not simply the addition of one hydrogen atom per atom plane in the dislocation core. Rather, the enhancement ratio of 1.23 requires that about 100 hydrogen atoms accumulate around a dislocation for each atom

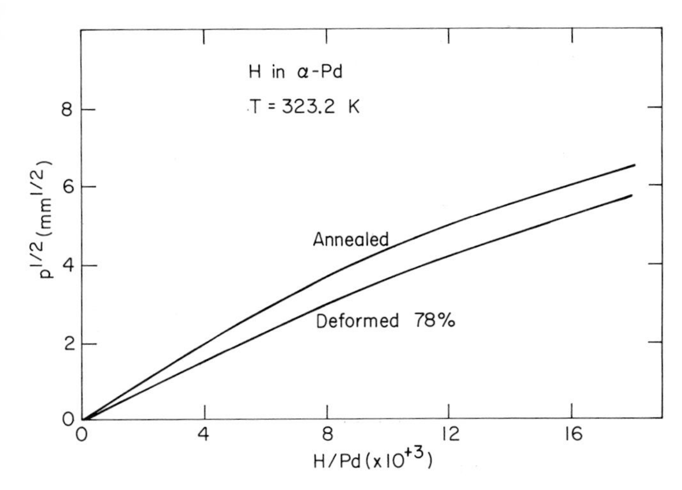

Fig. 8.3. Enhancement of solubility of hydrogen in palladium by mechanical deformation [8.39]

plane, for reasonable dislocation densities produced by the deformation. The average interaction of the interstitial hydrogen with the stress field of the dislocation is about 500 cal/mole, i.e., about 0.02 eV per atom of hydrogen. This work also showed that vacancies produced by the deformation could not be important contributors to traps for hydrogen (see [Ref. 8.39, p. 21]).

The effect described is almost solely that of absorption of hydrogen by dislocations, so the hydrogen solubility can be used as a measure of dislocation content. *Flanagan* and collaborators described this possibility in a later paper [8.40].

Changes in permeability of hydrogen through deformed Armco iron measured by *Kumnick* and *Johnson* were also interpreted in terms of trapping of hydrogen by dislocations in iron [8.16]. However, they apparently found the trapping efficiency of a dislocation to be about one or two atoms of hydrogen per atom plane of dislocation length, more like the usual Cottrell atmosphere. Is it possible that differences in crystal type account for the differences between Pd and Fe? If so, an additional fact needs explanation: How can it be that an average binding enthalpy of only 500 cal/mole produces such extensive trapping (some 100 atoms per Burgers vector) for Pd whereas a binding enthalpy more than ten times as large for Fe produces only about one or two atoms trapped per Burgers vector?

8.6 Grain Boundary Traps

Grain boundaries in metals have long been known to be sinks for impurities. For substitional impurities, an enhanced diffusion rate both of the effect of trapping and the higher mobility in the grain boundary may result. For

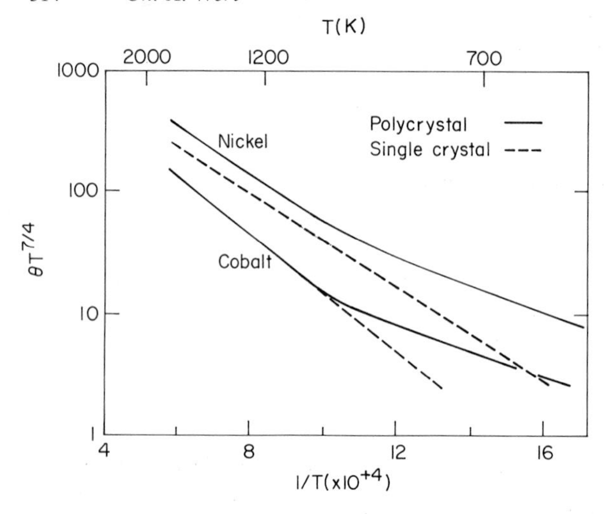

Fig. 8.4. Enhancement of solubility at low temperature in polycrystalline Ni and Co [8.41]

interstitials, the extent of trapping of carbon, nitrogen, and oxygen is not so certain, and the same is true of hydrogen. To be sure, methane-containing bubbles in grain boundaries in steels and water-vapor-containing bubbles in copper have long been known. But the trapping of individual hydrogen atoms in grain boundaries has not been so well established. Indeed, as the following examples show, the situation is not clear.

The solubility of hydrogen in polycrystalline nickel and cobalt measured by *Stafford* and *McLellan* shows the normal linear behavior at high temperatures [8.41] (see Fig. 8.4). Here the solubility θ is plotted according to the equation

$$\theta T^{7/4} = p^{1/2} \lambda \exp[-(H_u - E_D^0/2)/kT] \exp(\bar{S}_u^{xs}/k). \tag{8.5}$$

In this expression, p is the pressure of the surrounding gas, H_u is the partial molar enthalpy of solution, E_D^0 is the dissociation energy of the H_2 molecule at 0 K, $\bar{S}_u^{xs}$ is the excess partial molar entropy of solution, and λ is a constant. At about 1000 K the high temperature linearity breaks down, and the line slopes off less steeply to lower temperatures. Single crystals of nickel showed good linearity of $\theta T^{7/4}$ with l/T over the entire range. Clearly the grain boundaries in nickel produce a large excess solubility at low temperature. The same may be true for cobalt, but it is not possible to make a single crystal of β-cobalt and test the single-crystal linearity.

Similar data for iron do not show the difference between single-crystal and polycrystalline iron, even though some curvature exists over the whole temperature range (see Fig. 8.5 taken from *Da Silva* et al. [8.42]). Thus, the grain boundaries in iron do not produce excess solubility as they do for nickel. *Da Silva* et al. interpreted the curvature of the solubility plot with joint occupancy of the tetrahedral and octahedral sites in the bcc iron. They calculated an energy difference between tetrahedral and octahedral sites of

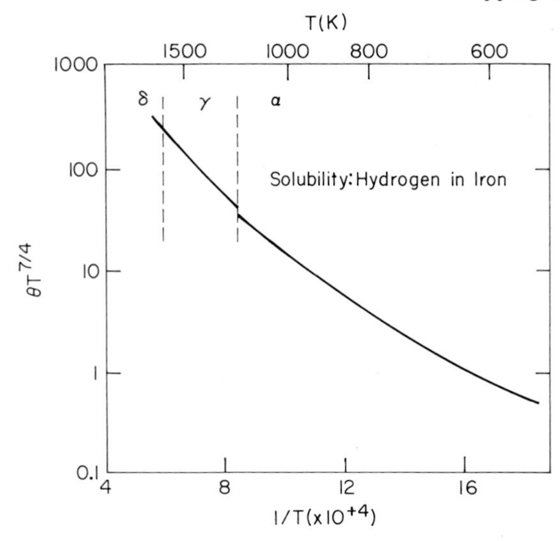

Fig. 8.5. Solubility of hydrogen in iron [8.42]

about 7000 cal/mole, the tetrahedral sites being more highly populated at low temperatures.

Several other features of the data for nickel should be described. *Stafford* and *McLellan* assumed that the solubility in polycrystalline nickel is independent of the actual grain size, provided that it is small. Assuming octahedral interstitial site occupancy in nickel, these authors calculated that about 25% of the sites in the solid are low energy, i.e., trapping sites. They also calculate that the energy of trapping is about 4.3 kcal/mole. If these calculations are true, then the number of actual grain boundary sites is nowhere near plentiful enough. Perhaps the polycrystalline material has local strains large enough to trap hydrogen in the elastic strain fields. Yet, if that is so, another difficulty of interpretation exists, for the partial molar volumes of solution of hydrogen in nickel and iron are about the same. The expression of *Wriedt* and *Oriani*, (8.4), would, therefore, predict excess trapping in iron, also. Figure 8.5 shows that does not occur. Altogether, something of a dilemma.

Hydrogen may be absorbed on the interfaces between precipitates and the metal lattice. *Podgurski* and *Oriani* showed that hydrogen is absorbed on the surfaces AlN particles in iron-aluminum alloys [8.43]. The interface can reversibly absorb hydrogen at room temperature, but prior absorption of nitrogen at the interface at somewhat higher temperatures (500 °C) appears to block surface sites from subsequent hydrogen absorption. Other examples of hydrogen absorption at interfaces between iron and metallic carbides in steels are described at length in the review paper of *Bernstein* and *Thompson* [8.44]. Trapping by carbides formed at low temperatures in steels was claimed by *Mindyuk* and *Svist* to be responsible for reduction in the rate of permeation of hydrogen [8.45]. This explanation cannot be solely responsible for the often-measured reduction in permeability and diffusion coefficient in irons and steels at low temperatures, though.

8.7 Dislocation Trapping

Dislocation trapping of interstitials has been examined both theoretically and experimentally for decades. Some of the first successful correlations of theory and experiment were made for carbon and nitrogen in iron. The papers of *Cottrell* and colleagues were followed by a rash of studies on these interstitials in iron. Many refinements of the earlier calculations were made, but the biggest gains in recent years have been made in experimental determinations. Calculation of the quantity of interstitial which can be absorbed in a material containing dislocations has been treated by *Li* and *Chou* [8.46] in an appendix to a paper of *Podgurski*, *Oriani*, and *Davis*. The calculation of *Li* and *Chou* is general for any interstitial, although they applied the expressions to nitrogen in iron.

Dislocation-interstitial interactions may be examined in many ways, and a brief review does not permit detailed examination of many of the plethora of papers which exist. Consequently, only a few examples are discussed to show the nature of the effects. To provide as complete a picture as possible for a specific material, most of the following discussion is related to iron.

A standard method of detecting trapping of hydrogen by dislocations in iron is observation of the absorption and evolution of hydrogen from a gaseous medium. The study of *Kumnick* and *Johnson* was noted earlier. To eliminate certain surface problems, they coated the exit surfaces of their specimens with a thin palladium plate. They estimated that about one atom of hydrogen was trapped per atom plane of dislocation length. They were not, however, able to deduce the trapping energy from their measurements. Transients in hydrogen permeability were assumed, in fact, to be caused by two types of traps; one deep trap—microvoids—which were considered to be similar to the voids postulated by *von Ellerbrock* et al. [8.15]. The second type of trap was assumed to be dislocations, which they deduced from annealing studies were probably not isolated, but were in the form of cell walls.

One should note that steady-state permeation of hydrogen through materials is unaffected by the presence of traps, providing that the traps are filled to a fixed level. Permeation of hydrogen in alpha-iron was reviewed by *Gonzales*, whose collection of data shows remarkable consistency of values of permeation rate from different investigators even though iron must have had widely varying density of traps of various kinds [8.47].

Transport of trapped hydrogen by moving dislocations can easily be observed. Many studies have shown the growth of microvoids and grain boundary cracks during plastic deformation; it is usually assumed that the moving dislocations transport hydrogen from a surrounding gas to these voids and cracks. A recent study of hydrogen transport by moving dislocations was reported by *Tien* et al. [8.48]. They developed a model for such transport which yields a rate appreciably higher than that for lattice diffusion of hydrogen. They described the deposition of hydrogen at inclusions and voids within an alloy of copper-silicon. They showed, by electron microscopy, that such voids can nucleate on particles of silica in an oxidized copper-silicon alloy. They further described both the rate of transport of hydrogen by moving dislocations and the subsequent leakage of hydrogen from voids in which it has been deposited. A similar calculation for hydrogen transport was made by *Johnson* and *Hirth*, who attempted also to relate the kinetic supersaturation to known technological facts [8.49]. Although expressions for rate of transport and for maximum supersaturation were goals of both of the papers, it is difficult to compare the results of the two calculations. One important difference appears to be the following: *Johnson* and *Hirth* found small supersaturation in all cases they studied; *Tien* et al. found large supersaturations—and large pressure increases—in voids which nucleate on the surface of precipitates.

Accelerated evolution of hydrogen from metals during plastic flow has also been examined. An excellent recent study is that of *Donovan* of evolution of tritium from iron, nickel, and several alloys [8.50]. Tritium was used as a radioactive tracer to permit use of low release rates. By noting release rate as a function of strain, *Donovan* showed that excellent correlations could be made with yield phenomena, uniform flow, fracture, serrated stress-strain regions, etc. Furthermore, measurements made as a function of temperature permitted activation energies to be calculated for tritium diffusion. Good correlation of the measured diffusion energies was obtained for iron and nickel, the only materials for which *Donovan* had good data. Again, though, these measurements did not permit an independent evaluation of the binding energy. Neither did they give evidence for high supersaturation of hydrogen in these metals.

Perhaps the most fruitful technique for measuring many features of hydrogen in metals has been internal friction. An excellent review of the several phenomena associated with internal friction due to hydrogen was given by *Schiller* [8.51]. A related technique, magnetic disaccommodation, was described by *Kronmüller* et al. [8.52 and Ref. 8.53, Chap. 11]. The work noted in this section is concerned only with interaction of hydrogen with dislocations.

Several internal friction maxima are associated with presence of hydrogen in iron. *Gibala* examined some of these peaks in some detail and gave an explanation for them [8.54]. A typical curve of *Gibala* is shown in Fig. 8.6. Here, for a frequency of 80 kHz, two damping peaks are observed. The lower temperature peak, shown in the inset extracted from a broad peak at even lower temperature, was described by *Gibala* as the Snoek peak. The peak at higher temperatures is termed the cold-work peak; it is observed only in deformed iron. One of the reasons *Gibala* was led to correlate the lower temperature peak with the Snoek relaxation was the excellent fit it gave to an extrapolation of the high temperature diffusion data of hydrogen in iron (see Fig. 11 of the review by *Birnbaum* and *Wert* as well as *Gibala's* Fig. 14 [8.55]). Although *Gibala* may be correct, the chances are that this is not a simple hydrogen Snoek peak; if it is caused by motion of hydrogen atoms, they may be trapped impurities at this low temperature. Similar peaks have been observed in many other metals, niobium and tantalum in particular.

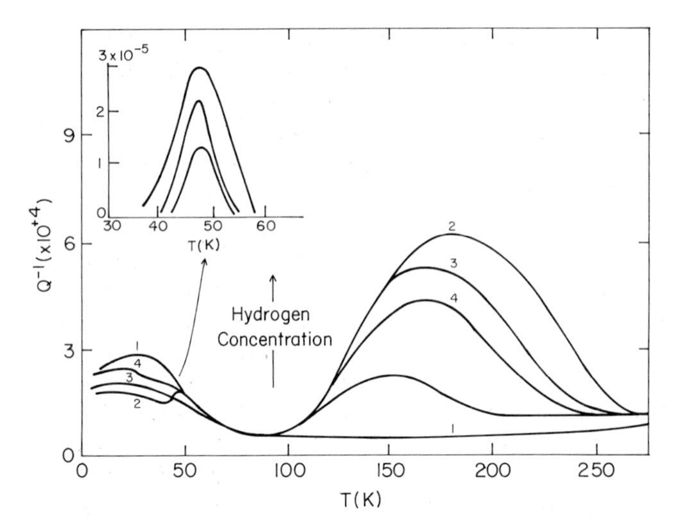

Fig. 8.6. Internal friction of iron containing varying amounts of hydrogen. $f = 80$ kHz. The sequence of changes in heights of the several peaks is indicated by the numbers on the curves. Three peaks were presumed by *Gibala* to be present. The cold-work peak near 200 K requires the presence of both hydrogen and dislocations. The broad peak near 30 K is presumably an intrinsic dislocation peak—hydrogen additions reduce its height. The peak at 50 K, extracted in the inset, was described by *Gibala* as the Snoek peak. It is probably not; *Gibala* now believes that it is a defect cluster peak—perhaps C–H, N–H, or O–H

Gibala observed that the cold-work peak and his "Snoek peak" showed a partitioning of hydrogen; their heights had a reciprocal relationship to each other. He was able to deduce a binding energy of hydrogen to dislocations in iron of 6400 cal/mole.

An excellent correlation of the height of the cold-work peak with the quantity of hydrogen involved was made by *Sturges* and *Miodownik* [8.56]. They made many measurements of the height of the cold-work peak and correlated that height with the amount of hydrogen released from the specimen during outgassing. They found a linear of variation of peak height and hydrogen effused from the specimen. They also measured an activation energy of the cold-work peak and compared that with the activation energy for diffusion of hydrogen in deformed iron. From their data they calculated a binding enthalpy of hydrogen to the dislocation trap of about 5000 cal/mole. In doing this, they used a value for diffusion of hydrogen in the perfect iron lattice of 3000 cal/mole. Had they used the presently accepted value around 2000 cal/mole, they would have obtained a value of about 6000 cal/mole for the binding enthalpy, a number very close to that of *Gibala*.

This entire problem was reexamined by *Takita* and *Sakamoto* [8.57]. They again studied the 50 K peak and the hydrogen cold-work peak, as *Gibala* had done. These investigators used a frequency of 70 kHz and showed that the two peaks go up and down oppositely, as *Gibala* had shown. They attempted to show that the bowing of dislocations provides the anelastic strain for both of these peaks. Thus, in their view, the low temperature peak would not be caused by hydrogen diffusion around an impurity, but would be caused by a dislocation loss mechanism. However, their peaks (or peak) around 50 K are much larger than those of *Gibala*, so their conclusions may not be applicable to his measurements.

A complex dilemma can be seen from the values of binding energy just quoted. The relative partial molar heat of solution of hydrogen in iron is about 6000 cal/mole. The binding energy of hydrogen to dislocation traps is also about 6000 cal/mole. Thus, in any experiment involving effusion of hydrogen from a solid, differentiation between dislocation traps and microvoids will be difficult; if the surface of a microvoid is clean, it would act like a free surface.

The correlation between presence of hydrogen at dislocations and embrittlement, as measured by the notch tensile strength, was determined by *Kikuta* et al. [8.58]. They found the same development of the hydrogen cold-work peak with aging time as has been seen by others, but they also saw a decrease in the maximum height of the peak as aging time was continued. They observed, also, that the notch tensile strength decreased as the cold work developed, then increased again as the cold-work peak aged away. They did not, though, develop a detailed model. Embrittlement of iron and steel was described in the review by *Bernstein* [8.7]. In particular, he discussed the concentration of hydrogen in the highly strained region around the tip of a crack.

The elusive Snoek peak of H in bcc metals may never be seen. First, the low temperature at which it would be found at any reasonable frequency may be so low that nearly all dissolved hydrogen might be trapped or clustered in pairs. Furthermore, if the displacement field around the dissolved H atoms has nearly cubic symmetry, as the observations seem to show, [8.34, 35], then it ought not to exist in principle.

8.8 Impurity Trapping

The previous sections have all been concerned with trapping of hydrogen on a large scale—by voids, dislocations, boundaries, strained regions. Because the geometry cannot be specified completely, and because hydrogen can be trapped in a variety of sites with different binding energies, it has been difficult to specify the trapping in quantitative terms. However, impurity trapping, to be discussed in this section, has simpler and more easily identifiable geometry, so the thermodynamics and kinetics of hydrogen bound to impurities are much easier to specify.

The concepts of trapping of hydrogen by impurities go back to the first studies of *Powers* and *Doyle* on clustering of oxygen in tantalum [8.59]. They showed that O–O pairs existed in Ta along with single O atoms, the two being in equilibrium if observations were made at high enough temperatures. The trapping of oxygen, nitrogen, and carbon by substitutional atoms was examined by a number of investigators as well [8.60–66]. The same principles have been carried over to the solutions of hydrogen in metals. Although measurements for a number of metals have been reported, the most complete study has been made for solutions of hydrogen in niobium; therefore, these are described in some detail.

Thermodynamic equilibrium of hydrogen which exists as single atoms, H–H clusters, and cluster of hydrogen with an immobile impurity, say oxygen, can be written as follows:

$$C_{\mathrm{H}}(\mathrm{total}) = C_{\mathrm{H}}(\mathrm{singles}) + 2C(\mathrm{pairs}) + C(\mathrm{O\text{–}H\ clusters}). \quad (8.6)$$

The thermodynamic equilibrium is expressed in detail according to the following equation:

$$C_{\mathrm{H}}^{\mathrm{T}} = C_{\mathrm{H}} + 2A_2 C_{\mathrm{H}}^2 \exp(B_{\mathrm{HH}}/RT) + A_3 C_{\mathrm{O}} C_{\mathrm{H}} \exp(B_{\mathrm{OH}}/RT). \quad (8.7)$$

The reader may find more general equations for clusters of higher order in the literature. Since the examples shown here are concerned only with simple pairwise clusters, the thermodynamic expression is written in this brief form. In this expression C_{H} is the concentration of single hydrogen atoms, C_{O} of single oxygen atoms. B_{HH} and B_{OH} are binding energies of H–H and O–H clusters; A_2 and A_3 are geometrical constants.

The extent of partitioning of hydrogen among these three possible configurations might be studied in a variety of ways, but the most satisfying and easily interpreted is the internal friction of these alloys. For metals for which each of the three configurations has a strain field of lower symmetry than the crystal itself—cubic, of course, for niobium—well-defined anelastic peaks exist. The position of these peaks depends on the frequency of observation and the mean-jump time of the configuration, and the amplitudes depend on the dipole moments of the configurations and their concentration.

The first detailed study of hydrogen trapping in the simple configurations was made by *Baker* and *Birnbaum* for hydrogen in niobium containing an oxygen or nitrogen impurity. At that time, they were not certain which of these gases was responsible for the effect they observed. Later studies showed it to be oxygen [8.67]. They saw two damping peaks at low temperature for frequencies around 40 MHz. Some of their observations are shown in Fig. 8.7. Peak B is caused by reorientation of the O–H pair and peak A, the H–H pair. By changing frequency the authors were able to deduce that the reorientation energy of the O–H peak is 0.13 eV, of the H–H peak, 0.05 eV. By changing frequency and observing the variation of peak height with temperature, they were able to deduce a binding energy of the O–H pair of 0.09 eV and for the H–H pair of 0.06 eV. At a later time, strain relaxation methods were employed by

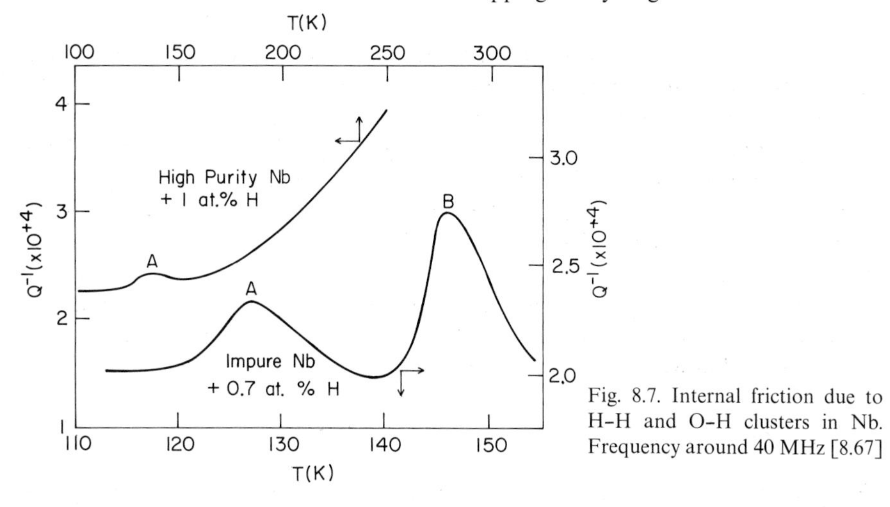

Fig. 8.7. Internal friction due to H–H and O–H clusters in Nb. Frequency around 40 MHz [8.67]

Table 8.1. Thermodynamic parameters of clusters

Metal	Complex	Binding enthalpy [eV]	Motion enthalpy [eV]
V	O–D		0.17
Nb	N–H	0.1	
	O–H	0.09	0.14
	O–D	0.13	0.19
	H–H	0.06	0.05
Ta	N–H	0.06	
	O–H		0.12
	O–D		0.12

Chen and *Birnbaum* to extend these measurements to much lower temperatures [8.68]. They were thereby able to test theories of diffusion by tunneling in a way which had not been possible before.

Measurements of thermodynamic and kinetic parameters have been made for V, Nb, and Ta; values are listed in Table 8.1. These data were assembled from [8.51, 67, 69].

The next important factor is the detailed geometry of the cluster. This can be measured by determining the magnitude of the internal friction peak as a function of crystal orientation for specimens of the same chemical composition of all constituents. This is a difficult measurement, but two determinations have been made for the O–H pair in niobium.

Measurement of the orientation dependence of the internal friction peak was reported by *Schiller* and *Nijman* [8.70]. Their results are sketched in Fig. 8.8. The low temperature peak at 70 K, the O–H peak, shows a much higher

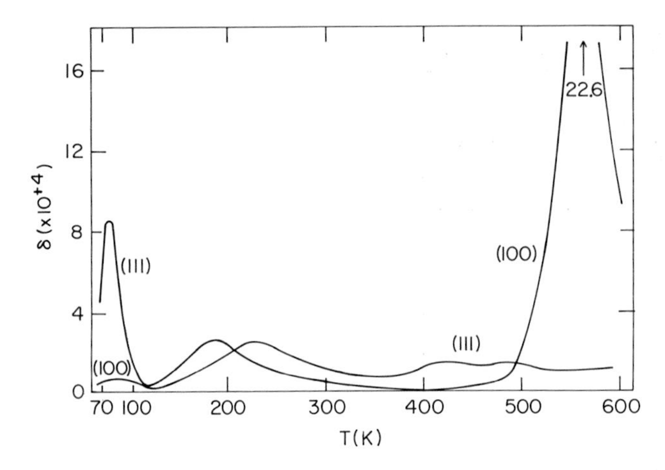

Fig. 8.8. Internal friction in single crystals of Nb showing orientation dependence of O–H cluster [8.70]

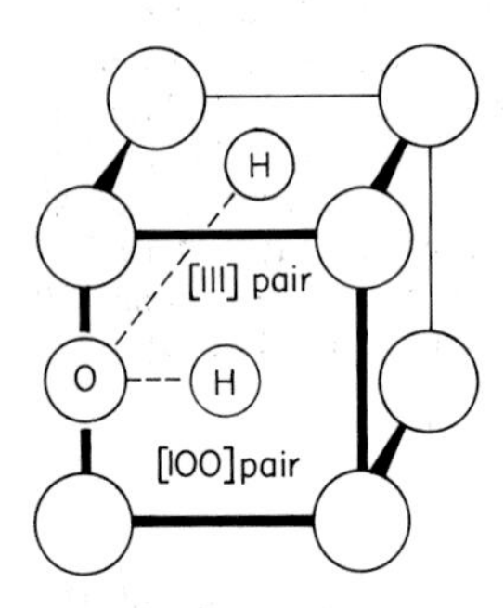

Fig. 8.9. Geometry of O–H pairs in [100] and [111] directions. Octahedral site occupancy of both interstitials is assumed

amplitude for a (111) crystal than for a (100) crystal. Although one cannot know whether the (111) data show a true maximum, one would conclude that the O–H pair may have trigonal symmetry, i.e., lie along the [111] direction. The "maximum" for the oxygen peak in (100) at 550 K is consistent with octahedral site occupancy of oxygen atoms in Nb. A trigonal configuration of the pair for nearest neighbor site occupancy is shown in Fig. 8.9.

An alternate view was expressed by *Mattas* and *Birnbaum*, who proposed two pair configurations [100] and [111] (Fig. 8.9 [8.71]). Later work by *Zapp* and *Birnbaum* [8.72] on crystals of three orientations, (100), (111), and (110), shows a null in *no* direction, contrary to the data reported by *Schiller* and *Nijman*. Thus, both agree that a [111] pair may be possible, but they do not agree on existence of a [100] pair. The reader should note that these pairs have been sketched as though the H atom occupied octahedral positions. Occupancy of tetrahedral sites poses no problem in concept for the [100] pair, but a true [111] trigonal pair is not possible for tetrahedral site occupancy of the H atom.

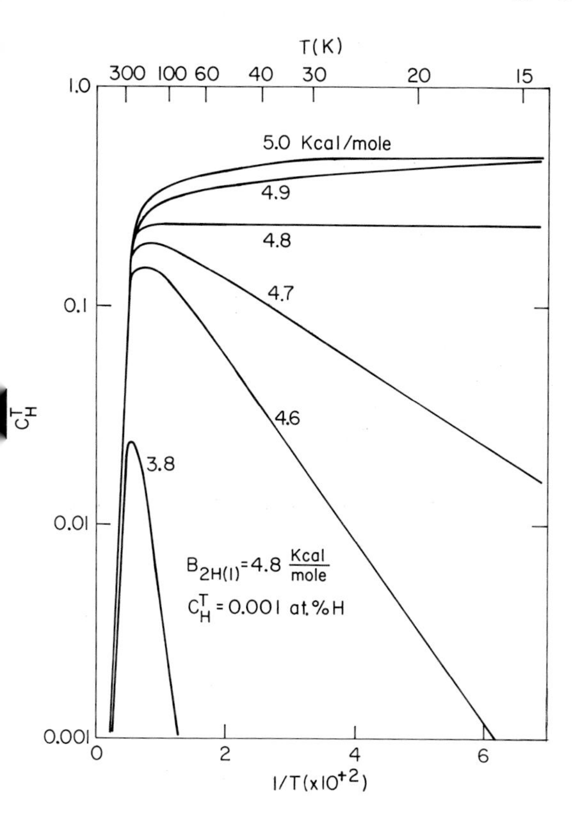

Fig. 8.10. Concentration of H–H clusters of one type in equilibrium with H–H clusters of another type. The binding energy of type 1 is held at 4.8 kcal/mole; that of type 2 is assumed to vary from 5.0 kcal/mole to 3.8 [8.73]

One possible reconciling feature of these measurements is found in the differences in frequency of measurement, ~2000 Hz and 50 MHz; the O–H peak thus was found near 75 K by *Schiller* and *Nijman* and near 250 K by *Zapp* and *Birnbaum.* Clusters with different binding energies, as would almost certainly be true for the two orientations, would have different relative populations at the two temperatures. Such an effect was calculated for H–H pairs of two possible configurations and binding energies by *Au* [8.73]. One of his plots is reproduced as Fig. 8.10. Pairs of type 1 are assumed to have a binding energy constant at 4800 cal/mole, pairs of type 2 of variable binding energy, 5000 cal/mole or less. For a total concentration of hydrogen of C_H^T, the concentration of type 2 pairs is plotted for various binding energies. One sees that, for these conditions, a difference in binding energy of the pairs produces differences in concentration of a factor of 10 for the most favorable temperature and of factors of many orders of magnitude as the temperature decreases.

Thus, it is not possible to conclude that the observations of *Schiller* and *Nijman* and those of *Zapp* and *Birnbaum* are at variance. If the concentrations of oxygen and hydrogen differ appropriately, the difference in temperature of measurement may account for the alternate conclusions.

Most internal friction measurements have been assumed to involve no mutual interaction of adjacent clusters of hydrogen. When the hydrogen concentrations become so large that large interactions exist between hydrogen atoms, then additional effects—blocking or critical point phenomena—exist and the entire system must be considered thermodynamically. For the discussion of such effects, see the paper of *Völkl* and *Alefeld* [8.74].

8.8.1 The Gorsky Effect

The anelastic measurements just described for O–H and H–H pairs do not provide information about effects of trapping on long-range diffusion of hydrogen, since the anelastic effect is the local tumbling of hydrogen atoms about a fixed impurity. Tumbling of the H–H pair does produce long-range diffusion, of course. An anelastic method does exist, however, for measurement of long-range diffusion in the presence of fixed traps, the Gorsky effect. That technique measures the energy loss of a specimen in flecture as hydrogen diffuses back and forth across the width of a specimen. This effect, a diffusion analogue of thermomechanical damping, permits evaluation of the effective diffusion coefficient. For example, the following expression may be written for a rectangular bar of thickness d vibrating in flecture:

$$D_{\mathrm{eff}}^{\tau}=(d/\pi)^{2}\,. \tag{8.8}$$

For diffusion of single H atoms in a perfect crystal, D_{eff} is the true lattice diffusion coefficient. In the presence of traps, the relaxation time τ will be lengthened and the measured diffusion coefficient will then be reduced. One can also measure the relaxation strength Δ and determine additional material parameters relating to changes in dipole strength resulting from trapping of hydrogen.

Trapping alters the diffusion coefficient as was expressed by *Oriani* in (8.3). Since θ_{x} varies with temperature, and since the trapped concentration c_{x} varies both with temperature and trap density N, the effective activation energy ΔH_{eff} in the presence of traps, varies with trap concentration. Unfortunately, the expression is not a simple one, but the value of ΔH_{eff} should increase as the number of trapping increases (see the paper of *Matusiewicz* et al. [8.75]). Figure 8.11, sketched from their paper, shows that ΔH_{eff} increases linearly with concentration of trapping sites, oxygen atoms in this case.

Thermodynamic and kinetic parameters of complexes formed in ferromagnetic alloys can be studied by the magnetic aftereffects. These measurements were summarized by *Kronmüller* [Ref. 8.53, Chap. 11]. Furthermore, diffusion of hydrogen, both in the free state and in the trapped configurations, can be studied by quasi-elastic neutron scattering. For a discussion of this technique and measurements of trapping, see Chapter 10.

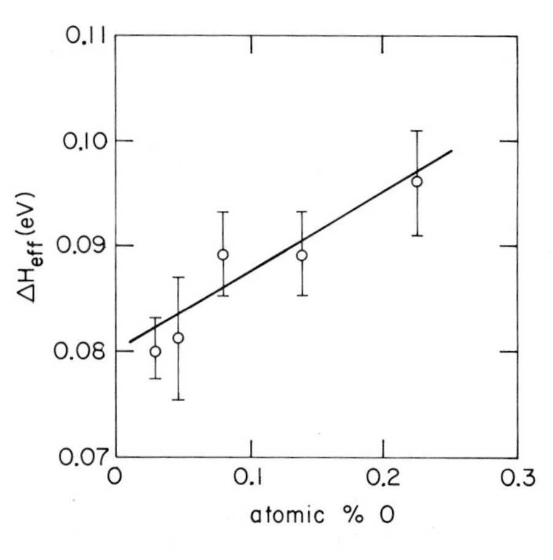

Fig. 8.11. Increase in ΔH_{eff} for hydrogen diffusion as trapping centers are increased in concentration [8.75]

8.8.2 Measurement of Trapping by Study of the Kinetics of Phase Changes

Interstitial impurities have been shown to retard decomposition of a supersaturated solution of hydrogen into its equilibrium constituents. Many papers have been published on this subject, in particular for V, Nb, and Ta [8.76–82]. An enormous retardation is found to exist, much larger than seems warranted by the retardation of mobility, as determined by anelastic measurements. Very likely the retardation is an effect both of trapping directly and of reduction of nucleation rate of the β phase. One essential feature of the studies is a demonstration that the formation of the β phase is retarded by gas atoms which trap hydrogen, but is not prevented, providing that the system is supersaturated. Thus, the binding enthalpies of hydrogen to oxygen and nitrogen impurities in V, Nb, and Ta must be less than the enthalpies of solution of the hydrides.

8.8.3 Trapping by Substitutional Alloying Elements

Trapping of hydrogen by other interstitials occurs because such interstitials alter the local electronic and strain field environment. The same ought to be true for substitutional alloying additions, but few observations of such trapping have been reported. Some of the myriad internal friction phenomena at low temperatures may be caused by such trapping, but conclusive studies are difficult, in spite of the fact that many instances of trapping of *C* and *N* in alloys of iron have been reported.

A conclusive study of trapping of hydrogen by V in Nb was made by *Matsumoto* [8.82]. Using the NMR line of ^{51}V, ^{93}Nb, and ^{1}H, he identified and determined properties of a close-pair H–V complex between interstitial hydrogen and substitutional vanadium. He assumed that the trapped hydrogen was located in the tetrahedral sites about the vanadium. The binding energy is about 0.09 ± 0.05 eV, not much different from that of the H–i pairs reported in Table 8.1. Line narrowing of the ^{1}H line occurs around 150 K, indicating low energy motion of the hydrogen among the 24 equivalent tetrahedral sites.

8.8.4 Low Temperature Resistivity Measurements

Measurement of resistivity changes as a function of temperature has long been used to deduce changes in defect states in materials. A report of such changes for H quenched into V, Nb, and Ta of varying purity was made by *Hanada* [8.83]. Recovery of quenched-in resistivity occurs in the range 40 to 100 K. Isochronal measurement showed binding energies and motion energies in the range 0.01 to 0.1 eV, on the lower edge of the values given in Table 8.1. *Hanada* believes that the resistivity recovery is related to interstitial impurity-hydrogen complexes, but was unable to specify their properties other than the measured motion and binding energies.

8.9 Interstitial Location by Nuclear Microanalysis

Precise location of light interstitials in metals has been made utilizing nuclear microanalysis combined with ion channeling. This technique has been demonstrated to have extensive application to He and H in metals and to impurity elements in semiconductors [8.84, 85]. Although it has been applied mainly to solid solutions produced by ion implantation of the element under study, it is amenable to analysis of specimens prepared by usual thermal or electrochemical methods.

The location of hydrogen in metals is made by measurement of reaction products which result from a nuclear reaction between the implanted hydrogen (or deuterium or tritium) and an appropriate incident ion which is injected into the material. For example, the presence of solutions of deuterium in metals can be measured by utilizing an incident beam of ^{3}He ions, which produce protons (and alpha particles) by the reaction D(^{3}He, p)^{4}He; the high energy protons (or alphas) can be counted by an appropriate detector. Other reactions of H, D, or T with appropriate incident ions yield α particles, x-rays, neutrons (see, for example, [Ref. 8.86, p. 813]. The simplest application of this technique is for depth profiling, i.e., atomic (or isotopic) composition as a function of distance from the surface. This capability results from the fact that the reaction cross section can have a sharp energy resonance. Thus, injection of an incident ion with an energy higher than that of the resonance permits detection of an implanted ion at a depth from the surface at which the injected ion has slowed to the critical energy. Even for reactions with constant cross section, depth profiling can be achieved from kinetics and the known stopping power of the reaction products. For examples of such depth profiling, see the paper of *Picraux* et al. [8.87].

Use of the technique of channeling of the incident beam of ions in combination with the nuclear reaction permits another significant measurement to be made. If the incident beam is highly collimated, the yield around directions of intense channeling permits rather precise location of the interstitial hydrogen to be determined. Examination of yield curves as a function of angle near [100] directions for D in Cr and W was performed by *Picraux* and *Vook* [8.85, 88]. They found that D ions occupy octahedral sites in Cr and tetrahedral sites in W at room temperature. Furthermore, the precision of the site assignment is rather exact: for example, D occupies the tetrahedral site in Mo when it is implanted at 90 K, but it moves to a position near (but not exactly identical to) the octahedral site when the Mo is heated to room temperature [8.89]. It is this precision of determination of interstitial position which makes the ion-channeling-nuclear-reaction technique so promising; if the formation of clusters causes change in interstitial position of more than a few tenths of an Angstrom from its regular interstitial location, the change will be observable and the new location can be determined. Thus, the technique offers the possibility of determination of cluster geometry in a way not feasible by other methods.

The method has its weaknesses. Implantation of the interstitials themselves produces defects which were shown by *Picraux* et al. to trap hydrogen [8.90]. This problem might be alleviated by production of specimens by usual thermal methods, although the measurement technique requires careful control of the near surface region, since the measurement depth is only about 1000 Å. Another radiation influence is the near-certainty that the injected ions used for measurement will produce defects. (They are typically of energy 0.2 to several MeV.) This difficulty can only be avoided by using small fluence of the ions used for measurement.

Most measurements of site occupancy reported so far are for the normal untrapped interstitial (see [Ref. 8.89, Table I]). An additional group of investigations was made for He implanted in metals (see, for example, the measurements of helium trapped in clusters in *W* [8.25]).

Two metals for which hydrogen has been found to exist in traps are *W* and Mo. For D in *W*, measurements at small fluence showed D to occupy tetrahedral sites. However, continued injection of ^{3}He ions showed that the D ions moved away from that site, possibly toward the octahedral site. This was presumed to be caused by trapping of D in radiation-induced traps. Finally, a series of measurements on Mo, both annealed and cold worked, showed existence of a deep trap of unknown type [8.87]. Persistence of this trapped state to high temperatures was interpreted to imply a large energy of trapping, perhaps 1.5 eV. A trap of this energy is not consistent with any other described to now, and probably implies a chemical reaction analogous to methane formation in hydrogen attack of steels.

Application of this technique has not yet developed to the point that a systematic study has been made on any system—rather, the studies have shown the promise and the limitations of the technique.

8.10 Summary

1) Hydrogen can be trapped in many kinds of defects in solids. The binding energy of trapping is of the same order as that for other interstitials (C, N, and O) to defects—a few tenths of an eV. However, the high mobility of hydrogen in metals permits equilibrium to be established at much lower temperatures than is true of the other interstitials; thus, the effects of trapping are more pronounced.

2) Hydrogen can be trapped in voids of macroscopic size. For small voids—tetrahedra observed microscopically—the nature of the trapping is less certain. For the smallest possible void—a lattice vacancy—the details of trapping are even less clear.

3) Dislocations trap hydrogen by forming atmospheres. Since the dilatational strain field about a dissolved hydrogen atom may have nearly cubic symmetry in cubic metals, the interaction may mainly be through the dilatational strain field of a dislocation. Thus, edge dislocations appear to have strong binding (of order 0.25 eV) but screw dislocations may have little interaction.

4) Strain field interactions must be very large, both externally produced strains and internal strains of boundaries and dislocations.

5) Enhancement of diffusion along dislocations cores seems uncertain, but moving dislocations transport hydrogen effectively.

6) Trapping of hydrogen by impurities is pronounced, especially by interstitials O, N, and C in bcc metals. Trapping energies are typically a few tenths of an eV. The geometry of the clusters is uncertain, and the relative binding energies of clusters of various geometries are not known.

References

8.1 J.D.Fast: *Interaction of Metals and Gases*, Vol. 1; Thermodynamics and Phase Relations (Academic Press, New York 1965)
8.2 J.D.Fast: *Interaction of Metals and Gases*, Vol. 2: Kinetics and Mechanisms (Academic Press, New York 1971)
8.3 E.Fromm, E.Gebhardt: *Gase und Kohlenstoff in Metallen* (Springer, Berlin, Heidelberg, New York 1976)
8.4 R.B.McLellan, W.A.Oates: Acta Met. **21**, 181 (1973)
8.5 G.Boureau, O.J.Kleppa, P.Dantzer: J. Chem. Phys. **64**, 5247 (1976)
8.6 G.Boureau, O.J.Kleppa: J. Chem. Phys. **65**, 3915 (1976)
8.7 I.M.Bernstein: Mater. Sci. Eng. **6**, 1 (1970)
8.8 A.McNabb, P.K.Foster: Trans. AIME **227**, 619 (1963)
8.9 R.A.Oriani: Acta Met. **18**, 147 (1970)
8.10 M.Koiwa: Acta Met. **22**, 1259 (1974)
8.11 L.M.Clarebrough, P.Humble, M.H.Loretto: Acta Met. **15**, 1007 (1967)
8.12 I.A.Johnston, P.S.Dobson, R.E.Smallman: Proc. Roy. Soc. A **315**, 231 (1970)
8.13 W.R.Wampler, T.Schober, B.Lengeler: Phil. Mag. **34**, 129 (1976)
8.14 A.A.Baranov, T.A.Goncharova: Fiz.-Khim. Mekh. Mat. **10**, 91 (1974)

8.15 H.-G. von Ellerbrock, G. Vibrans, H.-P. Stüwe: Acta Met. **20**, 53 (1972)
8.16 A.J. Kumnick, H.H. Johnson: Met. Trans. **5**, 1199 (1974)
8.17 C.E. Ells, W. Evans: Trans. AIME **227**, 438 (1963)
8.18 P.G. Shewmon: "Hydrogen Attack in Carbon Steel", in *Effect of Hydrogen on Behavior of Materials*, ed. by A.W. Thompson, I.M. Bernstein (The Metallurgical Society of AIME, New York 1976) pp. 59–69
8.19 H.M. Shih, H.H. Johnson: Scripta Met. **11**, 151 (1977)
8.20 D.E.J. Talbot: Int. Met. Rev. **20**, 166 (1975)
8.21 C. Budin, A. Lucasson, P. Lucasson: J. Phys. (Paris) **25**, 751 (1964)
8.22 G.R. Caskey, Jr., R.G. Derrick: Scripta Met. **10**, 377 (1976)
8.23 C. Budin, P. Lucasson, A. Lucasson: J. Phys. (Paris) **26**, 9 (1965)
8.24 V.A. Gol'tzov, T.A. Podolinskaya: Fiz.-Khim. Mekh. Mat. **10**, 607 (1974)
8.25 S.T. Picraux, F.L. Vook: "Lattice Location Studies of ^{2}D and ^{3}He in *W*", in *Applications of Ion Beams to Metals*, ed. by S.T. Picraux, E.P. EerNirse, F.L. Vook (Plenum Press, New York 1974) pp. 407–421
8.26 A.H. Cottrell, B.A. Bilby: Proc. Phys. Soc. A **62**, 49 (1949)
8.27 A.W. Cochardt, G. Schoek, H. Wiedersich: Acta Met. **3**, 533 (1955)
8.28 J.C.M. Li: "The Interaction of Mobile Solutes with Microstructural Defects", in *Physical Chemistry in Metallurgy*, ed. by R.M. Fisher, R.A. Oriani, E.T. Turkdogan (United States Steel Corporation, Pittsburgh 1976) pp. 405–420
8.29 J.P. Hirth, M. Cohen: Scripta Met. **3**, 107 (1969)
8.30 J.P. Hirth, M. Cohen: Scripta Met. **3**, 311 (1969)
8.31 J.P. Hirth, M. Cohen: Scripta Met. **4**, 167 (1970)
8.32 G. Schoeck: Scripta Met. **3**, 239 (1969)
8.33 D.J. Bacon: Scripta Met. **3**, 735 (1969)
8.34 D.J. Bacon: Scripta Met. **4**, 267 (1970)
8.35 G. Bauer, E. Seitz, H. Horner, W. Schmatz: Solid State Commun. **17**, 161 (1975)
8.36 H. Metzger, J. Peisl, J. Wanagel: J. Phys. F **6**, 2195 (1976)
8.37 H.A. Wriedt, R.A. Oriani: Acta Met. **18**, 753 (1970)
8.38 T.B. Flanagan, J.F. Lynch, J.D. Clewley, B. von Turkovich: J. Less-Common Metals **49**, 13 (1976)
8.39 T.B. Flanagan, J.F. Lynch: J. Less-Common Metals **49**, 25 (1976)
8.40 T.B. Flanagan, J.F. Lynch, J.D. Clewley, B. von Turkovich: Scripta Met. **9**, 1063 (1975)
8.41 S.W. Stafford, R.B. McLellan: Acta Met. **22**, 1463 (1974)
8.42 J.R.G. da Silva, S.W. Stafford, R.B. McLellan: J. Less-Common Metals **49**, 407 (1976)
8.43 H.H. Podgurski, R.A. Oriani: Met. Trans. **3**, 2055 (1972)
8.44 I.M. Bernstein, A.W. Thompson: Int. Met. Rev. **21**, 269 (1976)
8.45 A.K. Mindyuk, E.I. Svist: Fiz.-Khim. Mekh. Mat. **9**, 36 (1973)
8.46 J.C.M. Li, Y.T. Chou: Trans. AIME **245**, 1606 (1969)
8.47 O.D. Gonzales: Trans. AIME **245**, 607 (1969)
8.48 J.K. Tien, A.W. Thompson, I.M. Bernstein, R.J. Richards: Met. Trans. **7** A, 821 (1976)
8.49 H.H. Johnson, J.P. Hirth: Met. Trans. **7** A, 1543 (1976)
8.50 J.A. Donovan: Met. Trans. **7** A, 1677 (1975)
8.51 P. Schiller: Nuovo Cimento **33** B, 226 (1976)
8.52 H. Kronmüller, H. Steeb, N. König: Nuovo Cimento **33** B, 205 (1976)
8.53 G. Alefeld, J. Völkl (eds.): *Hydrogen in Metals I. Basic Properties*, Topics in Applied Physics, Vol. 28 (Springer, Berlin, Heidelberg, New York 1978)
8.54 R. Gibala: Trans. AIME **239**, 1574 (1967)
8.55 H.K. Birnbaum, C.A. Wert: Ber. Bunsenges. Physik. Chem. **76**, 206 (1972)
8.56 C.M. Sturges, A.P. Miodownik: Acta Met. **17**, 1197 (1969)
8.57 K. Takita, K. Sakamoto: Scripta Met. **10**, 399 (1976)
8.58 Y. Kikuta, K. Sugimoto, S. Ochiai, K. Iwata: Trans. Iron and Steel Inst. Japan **15**, 87 (1975)
8.59 R.W. Powers, M.V. Doyle: Trans. AIME **215**, 665 (1959)

8.60 A.A.Sauges, R.Gibala: "Crystallographic Orientation Dependence of Interstitial Anelasticity in Ta–Re–N and Ta–Re–O Alloys", in *Internal Friction and Ultrasonic Attenuation in Crystalline Solids*, ed. by D.Lenz, K.Lücke (Springer, Berlin, Heidelberg, New York 1975) pp. 289–297
8.61 H.Hashizume, T.Sugeno: J. Appl. Phys. (Japan) **6**, 567 (1967)
8.62 J.D.Fast, J.L.Meijering: Philips Res. Rep. **8**, 1 (1953)
8.63 L.J.Dijkstra, R.J.Sladek: Trans. AIME **197**, 69 (1953)
8.64 D.Mosher, C.Dollins, C.Wert: Acta Met. **18**, 797 (1970)
8.65 R.Gibala: "Hydrogen-Defect Interactions in Iron-Base Alloys", in *Stress Corrosion Cracking and Hydrogen Embrittlement in Iron-Base Alloys*, ed. by R.W.Staehle (NACE, Houston) in press
8.66 M.Pope, D.M.Jones, K.H.Jack: "Internal Friction Associated with Substitutional-Interstitial Solute-Atom Clusters in Iron-Vanadium-Nitrogen Alloys and Other Ternary Nitrogen Ferrites", in *Internal Friction and Ultrasonic Attenuation in Crystalline Solids*, ed. by D.Lenz, K.Lücke (Springer, Berlin, Heidelberg, New York 1975) pp. 266–275
8.67 C.C.Baker, H.K.Birnbaum: Acta Met. **21**, 865 (1973)
8.68 C.G.Chen, H.K.Birnbaum: Phys. Stat. Sol. (a) **36**, 687 (1976)
8.69 K.Rosan, H.Wipf: Phys. Stat. Sol. (a) **38**, 611 (1976)
8.70 P.Schiller, H.Nijman: Phys. Stat. Sol. (a) **31**, K 77 (1975)
8.71 R.F.Mattas, H.K.Birnbaum: Acta Met. **23**, 973 (1975)
8.72 P.Zapp, H.K.Birnbaum: Private communication
8.73 J.J.-S.Au: Thesis, University of Illinois at Urbana-Champaign (1976)
8.74 J.Völkl, G.Alefeld: Nuovo Cimento **33**B, 190 (1976)
8.75 G.Matusiewicz, R.Booker, J.Keiser, H.K.Birnbaum: Scripta Met. **8**, 1419 (1974)
8.76 D.Westlake: Trans. AIME **245**, 287 (1969)
8.77 D.G.Westlake, S.T.Ockers: J. Less-Common Metals **42**, 255 (1975)
8.78 D.G.Westlake, S.T.Ockers: Met. Trans. **6**A, 399 (1975)
8.79 D.Sherman, C.Owen, T.Scott: Trans. AIME **242**, 1775 (1968)
8.80 G.Cannelli, R.Cantelli: Appl. Phys. **3**, 325 (1974)
8.81 G.Cannelli, F.Mazzolai: J. Phys. Chem. Sol. **31**, 1913 (1970)
8.82 T.Matsumoto: J. Phys. Soc. Japan **42**, 1583 (1977)
8.83 R.Hanada: "A Resistometric Study of Hydrogen and Deuterium in Group Va Metals at Low Temperatures", in *Effect of Hydrogen on Behavior of Materials*, ed. by A.W.Thompson, I.M.Bernstein (The Metallurigical Society of AIME, New York 1976) pp. 676–685
8.84 H.D.Carstanjen, R.Sizmann: Ber. Bunsenges. Physik. Chem. **76**, 1223 (1972)
8.85 S.T.Picraux, F.L.Vook: "Lattice Location of Deuterium Implanted into W and Cr", in *Ion Implantation in Semiconductors*, ed. by S.Namba (Plenum Press, New York 1974) pp. 355–360
8.86 J.Bøttinger, S.T.Picraux, N.Rud: "Depth Profiling of Hydrogen and Helium Isotopes in Solids by Nuclear Reaction Analysis", in *Ion Beam Surface Layer Analysis*, ed. by O.Meyer, G.Linker, F.Käppelar (Plenum Press, New York 1976) pp. 811–819
8.87 S.T.Picraux, J.Bøttinger, N.Rud: Appl. Phys. Lett. **28**, 179 (1976)
8.88 S.T.Picraux, F.L.Vook: Phys. Rev. Lett. **33**, 1216 (1974)
8.89 F.L.Vook, S.T.Picraux: "Lattice Location Studies of Gases in Metals", in *Radiation Effects on Solid Surface*, ed. by M.Kaminski, Advances in Chemistry Series, No. 158 (American Chemical Society, Washington D.C. 1976) pp. 308–324
8.90 S.T.Picraux, J.Bøttinger, N.Rud: J. Nucl. Mat. **63**, 110 (1976)

Additional References with Titles

Chapter 2

Y. Fukai, S. Kazama: NMR studies of anomalous diffusion of hydrogen and phase transition in vanadium-hydrogen alloys. Acta Metall. **25**, 59 (1977)

O. Yoshinari, M. Koiwa, H. Asano, M. Hirabayashi: Low frequency internal friction study on vanadium-deuterium alloys. Trans. Jpn. Inst. Met. **19**, 171 (1978)

B.J. Makenas: "Precipitation and Ordering in the Niobium-Hydrogen System": Ph.D. Thesis, University of Illinois, Urbana (1978)

M. Amano, Y. Sasaki: Internal friction in niobium hydride. Trans. Nat. Res. Inst. Met. **19** (4) 155 (1977)

W. Pesch, T. Schober, H. Wenzl: A TEM investigation of anisotropic lattice distortions in ordered NbH and NbD alloys. Scr. Metall. **12**, 815 (1978)

G. Ferron, M. Quintard: Evidence of an internal friction phase transformation peak in niobium-hydrogen alloys. Phys. Status Solidi (a) **46**, K 43 (1978)

H. Metzger, H. Peisl: Huang diffuse x-ray scattering from lattice strains in high-concentration Ta-H alloys. J. Phys. F: Metal Phys. **8**, 391 (1978)

C.E. Laciana, A.J. Pedraza, E.J. Savino: Lattice distortion and migration energy of oxygen in vanadium and hydrogen in niobium, tantalum and vanadium. Phys. Status Solidi (a) **45**, 315 (1978)

J. Hauck: Structural relations between vanadium, niobium, tantalum hydrides and deuterides. Acta Crystallogr. A **34**, 389 (1978)

H. Metzger, H. Jo, S.C. Moss, D.G. Westlake: Single crystal x-ray study of the superstructure modulation and long-range order in V_2D. Phys. Status Solidi (a) **17**, 631 (1978)

Chapter 5

R.A. Guidotti, G.B. Atkinson, M.M. Wong: Hydrogen absorption by rare-earth transition metal alloys. J. Less-Common Metals **52** (1), 13 (1977)

S. Yajima, H. Kayano: Hydrogen sorption in La_2Mg_{17}. J. Less-Common Metals **55** (1), 139 (1977)

T. Takeshita, W.E. Wallace: Hydrogen absorption in $Th(Ni, Al)_5$ ternaries. J. Less-Common Metals **55** (1), 61 (1977)

E.C. Ashby, K.C. Nainan, H.S. Prasad: Preparation of magnesium zinc hydrides, $MgZnH_4$ and $Mg(ZnH_3)_2$. Inorg. Chem. **16** (2), 348 (1977)

Proc., 2nd World Hydrogen Energy Conf., Zürich, 1978, ed. by T.N. Veziroglu, W. Seifritz (Pergamon Press, Oxford, 1978). Has 23 papers on metal hydrides and their uses

Proc. Int. Symp. on Energy Storage, Geilo, 1977, (Pergamon Press, Oxford, 1978). Has numerous review articles by authorities in the field

Chapter 6

H. Oesterreicher, J. Clinton, H. Bittner: Hydrides of La–Ni compounds. Mater. Res. Bull. **11**, 1241 (1976)

W.A. Lanford, P.H. Schmidt, J.M. Rowell, J.M. Poate, R.C. Dynes, P.D. Dernier: Sensitivity of the T_c of Nb_3Ge to hydrogen content. Appl. Phys. Lett. **32**, 339 (1978)

L. Ya. Vinnikov, O.V. Zharikov, Ch.V. Kopetskii, V.M. Polovov: Pinning in single-crystal Nb and V containing H. Sov. J. Low. Temp. Phys. **3**, 11 (1977)

E. F. Khodosov, V. F. Shkav: Low-temperatures (up to 15 K) conductivity of Nb–H systems. JETP Lett. **25**, 313 (1977)
B. M. Klein, E. N. Enconomou, D. A. Papaconstantopoulos: Inverse isotope effect and the x dependence of the superconducting transition temperature in PdH_x and PdD_x. Phys. Rev. Lett. **39**, 574 (1977)
A. I. Morozov: Superconductivity in two-component compounds. The isotope effect in PdH. Sov. J. Low. Temp. Phys. **3**, 404 (1977)
P. Jena, C. L. Wiley, F. Y. Fradin: Isotope effect on the electronic spin density in PdH superconductors. Phys. Rev. Lett. **40**, 578 (1978)
C. B. Satterthwaite: On isotope effects in $PdH(D)_x$. Phys. Status Solidi A **43**, K 147 (1978)
B. N. Ganguly: "Anharmonicity and superconductivity in metal-hydrides". Proc. 2nd Int. Congr. H in Metals, Paris (1977)
M. Gupta, A. J. Freeman: Electronic structure and proton spin-lattice relaxation in PdH. Phys. Rev. B **17**, 3029 (1978)
A. Gorska, A. M. Gorski, J. Igalson, A. J. Pindor, L. Sniadower: "Ideal resistivity and electron-phonon interaction in $Pd–H_x$". Proc. 2nd Int. Congr. H in Metals, Paris (1977)
T. F. Smith, G. K. White: Grüneisen parameters, electron-phonon enhancement and superconductivity for Pd–H alloys. J. Phys. F.: Metal Physics **7**, 1029 (1977)
J. C. H. Chiu, R. A. B. Devine: Electrical resistivity evidence for the correlation between superconductivity and optical phonons in Pd–H. Solid State Commun. **22**, 631 (1977)
N. Jacobi, L. G. Caron: The role of optic phonons in the superconductivity of PdH and PdD. J. Low Temp. Phys. **30**, 51 (1978)
L. Dumoulin, P. Nédellec, C. Arzoumanian, J. P. Burger: Optical phonons in PdH_x and PdD_x by superconducting tunneling measurements. To be published
D. A. Papaconstantopoulos, E. N. Economou: Superconductivity in the Pd–Ag–H system. Inst. Phys. Conf. Ser. No. **39**, 489 (1978)
D. A. Papaconstantopoulos, B. M. Klein, E. N. Economou, L. L. Boyer: Band structure and superconductivity of PdH_x and PdD_x. Phys. Rev. B **17**, 141 (1978)
A. W. Szafranski, T. Skoskiewicz, B. Baranowski: Superconductivity in the $Pd_{1-x}M_xH_c$ (M = noble metal) and PdB_yH_c alloy systems. Phys. Status Solidi A **37**, K 163 (1976)
G. Wolf, M. Zimmermann: The molar heat capacity of superconducting PdD_x in the temperature range from 2 to 12 K. Phys. Status Solidi (a) **37**, 485 (1976)
O. J. Kleppa, C. Picard: H–H interaction in Pd–Ag alloys. Solid State Commun. **26**, 421 (1978)
R. J. Miller, T. O. Brun, C. B. Satterthwaite: Magnetic susceptibility of Pd–H and Pd–D at temperatures between 6 and 150 K. To be published
M. Horobiowski, T. Skoskiewicz, E. Trojnar: Magnetization of the superconducting PdH. Phys. Status Solidi (b) **79**, K 147 (1977)

Author Index

Page numbers for this volume are indicated by *italics*. All other page numbers refer to **Hydrogen in Metals I: Basic Properties,** Topics in Applied Physics, Vol. 28, ed. by G. Alefeld, J. Völkl (Springer, Berlin, Heidelberg, New York 1978)

Subject Index

Page numbers for this volume are indicated by *italics*. All other page numbers refer to **Hydrogen in Metals I: Basic Properties**, Topics in Applied Physics, Vol. 28, ed. by G. Alefeld, J. Völkl (Springer, Berlin, Heidelberg, New York 1978)

G. Leibfried, N. Breuer

Point Defects in Metals I

Introduction to the Theory
1978. 138 figures, 22 tables.
XIV, 342 pages
(Springer Tracts in Modern Physics, Volume 81)
ISBN 3-540-08375-8

Contents: Introduction and survey. – Harmonic approximation and linerar response (Green's function) of an arbitrary system. – Lattice theory. – Continuum Theory. Transition from lattice to continuum theory.– Statics and dynamics of simple single point defects. – Scattering of neutrons and X-rays by crystals.– Probability, distributions and statistics. – Properties of crystals with defects in small concentration. – Appendix.

C. P. Slichter

Principles of Magnetic Resonance

2nd revised and expanded edition. 1978.
115 figures. X, 397 pages
(Springer Series in Solid State Sciences, Volume 1)
ISBN 3-540-08476-2
(Originally published by Harper & Row Publ., New York)

Contents: Elements of Resonance. – Basic Theory. – Magnetic Dipolar Broadening of Rigid Lattices. – Magnetic Interactions of Nuclei with Electrons. – Spin-Lattice Relaxation and Motional Narrowing of Resonance Lines. – Spin Temperature in Magnetism and in Magnetic Resonance. – Double Resonance. – Advanced Concepts in Pulsed Magnetic Resonance. – Electric Quadrupole Effects. – Electron Spin Resonance. Summary. – Appendices.

O. Madelung

Introduction to Solid-State Theory

Translated from the German by B. C. Taylor
1978. 144 figures. XI, 486 pages
(Springer Series in Solid-State Sciences, Volume 2)
ISBN 3-540-08516-5

Contents: Fundamentals. – The One-Electron Approximation. – Elementary Excitations. Electron-Phonon Interaction: Transport Phenomena.– Electron-Electron Interaction by Exchange of Virtual Phonons: Superconductivity. – Interaction with Photons: Optics. – Phonon-Phonon Interaction: Thermal Properties. – Local Description of Solid-State Properties. – Localized States. – Disorder. – Appendix: The Occupation Number Representation.

Z. G. Pinsker

Dynamical Scattering of X-Rays in Crystals

1978. 124 figures, 12 tables.
XII, 511 pages
(Springer Series in Solid State Sciences, Volume 3)
ISBN 3-540-08564-5

Contents: Wave Equation and Its Solution for Transparent Infinite Crystal. – Transmission of X-Rays Through a Transparent Crystal Plate. Laue Reflection. – X-Ray Scattering in Absorbing Crystal. Laue Reflection. – Poynting's Vectors and the Propagation of X-Ray Wave Energy. – Dynamical Theory in Incident-Spherical-Wave Approximation. – Bragg Reflection of X-Rays. I. Basic Definitions. Coefficients of Absorption; Diffraction in Finite Crystal. – Bragg Reflection of X-Rays. II. Reflection and Transmission Coefficients and Their Integradet Values. – X-Ray Spectrometers Used in Dynamical Scattering Investigations. Some Results of Experimental Verification of the Theory. – X-Ray Interferometry. Moiré Patterns in X-Ray Diffraction. – Generalized Dynamical Theory of X-Ray Scattering in Perfect and Deformed Crystals. Dynamical Scattering in the Case of Three Strong Waves and More. – Appendices.

Springer-Verlag
Berlin Heidelberg New York

1-MONTH